Lecture Notes in Computer Science   12000

More information about this series at http://www.springer.com/series/7407

Ding-Zhu Du · Jie Wang (Eds.)

Complexity and Approximation

In Memory of Ker-I Ko

 Springer

Editors
Ding-Zhu Du (ID)
The University of Texas at Dallas
Richardson, TX, USA

Jie Wang (ID)
University of Massachusetts Lowell
Lowell, MA, USA

ISSN 0302-9743   ISSN 1611-3349 (electronic)
Lecture Notes in Computer Science
ISBN 978-3-030-41671-3   ISBN 978-3-030-41672-0 (eBook)
https://doi.org/10.1007/978-3-030-41672-0

LNCS Sublibrary: SL1 – Theoretical Computer Science and General Issues

Cover illustration: The cover illustration was painted by Ker-I Ko and is an artistic interpretation of turing machines.

This Springer imprint is published by the registered company Springer Nature Switzerland AG
The registered company address is: Gewerbestrasse 11, 6330 Cham, Switzerland

Ker-I Ko (1950–2018)

Preface

An ad hoc International Workshop on Complexity and Approximation was held during April 27–28, 2019, in Qingdao, hosted by Ocean University of China, for the purpose of honoring Ker-I Ko, who passed away in December 2018. A talented computer scientist, Ko was one of the key players in the areas of Computational Complexity Theory, Complexity Theory of Real Functions, and Combinatorial Optimization. Colleagues, friends, and family members gathered together at the workshop to celebrate his life, present advanced progress in the aforementioned areas, and highlight Ko's significant contributions. This book is a collection of articles selected from presentations at the workshop. It also includes invited articles not presented at the workshop by authors unable to travel to China at that time.

We are grateful to Ocean University of China for sponsoring and hosting the workshop and members of the Organization and Local Arrangement Committee for their hard and efficient work, which helped make the workshop a great success. We thank all authors for their contributions and to all reviewers for their constructive comments.

December 2019

Ding-Zhu Du
Jie Wang

Contents

In Memoriam: Ker-I Ko (1950–2018)

Ding-Zhu Du[1]($\boxtimes$) and Jie Wang[2]

[1] University of Texas at Dallas, Richardson, TX 75080, USA
dzdu@utdallas.edu
[2] University of Massachusetts, Lowell, MA 01854, USA
wang@cs.uml.edu

> Ker-I Ko is a talented scientist, novelist,
> and a warm, sincere long time friend. We
> will miss him profoundly.
>
> _____
>
> Andrew Chi-Chih Yao
> Turing Award recipient, 2000

Ker-I Ko was a colleague and a friend, and our friendships began from the mid 1980's. Ker-I passed away peacefully due to lung failure on the 13th of December in 2018 at a hospital in New York, with his wife Mindy Pien and all three children by his side. His passing is a great loss to the theoretical computer science community.

He received his BS in mathematics from National Tsing Hua University in 1972, his MS in mathematics and his PhD in computer science both from the Ohio State University in 1974 and 1979, respectively. He started his academic career as a faculty member at the University of Houston in 1979 and moved to New York in 1986 as Full Professor at SUNY Stony Brook. He remained in that position until his retirement in 2012. After that he taught at National Chiao Tung University in Taiwan.

A versatile and productive researcher, Ker-I had published a total of 66 journal papers [1–66], 29 conference papers [67–85], 5 book chapters [86–90], and 7 books [91–97]. In his homepage at SUNY Stony Brook, he described his research interests in five areas and selected 2–3 publications as representatives for each area: Computational Complexity Theory [28,38,51], Complexity Theory of Real Functions [20,65,72], Combinatorics [42,44], Combinatorial Optimization [25,90], and Computational Learning Theory [76,77].

Ker-I was one of the founding fathers of computational complexity over real numbers and analysis. He and Harvey Friedman devised a theoretical model for real number computations by extending the computation of Turing machines [65], based on which he established the computational complexity of several important real number operations, such as NP-hard for optimization and $\#P$-hard for integration [97], and introduced computational aspect into several real number mathematical subjects [20,72].

He contributed significantly to advancing the theory of structural complexity. Ker-I's best known work was perhaps his brilliant construction of an infinite series of oracles $A_1, A_2, \ldots$ that collapses the polynomial-time hierarchy, relative to A_k, to

© Springer Nature Switzerland AG 2020
D.-Z. Du and J. Wang (Eds.): Ko Festschrift, LNCS 12000, pp. 1–7, 2020.
https://doi.org/10.1007/978-3-030-41672-0_1

exactly the k-th level [38]. In addition, to help understand polynomial-time isomor-phism, Ker-I, with Tim Long and Ding-Zhu Du, showed that $P \neq UP$ only if there exist two sets that are one-to-one length-increasing polynomial-time reducible to each other but not polynomial-time isomorphic [51]. Moreover, Ker-I, with Pekka Orponen, Uwe Schöning, and Osamu Watanabe, showed how to use Kolmogorov complexity to measure the complexity of an instance of a problem [28].

Fig. 1. Ker-I Ko's artistic interpretation of Turing machines

Ker-I also studied a number of research problems in combinatorics, combina-torial optimization, and computational learning theory. In particular, he studied the complexity of group testing [44] and min-max optimization problems [90], and designed efficient algorithms for searching and learning problems, such as searching two objects by underweight feedbacks [42] and learning string patterns and tree patterns from examples [77].

In addition to being an outstanding researcher and educator in computer science, Ker-I was also a skilled artist and writer. His work includes over 100 oil and watercolor paintings, and a book of short fiction titled A Narrow Alley that was released in Chinese by a highly reputable publisher, the Commercial Press in Taiwan. Figure 1 is an example of his oil painting, reflecting his artistic interpretation of Turing machines.

Ker-I was quiet, but he always had a great deal to say about mathematics, computation theory, algorithms, art, literature, and life itself. We value the intelligence and insight he brought to our many discussions. We will miss him deeply.

References

1. Chen, X., et al.: Centralized and decentralized rumor blocking problems. J. Comb. Optim. **34**(1), 314–329 (2017)
2. Li, W., Liu, W., Chen, T., Qu, X., Fang, Q., Ko, K.-I.: Competitive profit maximization in social networks. Theor. Comput. Sci. **694**, 1–9 (2017)
3. Ran, Y., Zhang, Z., Ko, K.-I., Liang, J.: An approximation algorithm for maximum weight budgeted connected set cover. J. Comb. Optim. **31**(4), 1505–1517 (2016)
4. Ko, K.-I.: On the complexity of computing the Hausdorff distance. J. Complex. **29**(3–4), 248–262 (2013)
5. Fuxiang, Y., Ko, K.-I.: On logarithmic-space computable real numbers. Theor. Comput. Sci. **469**, 127–133 (2013)
6. Fuxiang, Y., Ko, K.-I.: On parallel complexity of analytic functions. Theor. Comput. Sci. **489–490**, 48–57 (2013)
7. Bauer, A., Hertling, P., Ko, K.-I.: Computability and complexity in analysis. J. UCS **16**(18), 2495 (2010)
8. Cheng, Y., Ding-Zhu, D., Ko, K.-I., Lin, G.: On the parameterized complexity of pooling design. J. Comput. Biol. **16**(11), 1529–1537 (2009)
9. Ko, K.-I., Fuxiang, Y.: On the complexity of convex hulls of subsets of the two-dimensional plane. Electr. Notes Theor. Comput. Sci. **202**, 121–135 (2008)
10. Cheng, Y., Ko, K.-I., Weili, W.: On the complexity of non-unique probe selection. Theor. Comput. Sci. **390**(1), 120–125 (2008)
11. Ko, K.-I., Fuxiang, Y.: Jordan curves with polynomial inverse moduli of continuity. Electr. Notes Theor. Comput. Sci. **167**, 425–447 (2007)
12. Ko, K.-I., Fuxiang, Y.: On the complexity of computing the logarithm and square root functions on a complex domain. J. Complex. **23**(1), 2–24 (2007)
13. Ko, K.-I., Weihrauch, K., Zheng, X.: Editorial: Math. Log. Quart. 4–5/2007. Math. Log. Q. **53**(4–5), 325 (2007)
14. Ko, K.-I., Fuxiang, Y.: Jordan curves with polynomial inverse moduli of continuity. Theor. Comput. Sci. **381**(1–3), 148–161 (2007)
15. Brattka, V., Hertling, P., Ko, K.-I., Tsuiki, H.: Computability and complexity in analysis. J. Complex. **22**(6), 728 (2006)
16. Fuxiang, Y., Chou, A.W., Ko, K.-I.: On the complexity of finding circumscribed rectangles and squares for a two-dimensional domain. J. Complex. **22**(6), 803–817 (2006)
17. Chou, A.W., Ko, K.-I.: On the complexity of finding paths in a two-dimensional domain II: piecewise straight-line paths. Electr. Notes Theor. Comput. Sci. **120**, 45–57 (2005)

18. Chou, A.W., Ko, K.-I.: The computational complexity of distance functions of two-dimensional domains. Theor. Comput. Sci. **337**(1–3), 360–369 (2005)
19. Brattka, V., Hertling, P., Ko, K.-I., Zhong, N.: Preface: MLQ - Math. Log. Quart. 4–5/2004. Math. Log. Q. **50**(4–5), 327–328 (2004)
20. Chou, A.W., Ko, K.-I.: On the complexity of finding paths in a two-dimensional domain I: shortest paths. Math. Log. Q. **50**(6), 551–572 (2004)
21. Ruan, L., Hongwei, D., Jia, X., Weili, W., Li, Y., Ko, K.-I.: A greedy approximation for minimum connected dominating sets. Theor. Comput. Sci. **329**(1–3), 325–330 (2004)
22. Ko, K.-I., Nerode, A., Weihrauch, K.: Foreword. Theor. Comput. Sci. **284**(2), 197 (2002)
23. Ko, K.-I.: On the computability of fractal dimensions and hausdorff measure. Ann. Pure Appl. Logic **93**(1–3), 195–216 (1998)
24. Ko, K.-I.: Computational complexity of fixed points and intersection points. J. Complex. **11**(2), 265–292 (1995)
25. Ko, K.-I., Lin, C.-L.: On the longest circuit in an alterable digraph. J. Glob. Optim. **7**(3), 279–295 (1995)
26. Chou, A.W.: Computational complexity of two-dimensional regions. SIAM J. Comput. **24**(5), 923–947 (1995)
27. Ko, K.-I.: A polynomial-time computable curve whose interior has a nonrecursive measure. Theor. Comput. Sci. **145**(1&2), 241–270 (1995)
28. Orponen, P., Ko, K.-I., Schöning, U., Watanabe, O.: Instance complexity. J. ACM **41**(1), 96–121 (1994)
29. Ko, K.-I.: On the computational complexity of integral equations. Ann. Pure Appl. Logic **58**(3), 201–228 (1992)
30. Ding-Zhu, D., Ko, K.-I.: A note on best fractions of a computable real number. J. Complex. **8**(3), 216–229 (1992)
31. Ko, K.-I., Tzeng, W.-G.: Three Σ_2^p-complete problems in computational learning theory. Comput. Complex. **1**, 269–310 (1991)
32. Ko, K.-I.: Separating the low and high hierarchies by oracles. Inf. Comput. **90**(2), 156–177 (1991)
33. Ko, K.-I.: On the complexity of learning minimum time-bounded turing machines. SIAM J. Comput. **20**(5), 962–986 (1991)
34. Ko, K.-I.: On adaptive versus nonadaptive bounded query machines. Theor. Comput. Sci. **82**(1), 51–69 (1991)
35. Ko, K.-I.: A note on separating the relativized polynomial time hierarchy by immune sets. ITA **24**, 229–240 (1990)
36. Ko, K.-I.: Separating and collapsing results on the relativized probabilistic polynomial-time hierarchy. J. ACM **37**(2), 415–438 (1990)
37. Ko, K.-I.: Distinguishing conjunctive and disjunctive reducibilities by sparse sets. Inf. Comput. **81**(1), 62–87 (1989)
38. Ko, K.-I.: Relativized polynomial time hierarchies having exactly k levels. SIAM J. Comput. **18**(2), 392–408 (1989)
39. Du, D.-Z., Ko, K.-I.: On the complexity of an optimal routing tree problem. Acta Math. Appl. Sinica (Engl. Ser.) **5**, 68–80 (1989)
40. Du, D.-Z., Ko, K.-I.: Complexity of continuous problems on convex functions. Syst. Sci. Math. **2**, 70–79 (1989)
41. Book, R.V., Ko, K.-I.: On sets truth-table reducible to sparse sets. SIAM J. Comput. **17**(5), 903–919 (1988)
42. Ko, K.-I.: Searching for two objects by underweight feedback. SIAM J. Discrete Math. **1**(1), 65–70 (1988)

43. Marron, A., Ko, K.-I.: Identification of pattern languages from examples and queries. Inf. Comput. **74**(2), 91–112 (1987)

44. Du, D., Ko, K.: Some completeness results on decision trees and group testing. SIAM J. Algebraic Discrete Methods **8**, 762–777 (1987)

45. Ko, K.-I., Hua, C.-M.: A note on the two-variable pattern-finding problem. J. Comput. Syst. Sci. **34**(1), 75–86 (1987)

46. Ko, K.-I.: On helping by robust oracle machines. Theor. Comput. Sci. **52**, 15–36 (1987)

47. Ko, K.-I.: Corrigenda: on the continued fraction representation of computable real numbers. Theor. Comput. Sci. **54**, 341–343 (1987)

48. Ko, K.-I.: Approximation to measurable functions and its relation to probabilistic computation. Ann. Pure Appl. Logic **30**(2), 173–200 (1986)

49. Ko, K.-I., Teng, S.-C.: On the number of queries necessary to identify a permutation. J. Algorithms **7**(4), 449–462 (1986)

50. Ko, K.-I.: On the computational complexity of best Chebyshev approximations. J. Complex. **2**(2), 95–120 (1986)

51. Ko, K.-I., Long, T.J., Ding-Zhu, D.: On one-way functions and polynomial-time isomorphisms. Theor. Comput. Sci. **47**(3), 263–276 (1986)

52. Ko, K.-I.: On the continued fraction representation of computable real numbers. Theor. Comput. Sci. **47**(3), 299–313 (1986)

53. Ko, K.-I.: On the notion of infinite pseudorandom sequences. Theor. Comput. Sci. **48**(3), 9–33 (1986)

54. Ko, K.-I.: Continuous optimization problems and a polynomial hierarchy of real functions. J. Complex. **1**(2), 210–231 (1985)

55. Ko, K.-I.: Nonlevelable sets and immune sets in the accepting density hierarchy in NP. Math. Syst. Theory **18**(3), 189–205 (1985)

56. Ko, K.-I., Schöning, U.: On circuit-size complexity and the low hierarchy in NP. SIAM J. Comput. **14**(1), 41–51 (1985)

57. Ko, K.-I.: On some natural complete operators. Theor. Comput. Sci. **37**, 1–30 (1985)

58. Ko, K.-I.: Reducibilities on real numbers. Theor. Comput. Sci. **31**, 101–123 (1984)

59. Ko, K.-I.: On the computational complexity of ordinary differential equations. Inf. Control **58**(1–3), 157–194 (1983)

60. Ko, K.-I.: On self-reducibility and weak p-selectivity. J. Comput. Syst. Sci. **26**(2), 209–221 (1983)

61. Ko, K.-I.: On the definitions of some complexity classes of real numbers. Math. Syst. Theory **16**(2), 95–109 (1983)

62. Ko, K.-I.: Some negative results on the computational complexity of total variation and differentiation. Inf. Control **53**(1/2), 21–31 (1982)

63. Ko, K.-I.: Some observations on the probabilistic algorithms and NP-hard problems. Inf. Process. Lett. **14**(1), 39–43 (1982)

64. Ko, K.-I.: The maximum value problem and NP real numbers. J. Comput. Syst. Sci. **24**(1), 15–35 (1982)

65. Ko, K.-I., Friedman, H.: Computational complexity of real functions. Theor. Comput. Sci. **20**, 323–352 (1982)

66. Ko, K.-I., Moore, D.J.: Completeness, approximation and density. SIAM J. Comput. **10**(4), 787–796 (1981)

67. Bauer, A., Hertling, P., Ko, K.-I.: CCA 2009 front matter - proceedings of the sixth international conference on computability and complexity in analysis. In: CCA (2009)

68. Bauer, A., Hertling, P., Ko, K.-I.: CCA 2009 preface - proceedings of the sixth international conference on computability and complexity in analysis. In: CCA (2009)

69. Yu, F., Chou, A.W., Ko, K.-I.: On the complexity of finding circumscribed rectangles for a two-dimensional domain. In: CCA, pp. 341–355 (2005)

70. Ko, K.-I., Yu, F.: On the complexity of computing the logarithm and square root functions on a complex domain. In: Wang, L. (ed.) COCOON 2005. LNCS, vol. 3595, pp. 349–358. Springer, Heidelberg (2005). https://doi.org/10.1007/11533719_36

71. Ko, K.-I.: Fractals and complexity. In: CCA (1996)

72. Ko, K.-I., Weihrauch, K.: On the measure of two-dimensional regions with polynomial-time computables boundaries. In: Computational Complexity Conference, pp. 150–159 (1996)

73. Chou, A.W., Ko, K.-I.: Some complexity issues on the simply connected regions of the two-dimensional plane. In: STOC, pp. 1–10 (1993)

74. Ko, K.-I.: A note on the instance complexity of pseudorandom sets. In: Computational Complexity Conference, pp. 327–337 (1992)

75. Ko, K.-I.: Integral equations, systems of quadratic equations, and exponential-time completeness (extended abstract). In: STOC, pp. 10–20 (1991)

76. Ko, K.-I.: On the complexity of learning minimum time-bounded Turing machines. In: COLT, pp. 82–96 (1990)

77. Ko, K.-I., Marron, A., Tzeng, W.-G.: Learning string patterns and tree patterns from examples. In: ML, pp. 384–391 (1990)

78. Ko, K.-I.: Computational complexity of roots of real functions (extended abstract). In: FOCS, pp. 204–209 (1989)

79. Ko, K.-I.: Relativized polynomial time hierarchies having exactly K levels. In: Computational Complexity Conference (1988)

80. Ko, K.-I.: Distinguishing bounded reducibilities by sparse sets. In: Computational Complexity Conference, pp. 181–191 (1988)

81. Ko, K.-I: Relativized polynominal time hierarchies having exactly K levels. In: STOC, pp. 245–253 (1988)

82. Book, R.V., Ko, K.-I.: On sets reducible to sparse sets. In: Computational Complexity Conference (1987)

83. Ko, K.-I.: On helping by robust oracle machines. In: Computational Complexity Conference (1987)

84. Ko, K.-I., Orponen, P., Schöning, U., Watanabe, O.: What is a hard instance of a computational problem? In: Selman, A.L. (ed.) Structure in Complexity Theory. LNCS, vol. 223, pp. 197–217. Springer, Heidelberg (1986). https://doi.org/10.1007/3-540-16486-3_99

85. Ko, K.-I, Long, T.J., Du, D.-Z.: A note on one-way functions and polynomial-time isomorphisms (extended abstract). In: STOC, pp. 295–303 (1986)

86. Ko, K.: Applying techniques of discrete complexity theory to numerical computation. In: Book, R. (ed.) Studies in Complexity Theory, pp. 1–62. Research Notes in Theoretical Computer Science, Pitman (1986)

87. Ko, K.: Constructing oracles by lower bound techniques for circuits. In: Du, D., Hu, G. (eds.) Combinatorics, Computing and Complexity, pp. 30–76. Kluwer Academic Publishers and Science Press, Boston (1989)

88. Ko, K.: Polynomial-time computability in analysis. In: Ershov, Y.L., et al. (eds.) Handbook of Recursive Mathematics. Recursive Algebra, Analysis and Combinatorics, vol. 2, pp. 1271–1317 (1998)

89. Ko, K.: Computational complexity of fractals. In: Downey, R., et al. (eds.) Proceedings of the 7th and 8th Asian Logic Conferences, pp. 252–269. World Scientific, Singapore (2003)

90. Ko, K., Lin, C.-L.: On the complexity of min-max optimization problems and their approximation. In: Du, D.-Z., Pardalos, P.M. (eds.) Minimax and Applications, pp. 219–239. Kluwer (1995)

91. Du, D.-Z., Ko, K.-I., Hu, X.: Design and Analysis of Approximation Algorithms. Springer, New York (2012). https://doi.org/10.1007/978-1-4614-1701-9

92. Du, D.-Z., Ko, K., Hu, X.: Design and Analysis of Approximation Algorithms. Higher Education Press, Beijing (2011). (in Chinese)

93. Du, D.-Z., Ko, K., Wang, J.: Introduction to Computational Complexity. Higher Education Press, Beijing (2002). (in Chinese)

94. Du, D.-Z., Ko, K.: Problem Solving in Automata, Languages and Complexity. Wiley, New York (2001)

95. Du, D.-Z., Ko, K.: Theory of Computational Complexity. Wiley, New York (2000)

96. Du, D.-Z., Ko, K. (eds.): Advances in Algorithms, Languages, and Complexity. Kluwer, Dordrecht (1997)

97. Ko, K.: Computational Complexity of Real Functions. Birkhauser Boston, Boston (1991)

Ker-I Ko and the Study of Resource-Bounded Kolmogorov Complexity

Eric Allender[(✉)] [iD]

Rutgers University, New Brunswick, NJ 08854, USA
allender@cs.rutgers.edu
http://www.cs.rutgers.edu/~allender

Abstract. Ker-I Ko was among the first people to recognize the importance of resource-bounded Kolmogorov complexity as a tool for better understanding the structure of complexity classes. In this brief informal reminiscence, I review the milieu of the early 1980's that caused an up-welling of interest in resource-bounded Kolmogorov complexity, and then I discuss some more recent work that sheds additional light on the questions related to Kolmogorov complexity that Ko grappled with in the 1980's and 1990's.

In particular, I include a detailed discussion of Ko's work on the question of whether it is NP-hard to determine the time-bounded Kolmogorov complexity of a given string. This problem is closely connected with the Minimum Circuit Size Problem (MCSP), which is central to several contemporary investigations in computational complexity theory.

Keywords: Kolmogorov complexity · Complexity theory · Minimum Circuit Size Problem

1 Introduction: A Brief History of Time-Bounded Kolmogorov Complexity

In the beginning, there was Kolmogorov complexity, which provided a satisfying and mathematically precise definition of what it means for something to be "random", and gave a useful measure of the amount of information contained in a bitstring.[1] But the fact that the Kolmogorov complexity function is not computable does limit its application in several areas, and this provided some of the original motivation for the study of resource-bounded Kolmogorov complexity.

A version of time-bounded Kolmogorov complexity appears already in Kolmogorov's original 1965 paper [41]. However, for the purposes of the story being told here, the first significant development came with the work of Kolmogorov's

---

[1] If the reader is not familiar with Kolmogorov complexity, then we recommend some excellent books on this topic [25,44].

Supported in part by NSF Grants CCF-1514164 and CCF-1909216.

D.-Z. Du and J. Wang (Eds.): Ko Festschrift, LNCS 12000, pp. 8–18, 2020.
https://doi.org/10.1007/978-3-030-41672-0_2

doctoral student Leonid Levin.[2] Levin's fundamental work on NP-completeness [42] has, as its second theorem, a result that can easily be proved[3] by making use of a notion of time-bounded Kolmogorov complexity called Kt, which Levin developed in the early 1970's, but whose formal definition did not appear in a published article until 1984 [43]. Adleman acknowledges communication with Levin in a 1979 MIT technical report [1] that discusses a very similar notion, which he called "potential".[4] Since Kt will be discussed at greater length later on, let us give the definition here:

Definition 1. *For any Turing machine M and strings x and y, $\mathsf{Kt}_M(x|y)$ is the minimum, over all "descriptions" d such that $M(d, y) = x$ in t steps, of the sum $|d| + \log t$. (If no such d exists, then $\mathsf{Kt}_M(x|y)$ is undefined.) $\mathsf{Kt}_M(x)$ is defined to be $\mathsf{Kt}_M(x|\lambda)$, where λ is the empty string.*

If M is chosen to be a universal Turing machine, then $\mathsf{Kt}_M(x|y)$ is always defined. As is usual when discussing Kolmogorov complexity we select one such universal machine U, and define $\mathsf{Kt}(x|y)$ to be equal to $\mathsf{Kt}_U(x|y)$. Kt has the appealing property that it can be used to design optimal search algorithms for finding witnesses for problems in NP. For instance, P = NP iff every $\phi \in$ SAT has some assignment v such that $\mathsf{Kt}(v|\phi) = O(\log |\phi|)$ [1, 42]. See [44] for a discussion.

Li and Vitányi [44], in their discussion of the origins of time-bounded Kolmogorov complexity, highlight not only the work of Adleman and Levin discussed above, but also a 1977 paper by Daley [24], where time-bounded Kolmogorov complexity is studied in the context of inductive inference. Indeed, in Ko's first paper that deals with K-complexity [37], Daley's work [24] is one of the four papers that Ko mentions as containing prior work on resource-bounded Kolmogorov complexity. (The others are [42], and the papers of Hartmanis and of Sipser that are discussed below.) But I think that this is only part of the story.

Adleman's work [1] remains even today an unpublished MIT technical report, which did not circulate widely. Levin's work [42] was still not particularly

---

[2] Levin was Kolmogorov's student, but he did not receive his Ph.D. until after he emigrated to the US, and Albert Meyer was his advisor at MIT. The circumstances around Levin being denied his Ph.D. in Moscow are described in the excellent article by Trakhtenbrot [59].

[3] This result also appears as Exercise 13.20 in what was probably the most popular complexity theory textbook for the early 1980's [33], which credits Levin for that result, but *not* for what is now called the Cook-Levin theorem.

[4] In [1], in addition to Levin, Adleman also credits Meyer and McCreight [46] with developing similar ideas. I have been unable to detect any close similarity, although the final paragraph of [46] states "Our results are closely related to more general definitions of randomness proposed by Kolmogorov, Martin-Löf, and Chaitin" [and here the relevant literature is cited, before continuing] "A detailed discussion must be postponed because of space limitations" [and here Meyer and McCreight *include a citation to a letter from the vice-president of Academic Press* (which presumably communicated the space limitations to the authors).] Indeed, Meyer and McCreight were interested in when a decidable (and therefore very non-random) set can be said to "look random" and thereby deserve to be called pseudorandom. We will return to this topic later in the paper.

well-known in the early 1980's, and the published paper contains very little detail. Daley's work [24] was part of the inductive inference research community, which was then and remains today rather distinct from the complexity theory community. Thus I would also emphasize the impact that the 1980 STOC paper by Paul, Seiferas, and Simon [52] had, in bringing the tools and techniques of Kolmogorov complexity to the STOC/FOCS community in the context of proving lower bounds. At the following FOCS conference, Gary Peterson introduced a notion of resource-bounded Kolmogorov complexity [54]. Peterson's article has a very interesting and readable introduction, highlighting the many ways in which different notions of succinctness had arisen in various other work on complexity theory. Peterson's FOCS'80 paper also introduces a theme that echoes in more recent work, showing how various open problems in complexity theory can be restated in terms of the relationships among different notions of resource-bounded Kolmogorov complexity. However, the precise model of resource-bounded Kolmogorov complexity that is introduced in [54] is rather abstruse, and it seems that there has been no further work using that model in the following four decades.

Perhaps it was in part due to those very deficiencies, that researchers were inspired to find a better approach. At the 1983 STOC, Sipser introduced a notion of polynomial-time "distinguishing" Kolmogorov complexity, in the same paper in which he showed that BPP lies in the polynomial hierarchy [58]. At FOCS that same year, Hartmanis introduced what he termed "Generalized Kolmogorov Complexity", in part as a tool to investigate the question of whether all NP-complete sets are isomorphic. Both Sipser and Hartmanis cited Ko's work, which would eventually appear as [37], as presenting yet another approach to studying resource-bounded Kolmogorov complexity.

Ko's motivation for developing a different approach to resource-bounded Kolmogorov complexity arose primarily because of the groundbreaking work of Yao [62] and Blum and Micali [18], which gave a new approach to the study of pseudorandom generators. Ko sought to find a relationship between the new notion of pseudorandomness and the classical notions of Martin-Löf randomness for *infinite sequences*. Other notions of "pseudorandomness" had been proposed by Meyer and McCreight [46] and by Wilbur [61], and Ko succeeded in finding the relationships among these notions, and in presenting new definitions that provided a complexity-theoretic analog of Martin-Löf randomness. (This analog is more successful in the context of space-bounded Kolmogorov complexity, than for time).

One of the people who had a significant impact on the development on resource-bounded Kolmogorov complexity at this time was Ron Book. Book took an active interest in mentoring young complexity theoreticians, and he organized some informal workshops in Santa Barbara in the mid-to-late 1980's. That was where I first met Ker-I Ko. Some of the others who participated were José Balcázar, Richard Beigel, Lane Hemaspaandra, Jack Lutz, Uwe Schöning, Jacobo Torán, Jie Wang, and Osamu Watanabe. Resource-bounded Kolmogorov complexity was a frequent topic of discussion at these gatherings. Four members of that group (Ko, Orponen, Schöning, and Watanabe) incorporated time-bounded Kolmogorov complexity into their work investigating the question of what it means for certain instances of a computational problem to be hard, whereas other instances can be easy [40]; I first learned about this work at Book's 1986 Santa Barbara workshop, shortly before the paper was presented at the first Structure in Complexity Theory conference (which was the forerunner to the Computational Complexity Conference (CCC)). A partial list of other work on resource-bounded Kolmogorov complexity whose origin can be traced in one way or another to Book's series of workshops includes [2,13,16,20,26,28], as well as the volume edited by Watanabe [60].

Research in resource-bounded Kolmogorov complexity has continued at a brisk pace in the succeeding years. This article will not attempt to survey – or even briefly mention – all of this work. Instead, our goal in this section is to sketch the developments that influenced Ker-I Ko's work on resource-bounded Kolmogorov complexity. Ko's research focus shifted toward other topics after the early 1990's, and thus later work such as [5,15,22,23] does not pertain to this discussion.

But there is one more paper that Ko wrote that deals with resource-bounded Kolmogorov complexity [38], which constitutes an important milestone in a line of research that is very much an active research topic today. In the next section, we place Ko's 1990 COLT paper [38] in context, and discuss how it connects to the current frontier in computational complexity theory.

2 Time-Bounded Kolmogorov Complexity and NP-Completeness

Ko was not the first to see that there is a strong connection between resource-bounded Kolmogorov complexity and one of the central tasks of computational learning theory: namely, to find a succinct explanation that correctly describes observed phenomena. But he does appear to have been the first to obtain theorems that explain the obstacles that have thus far prevented a classification of the complexity of this problem, where "succinct explanation" is interpreted operationally in terms of an efficient algorithm with a short description. There had been earlier work [55,56] showing that it is NP-hard to find "succinct explanations" that have size at all close to the optimal size, if these "explanations" are required to be finite automata or various other restricted formalisms. But for general formalisms such as programs or circuits, this remains an open problem.[5]

Ko approached this problem by defining a complexity measure called LT for partially-specified Boolean functions (which now are more commonly referred to as "promise problems"). Given a list of "yes instances" Y and a list of "no instances" N, $LT(Y, N, t)$ is the length of the shortest description d such that $U(d, x) = 1$ in at most t steps for all $x \in Y$, and $U(d, x) = 0$ in at most t steps

---

[5] During the review and revision phase of preparing this paper, I was given a paper that settles this question! Ilango, Loff, and Oliveira have now shown that the "circuit" version of this problem (which they call Partial-MCSP) is NP-complete [35]. For additional discussion of this result and how it contrasts with Ko's work [38], see [4].

for all $x \in N$, where U is some fixed universal Turing machine (in the tradition of Kolmogorov complexity). Given any oracle A, one can define a relativized measure LT^A, merely by giving the machine U access to A; for any A, the set $\mathsf{MinLT}^A ::= \{(Y, N, 0^s, 0^t) : \mathsf{LT}^A(Y, N, t) \leq s\}$ is in NP^A. Ko showed that there are oracles A relative to which MinLT^A is not NP^A-complete under polynomial-time Turing reductions. In other words, the question of whether this version of the canonical learning theory problem is NP-complete cannot be answered via relativizing techniques.

Ko proves his results about MinLT by first proving the analogous results about a problem he calls $\mathsf{MinKT} ::= \{(x, 0^s, 0^t) : \exists d \ |d| \leq s \wedge U(d) = x$ in at most t steps$\}$. Note that MinKT is essentially MinLT restricted to the case where $Y \cup N$ is equal to the set of all strings of length n (in which case this information can be represented by a string x of length 2^n). Quoting from [39]: "Indeed, there seems to be a simple transformation of the proofs of the results about MinKT to the proofs of analogous results about MinLT. This observation supports our viewpoint of treating the problem MinKT as a simpler version of MinLT, and suggests an interesting link between program-size complexity and learning in the polynomial-time setting." One can see that Ko had been working for quite some time on the question of whether it is NP-hard to determine the time-bounded Kolmogorov complexity of a given string (i.e, the question of whether MinKT is NP-complete), because this question also appears in [37], where it is credited to some 1985 personal communication from Hartmanis.

Ko's question about the difficulty of computing time-bounded Kolmogorov complexity was also considered by Levin in the early 1970's, as related by Trakhtenbrot[6] [59]; see also the discussion in [12]. More precisely, Levin was especially interested in what is now called the Minimum Circuit Size Problem $\mathsf{MCSP} ::= \{(x, s)| x$ is a string of length 2^k representing the truth-table of a k-ary Boolean function that is computed by a circuit of size at most $s\}$. A small circuit for a Boolean function f can be viewed as a short description of f, and thus it was recognized that MCSP was similar in spirit to questions about time-bounded Kolmogorov complexity, although there are no theorems dating to this period that make the connection explicit. Trakhtenbrot [59] describes how MCSP had been the focus of much attention in the Soviet Union as early as the late 1950's; Levin had hoped to include a theorem about the complexity of MCSP (or of time-bounded Kolmogorov complexity) in [42], but these questions remain unresolved even today.

The modern study of the computational complexity of MCSP can really be said to have started with the STOC 2000 paper by Kabanets and Cai [36]. They were the first to show that MCSP must be intractable if cryptographically-secure one-way functions are to exist, and they were the first to initiate an investigation of the consequences that would follow if MCSP were NP-complete under various types of reducibilities.

---

[6] In particular, this is the problem that Trakhtenbrot calls "Task 5" in [59].

A tighter connection between MCSP and resource-bounded Kolmogorov complexity was established in [6]. Prior to [6] most studies of time-bounded Kolmogorov complexity either concentrated on Levin's measure Kt, or else on a measure (similar to what Ko studied) that we can denote K^t for some time bound t (typically where $t(n) = n^{O(1)}$), where $K^t(x)$ is the length of the shortest d such that $U(d) = x$ in at most $t(|x|)$ steps. Although both of these definitions are very useful in various contexts, there are some drawbacks to each. Computing Kt(x) does not seem to lie in NP (and in fact it is shown in [6] that computing Kt is complete for EXP under P/poly reductions). The value of $K^t(x)$ can vary quite a lot, depending on the choice of universal Turing machine U; the usual way of coping with this is to observe that $K^t(x)$, as defined using some machine U_1 is bounded above by $K^{t'}(x)$ as defined using a different machine U_2, for some time bound t' that is not too much larger than t. Both definitions yield measures that have no clear connection to circuit complexity.

The solution presented in [6] is to modify Levin's Kt measure, to obtain a new measure called KT, as follows. First, note that Levin's Kt measure remains essentially unchanged if Definition 1 is replaced by

Definition 2. *Let $x = x_1 x_2 \ldots x_n$ be a string of length n. Kt(x) is the minimum, over all "descriptions" d such that $U(d, i) = x_i$ in t steps, of the sum $|d| + \log t$.*

In other words, the description d still describes the string x, but the way that U obtains x from d is to compute $U(d, i)$ for each $i \in \{1, \ldots n\}$. The main thing that is gained from this modification, is that now the runtime of U can be much less than $|x|$. This gives us the flexibility to replace "$\log t$" in the definition of Kt, with "t", to obtain the definition of KT:

Definition 3. *Let $x = x_1 x_2 \ldots x_n$ be a string of length n. KT(x) is the minimum, over all "descriptions" d such that $U(d, i) = x_i$ in t steps, of the sum $|d| + t$. (A more formal and complete definition can be found in [6].)*

When x is a bit string of length 2^k representing a k-ary Boolean function f, the circuit size of f is polynomially-related to KT(x) [6]. Thus it has been productive to study MCSP (the problem of computing the circuit size function) in tandem with MKTP (the problem of computing the KT function) [6,7,9–11, 31,45,49,57]. This has led to improved hardness results for MCSP (and MKTP) [6,7,9,31,57] and some non-hardness results [9–11]. (The non-hardness results of [47] for MCSP apply equally well to MKTP, and should also be listed here.) We now know that MCSP and MKTP are hard for a complexity class known as SZK under BPP-Turing reductions [7], and they cannot be shown to be NP-complete under polynomial-time many-one reductions without first proving that EXP $\neq$ ZPP [47]. These hardness results also hold for Ko's languages MinKT and MinLT.

Somewhat surprisingly, some hardness proofs currently work only for MKTP and the corresponding hardness conditions for MCSP are either not known to hold [8,9] or seem to require different techniques [27].

Some researchers have begun to suspect that MCSP may be hard for NP under sufficiently powerful notions of reducibility, such as P/poly reductions.

Interestingly, Ko explicitly considered the possibility that MinKT is NP-complete under a powerful notion of reducibility known as SNP reducibility. (Informally, "A is SNP reducible to B" means that A is $(\mathsf{NP} \cap \mathsf{coNP})$-reducible to B.) More recently, Hitchcock and Pavan have shown that this indeed holds under a plausible hypothesis [32]. Interestingly, Ilango has shown that a variant of MCSP is NP-complete under (very restrictive) AC^0 reductions [34]. Hirahara has shown that, if a certain version of time-bounded Kolmogorov complexity is NP-hard to compute, then this implies strong worst-case-to-average-case reductions in NP [30].

One especially intriguing recent development involves what has been termed "hardness magnification". This refers to the phenomenon wherein a seemingly very modest and achievable lower bound can be "magnified" to yield truly dramatic lower bounds which would solve longstanding open questions about the relationships among complexity classes. The problems MCSP, MKTP, and even MKtP (the problem of computing Kt complexity) figure prominently in this line of work [45, 49, 50]. In particular, it is shown in [49] that if one were able to show a certain lower bound for MKtP that is *known* to hold for the apparently much easier problem of computing the inner product mod 2, then it would follow that $\mathsf{EXP} \not\subseteq \mathsf{NC}^1$.

3   Conclusions

Ker-I Ko has left us. But he has left us a rich legacy. This brief article has touched on only a small part of his scientific accomplishments, and how they continue to affect the scientific landscape. Even within the very limited focus of this paper, much has been left out. For instance, the connection between resource-bounded Kolmogorov complexity and learning theory could itself be the subject of a much longer article; as a sample of more recent work in this line, let us mention [48].

References

1. Adleman, L.M.: Time, space and randomness. Technical report, MIT/LCS/TM-131, MIT (1979)
2. Allender, E.: Some consequences of the existence of pseudorandom generators. In: Proceedings of the 19th Annual ACM Symposium on Theory of Computing (STOC), pp. 151–159 (1987). https://doi.org/10.1145/28395.28412, see also [3]
3. Allender, E.: Some consequences of the existence of pseudorandom generators. J. Comput. Syst. Sci. **39**(1), 101–124 (1989). https://doi.org/10.1016/0022-0000(89)90021-4
4. Allender, E.: The new complexity landscape around circuit minimization. In: Proceedings of the 14th International Conference on Language and Automata Theory and Applications (LATA) (2020, to appear)
5. Allender, E., Buhrman, H., Friedman, L., Loff, B.: Reductions to the set of random strings: the resource-bounded case. Logical Methods Comput. Sci. **10**(3) (2014). https://doi.org/10.2168/LMCS-10(3:5)2014
6. Allender, E., Buhrman, H., Koucky, M., van Melkebeek, D., Ronneburger, D.: Power from random strings. SIAM J. Comput. **35**, 1467–1493 (2006). https://doi.org/10.1137/050628994

7. Allender, E., Das, B.: Zero knowledge and circuit minimization. Inf. Comput. **256**, 2–8 (2017). https://doi.org/10.1016/j.ic.2017.04.004. Special issue for MFCS 2014

8. Allender, E., Grochow, J., van Melkebeek, D., Morgan, A., Moore, C.: Minimum circuit size, graph isomorphism and related problems. SIAM J. Comput. **47**, 1339–1372 (2018). https://doi.org/10.1137/17M1157970

9. Allender, E., Hirahara, S.: New insights on the (non)-hardness of circuit minimization and related problems. ACM Trans. Comput. Theory (ToCT) **11**(4), 27:1–27:27 (2019). https://doi.org/10.1145/3349616

10. Allender, E., Holden, D., Kabanets, V.: The minimum oracle circuit size problem. Comput. Complex. **26**(2), 469–496 (2017). https://doi.org/10.1007/s00037-016-0124-0

11. Allender, E., Ilango, R., Vafa, N.: The non-hardness of approximating circuit size. In: van Bevern, R., Kucherov, G. (eds.) CSR 2019. LNCS, vol. 11532, pp. 13–24. Springer, Cham (2019). https://doi.org/10.1007/978-3-030-19955-5_2

12. Allender, E., Koucky, M., Ronneburger, D., Roy, S.: The pervasive reach of resource-bounded Kolmogorov complexity in computational complexity theory. J. Comput. Syst. Sci. **77**, 14–40 (2010). https://doi.org/10.1016/j.jcss.2010.06.004

13. Allender, E., Watanabe, O.: Kolmogorov complexity and degrees of tally sets. In: Proceedings: Third Annual Structure in Complexity Theory Conference, pp. 102–111. IEEE Computer Society (1988). https://doi.org/10.1109/SCT.1988.5269, see also [14]

14. Allender, E., Watanabe, O.: Kolmogorov complexity and degrees of tally sets. Inf. Comput. **86**(2), 160–178 (1990). https://doi.org/10.1016/0890-5401(90)90052-J

15. Antunes, L., Fortnow, L., van Melkebeek, D., Vinodchandran, N.V.: Computational depth: concept and applications. Theor. Comput. Sci. **354**(3), 391–404 (2006). https://doi.org/10.1016/j.tcs.2005.11.033

16. Arvind, V., et al.: Reductions to sets of low information content. In: Kuich, W. (ed.) ICALP 1992. LNCS, vol. 623, pp. 162–173. Springer, Heidelberg (1992). https://doi.org/10.1007/3-540-55719-9_72. See also [17]

17. Arvind, V., et al.: Reductions to sets of low information content. In: Ambos-Spies, K., Homer, S., Schoning, U. (eds.) Complexity Theory: Current Research, pp. 1–46. Cambridge University Press (1993)

18. Blum, M., Micali, S.: How to generate cryptographically strong sequences of pseudo random bits. In: 23rd Annual Symposium on Foundations of Computer Science (FOCS), pp. 112–117 (1982). https://doi.org/10.1109/SFCS.1982.72, see also [19]

19. Blum, M., Micali, S.: How to generate cryptographically strong sequences of pseudo-random bits. SIAM J. Comput. **13**(4), 850–864 (1984). https://doi.org/10.1137/0213053

20. Book, R.V., Lutz, J.H.: On languages with very high information content. In: Proceedings of the Seventh Annual Structure in Complexity Theory Conference, pp. 255–259. IEEE Computer Society (1992). https://doi.org/10.1109/SCT.1992.215400, see also [21]

21. Book, R.V., Lutz, J.H.: On languages with very high space-bounded Kolmogorov complexity. SIAM J. Comput. **22**(2), 395–402 (1993). https://doi.org/10.1137/0222029

22. Buhrman, H., Fortnow, L., Laplante, S.: Resource-bounded Kolmogorov complexity revisited. SIAM J. Comput. **31**(3), 887–905 (2001). https://doi.org/10.1137/S009753979834388X

23. Buhrman, H., Mayordomo, E.: An excursion to the Kolmogorov random strings. JCSS **54**, 393–399 (1997). https://doi.org/10.1006/jcss.1997.1484

24. Daley, R.: On the inference of optimal descriptions. Theor. Comput. Sci. **4**(3), 301–319 (1977). https://doi.org/10.1016/0304-3975(77)90015-9

25. Downey, R., Hirschfeldt, D.: Algorithmic Randomness and Complexity. Springer, Heidelberg (2010). https://doi.org/10.1007/978-0-387-68441-3

26. Gavaldà, R., Torenvliet, L., Watanabe, O., Balcázar, J.L.: Generalized Kolmogorov complexity in relativized separations (extended abstract). In: Rovan, B. (ed.) MFCS 1990. LNCS, vol. 452, pp. 269–276. Springer, Heidelberg (1990). https://doi.org/10.1007/BFb0029618

27. Golovnev, A., Ilango, R., Impagliazzo, R., Kabanets, V., Kolokolova, A., Tal, A.: $AC^0[p]$ lower bounds against MCSP via the coin problem. In: 46th International Colloquium on Automata, Languages, and Programming, (ICALP). LIPIcs, vol. 132, pp. 66:1–66:15. Schloss Dagstuhl - Leibniz-Zentrum fuer Informatik (2019). https://doi.org/10.4230/LIPIcs.ICALP.2019.66

28. Hemachandra, L.A., Wechsung, G.: Using randomness to characterize the complexity of computation. In: Proceedings of the IFIP 11th World Computer Congress on Information Processing 1989, pp. 281–286. North-Holland/IFIP (1989), see also [29]

29. Hemachandra, L.A., Wechsung, G.: Kolmogorov characterizations of complexity classes. Theor. Comput. Sci. **83**(2), 313–322 (1991). https://doi.org/10.1016/0304-3975(91)90282-7

30. Hirahara, S.: Non-black-box worst-case to average-case reductions within NP. In: 59th IEEE Annual Symposium on Foundations of Computer Science (FOCS), pp. 247–258 (2018). https://doi.org/10.1109/FOCS.2018.00032

31. Hirahara, S., Santhanam, R.: On the average-case complexity of MCSP and its variants. In: 32nd Conference on Computational Complexity, CCC. LIPIcs, vol. 79, pp. 7:1–7:20. Schloss Dagstuhl - Leibniz-Zentrum fuer Informatik (2017). https://doi.org/10.4230/LIPIcs.CCC.2017.7

32. Hitchcock, J.M., Pavan, A.: On the NP-completeness of the minimum circuit size problem. In: Conference on Foundations of Software Technology and Theoretical Computer Science (FST&TCS). LIPIcs, vol. 45, pp. 236–245. Schloss Dagstuhl - Leibniz-Zentrum fuer Informatik (2015). https://doi.org/10.4230/LIPIcs.FSTTCS.2015.236

33. Hopcroft, J.E., Ullman, J.D.: Introduction to Automata Theory, Languages and Computation. Addison-Wesley, Boston (1979)

34. Ilango, R.: Approaching MCSP from above and below: Hardness for a conditional variant and $AC^0[p]$. In: 11th Innovations in Theoretical Computer Science Conference, ITCS. LIPIcs, vol. 151, pp. 34:1–34:26. Schloss Dagstuhl - Leibniz-Zentrum fuer Informatik (2020). https://doi.org/10.4230/LIPIcs.ITCS.2020.34

35. Ilango, R., Loff, B., Oliveira, I.C.: NP-hardness of minimizing circuits and communication (2019, manuscript)

36. Kabanets, V., Cai, J.Y.: Circuit minimization problem. In: ACM Symposium on Theory of Computing (STOC), pp. 73–79 (2000). https://doi.org/10.1145/335305.335314

37. Ko, K.: On the notion of infinite pseudorandom sequences. Theor. Comput. Sci. **48**(3), 9–33 (1986). https://doi.org/10.1016/0304-3975(86)90081-2

38. Ko, K.: On the complexity of learning minimum time-bounded Turing machines. In: Proceedings of the Third Annual Workshop on Computational Learning Theory, (COLT), pp. 82–96 (1990), see also [39]

39. Ko, K.: On the complexity of learning minimum time-bounded Turing machines. SIAM J. Comput. **20**(5), 962–986 (1991). https://doi.org/10.1137/0220059

40. Ko, K.-I., Orponen, P., Schöning, U., Watanabe, O.: What is a hard instance of a computational problem? In: Selman, A.L. (ed.) Structure in Complexity Theory. LNCS, vol. 223, pp. 197–217. Springer, Heidelberg (1986). https://doi.org/10.1007/3-540-16486-3_99. See also [51]

41. Kolmogorov, A.N.: Three approaches to the quantitative definition ofinformation'. Probl. Inf. Transm. **1**(1), 1–7 (1965)

42. Levin, L.: Universal search problems. Probl. Inf. Transm. **9**, 265–266 (1973)

43. Levin, L.A.: Randomness conservation inequalities; information and independence in mathematical theories. Inf. Control **61**(1), 15–37 (1984). https://doi.org/10.1016/S0019-9958(84)80060-1

44. Li, M., Vitanyi, P.M.B.: An Introduction to Kolmogorov Complexity and Its Applications. Texts in Computer Science, 4th edn. Springer, Heidelberg (2019). https://doi.org/10.1007/978-3-030-11298-1

45. McKay, D.M., Murray, C.D., Williams, R.R.: Weak lower bounds on resource-bounded compression imply strong separations of complexity classes. In: Proceedings of the 51st Annual ACM SIGACT Symposium on Theory of Computing (STOC), pp. 1215–1225 (2019). https://doi.org/10.1145/3313276.3316396

46. Meyer, A., McCreight, E.: Computationally complex and pseudo-random zero-one valued functions. In: Theory of Machines and Computations, pp. 19–42. Elsevier (1971)

47. Murray, C., Williams, R.: On the (non) NP-hardness of computing circuit complexity. Theory Comput. **13**(4), 1–22 (2017). https://doi.org/10.4086/toc.2017.v013a004

48. Oliveira, I., Santhanam, R.: Conspiracies between learning algorithms, circuit lower bounds and pseudorandomness. In: 32nd Conference on Computational Complexity, CCC. LIPIcs, vol. 79, pp. 18:1–18:49. Schloss Dagstuhl - Leibniz-Zentrum fuer Informatik (2017). https://doi.org/10.4230/LIPIcs.CCC.2017.18

49. Oliveira, I.C., Pich, J., Santhanam, R.: Hardness magnification near state-of-the-art lower bounds. In: 34th Computational Complexity Conference (CCC). LIPIcs, vol. 137, pp. 27:1–27:29. Schloss Dagstuhl - Leibniz-Zentrum fuer Informatik (2019). https://doi.org/10.4230/LIPIcs.CCC.2019.27

50. Oliveira, I.C., Santhanam, R.: Hardness magnification for natural problems. In: 59th IEEE Annual Symposium on Foundations of Computer Science (FOCS), pp. 65–76 (2018). https://doi.org/10.1109/FOCS.2018.00016

51. Orponen, P., Ko, K., Schoning, U., Watanabe, O.: Instance complexity. J. ACM **41**(1), 96–121 (1994). https://doi.org/10.1145/174644.174648

52. Paul, W.J., Seiferas, J.I., Simon, J.: An information-theoretic approach to time bounds for on-line computation (preliminary version). In: Proceedings of the Twelfth Annual ACM Symposium on Theory of Computing, STOC 1980, pp. 357–367. ACM, New York (1980). https://doi.org/10.1145/800141.804685, see also [53]

53. Paul, W.J., Seiferas, J.I., Simon, J.: An information-theoretic approach to time bounds for on-line computation. J. Comput. Syst. Sci. **23**(2), 108–126 (1981). https://doi.org/10.1016/0022-0000(81)90009-X

54. Peterson, G.L.: Succinct representation, random strings, and complexity classes. In: 21st Annual Symposium on Foundations of Computer Science (FOCS), pp. 86–95 (1980). https://doi.org/10.1109/SFCS.1980.42

55. Pitt, L., Valiant, L.G.: Computational limitations on learning from examples. J. ACM **35**(4), 965–984 (1988). https://doi.org/10.1145/48014.63140

56. Pitt, L., Warmuth, M.K.: The minimum consistent DFA problem cannot be approximated within any polynomial. J. ACM **40**(1), 95–142 (1993). https://doi.org/10.1145/138027.138042

57. Rudow, M.: Discréte logarithm and minimum circuit size. Inf. Process. Lett. **128**, 1–4 (2017). https://doi.org/10.1016/j.ipl.2017.07.005
58. Sipser, M.: A complexity theoretic approach to randomness. In: Proceedings of the 15th Annual ACM Symposium on Theory of Computing (STOC), pp. 330–335 (1983). https://doi.org/10.1145/800061.808762
59. Trakhtenbrot, B.A.: A survey of Russian approaches to perebor (brute-force searches) algorithms. IEEE Ann. Hist. Comput. **6**(4), 384–400 (1984)
60. Watanabe, O.: Kolmogorov Complexity and Computational Complexity, 1st edn. Springer, Heidelberg (2012)
61. Wilber, R.E.: Randomness and the density of hard problems. In: 24th Annual Symposium on Foundations of Computer Science (FOCS), pp. 335–342 (1983). https://doi.org/10.1109/SFCS.1983.49
62. Yao, A.C.: Theory and applications of trapdoor functions (extended abstract). In: 23rd Annual Symposium on Foundations of Computer Science (FOCS), pp. 80–91 (1982). https://doi.org/10.1109/SFCS.1982.45

The Power of Self-Reducibility: Selectivity, Information, and Approximation

Lane A. Hemaspaandra

Department of Computer Science, University of Rochester,
Rochester, NY 14627, USA
http://www.cs.rochester.edu/u/lane/

*In memory of Ker-I Ko, whose indelible
contributions to computational complexity
included important work (e.g., [24–27]) on
each of this chapter's topics:
self-reducibility, selectivity, information,
and approximation.*

Abstract. This chapter provides a hands-on tutorial on the important technique known as self-reducibility. Through a series of "Challenge
Problems" that are theorems that the reader will—after being given definitions and tools—try to prove, the tutorial will ask the reader not to read
proofs that use self-reducibility, but rather to *discover* proofs that use
self-reducibility. In particular, the chapter will seek to guide the reader to
the discovery of proofs of four interesting theorems—whose focus areas
range from selectivity to information to approximation—from the literature, whose proofs draw on self-reducibility.

The chapter's goal is to allow interested readers to add self-reducibility
to their collection of proof tools. The chapter simultaneously has a related
but different goal, namely, to provide a "lesson plan" (and a coordinated
set of slides is available online to support this use [13]) for a lecture to
a two-lecture series that can be given to undergraduate students—even
those with no background other than basic discrete mathematics and
an understanding of what polynomial-time computation is—to immerse
them in hands-on proving, and by doing that, to serve as an invitation to
them to take courses on Models of Computation or Complexity Theory.

Keywords: Computational and structural complexity theory ·
Enumerative counting · P-selectivity · Self-reducibility · Sparse sets

---

This chapter was written in part while on sabbatical at Heinrich Heine University
Düsseldorf, supported in part by a Renewed Research Stay grant from the Alexander
von Humboldt Foundation.

D.-Z. Du and J. Wang (Eds.): Ko Festschrift, LNCS 12000, pp. 19–47, 2020.
https://doi.org/10.1007/978-3-030-41672-0_3

1 Introduction

Section 1.1 explains the two quite different audiences that this chapter is intended for, and for each describes how that group might use the chapter. If you're not a computer science professor it would make sense to skip Sect. 1.1, and if you are a computer science professor you might at least on a first reading choose to skip Sect. 1.1.

Section 1.2 introduces the type of self-reducibility that this chapter will focus on, and the chapter's central set, SAT (the satisfiability problem for propositional Boolean formulas).

1.1 A Note on the Two Audiences, and How to Read This Chapter

This chapter is unusual in that it has two intended audiences, and those audiences differ dramatically in their amounts of background in theoretical computer science.

For Those Unfamiliar with Complexity Theory. The main intended audience is those—most especially young students—who are not yet familiar with complexity theory, or perhaps have not even yet taken a models of computation course. If that describes you, then this chapter is intended to be a few-hour tutorial immersion in—and invitation to—the world of theoretical computer science research. As you go through this tutorial, you'll try to solve—hands-on—research issues that are sufficiently important that their original solutions appeared in some of theoretical computer science's best conferences and journals.

You'll be given the definitions and some tools before being asked to try your hand at finding a solution to a problem. And with luck, for at least a few of our four challenge problems, you will find a solution to the problem. Even if you don't find a solution for the given problem—and the problems increase in difficulty and especially the later ones require bold, flexible exploration to find possible paths to the solution—the fact that you have spent time trying to solve the problem will give you more insight into the solution when the solution is then presented in this chapter.

A big-picture goal here is to make it clear that doing theoretical computer science research is often about playful, creative, flexible puzzle-solving. The underlying hope here is that many people who thought that theoretical computer science was intimidating and something that they could never do or even understand will realize that they can do theoretical computer science and perhaps even that they (gasp!) *enjoy* doing theoretical computer science.

The four problems also are tacitly bringing out a different issue, one more specifically about complexity. Most people, and even most computer science professors, think that complexity theory is extraordinarily abstract and hard to grasp. Yet in each of our four challenge problems, we'll see that doing complexity is often extremely concrete, and in fact is about building a program that solves a given problem. Building programs is something that many people already have done, e.g., anyone who has taken an introduction to programming course or a data structures course. The only difference in the programs one builds when doing proofs in complexity theory is that the programs one builds typically draw on some hypothesis that provides a piece of the program's action. For example, our fourth challenge problem will be to show that if a certain problem is easy to approximate, then it can be solved exactly. So your task, when solving it, will be to write a program that exactly solves the problem. But in writing your program you will assume that you have as a black box that you can draw on as a program (a subroutine) that given an instance of the problem gives an approximate solution.

This view that complexity is largely about something that is quite concrete, namely building programs, in fact is the basis of an entire graduate-level complexity-theory textbook [18], in which the situation is described as follows:

> Most people view complexity theory as an arcane realm populated by pointy-hatted (if not indeed pointy-headed) sorcerers stirring cauldrons of recursion theory with wands of combinatorics, while chanting incantations involving complexity classes whose very names contain hundreds of characters and sear the tongues of mere mortals. This stereotype has sprung up in part due to the small amount of esoteric research that fits this bill, but the stereotype is more strongly attributable to the failure of complexity theorists to communicate in expository forums the central role that algorithms play in complexity theory.

Expected Background. To keep this chapter as accessible as possible, the amount of expected background has been kept quite small. But there are some types of background that are being assumed here. The reader is assumed to know following material, which would typically be learned within about the first two courses of most computer science departments' introductory course sequences.

1. What a polynomial is.
   As an example, $p(n) = n^{1492} + 42n^{42} + 13$ is a polynomial; $e(n) = 2^n$ is not.
2. What it means for a set or function to be computable in polynomial time, i.e., to be computed in time polynomial in the number of bits in the input to the problem. The class of all sets that can be computed in polynomial time is denoted P, and is one of the most important classes in computer science.
   As an example, the set of all positive integers that are multiples of 10 is a set that belongs to P.
3. Some basics of logic such as the meaning of quantifiers ($\exists$ and $\forall$) and what a (propositional) Boolean formula is.
   As an example of the latter, the formula $x_1 \wedge (x_2 \vee \overline{x_3})$ is a such a formula, and evaluates as True—with each of x_1, x_2, and x_3 being variables whose potential values are True or False—exactly if x_1 is True and either x_2 is True or the negation of x_3 is True.

If you have that background in hand, wonderful! You have the background to tackle this chapter's puzzles and challenges, and please (unless you're a professor thinking of modeling a lecture series on this chapter) skip from here right on to Sect. 1.2.

For Computer Science Professors. Precisely because this chapter is designed to be accessible and fun for students who don't have a background in theoretical computer science, the chapter avoids—until Sect. 7—trying to abstract away from the focus on SAT. In particular, this chapter either avoids mentioning complexity class names such as NP, coNP, and PSPACE, or at least, when it does mention them, uses phrasings such as "classes known as" to make clear that the reader is not expected to possess that knowledge.

Despite that, computer science professors are very much an intended audience for this chapter, though in a way that is reflecting the fact that the real target audience for these challenges is young students. In particular, in addition to providing a tutorial introduction for students, of the flavor described in Sect. 1.1, this chapter also has as its goal to provide to you, as a teacher, a "lesson plan" to help you offer in your course a one- or two-day lecture (but really hands-on workshop) sequence[1] in which you present the definitions and tools of the first of these problems, and then ask the students to break into groups and in groups spend about 10–25 minutes working on solving the problem,[2] and then you ask whether some group has found a solution and would like to present it to the class, and if so you and the class listen to and if needed correct the solution (and if no group found a solution, you and the class will work together to reach a solution). And then you go on to similarly treat the other three problems, again with the class working in teams. This provides students with a hands-on immersion in team-based, on-the-spot theorem-proving—something that most students never get in class. I've done this in classes of size up to 79 students, and they love it. The approach does not treat them as receptors of information lectured at them, but rather shows them that they too can make discoveries—even ones that when first obtained appeared in such top forums as *CCC* (the yearly *Computational Complexity Conference*), *ICALP* (the yearly *International Colloquium on Automata, Languages, and Programming*), the journal *Information and Computation*, and *SIAM Journal on Computing*.

---

[1] To cover all four problems would take two class sessions. Covering just the first two or perhaps three of the problems could be done in a single 75-minute class session.

[2] In this chapter, since student readers of the chapter will be working as individuals, I suggest to the reader, for most of the problems, longer amounts of time. But in a classroom setting where students are working in groups, 10–25 minutes may be an appropriate amount of time; perhaps 10 minutes for the first challenge problem, 15 for the second, 15 for the third, and 25 for the fourth. You'll need to yourself judge the time amounts that are best, based on your knowledge of your students. For many classes, the just-mentioned times will not be enough. Myself, I try to keep track of whether the groups seem to have found an answer, and I will sometimes stretch out the time window if many groups seem to be still working intensely and with interest. Also, if TAs happen to be available who don't already know the answers, I may assign them to groups so that the class's groups will have more experienced members, though the TAs do know to guide rather than dominate a group's discussions.

To support this use of the chapter as a teaching tool in class, I have made publicly available a set of LaTeX/Beamer slides that can be used for a one- or two-class hands-on workshop series on this chapter. The slides are available online [13], both as pdf slides and, for teachers who might wish to modify the slides, as a zip archive of the source files.

Since the slides might be used in courses where students already do know of such classes as NP and coNP, the slides don't defer the use of those classes as aggressively as this chapter itself does. But the slides are designed so that the mention of the connections to those classes is parenthetical (literally so—typically a parenthetical, at the end of a theorem statement, noting the more general application of the claim to all of NP or all of coNP), and those parentheticals can be simply skipped over. Also, the slides define on the fly both NP and coNP, so that if you do wish to cover the more general versions of the theorems, the slides will support that too.

The slides don't themselves present the solutions to Challenge Problems 1, 2, or 3. Rather, they assume that one of the class's groups will present a solution on the board (or document camera) or will speak the solution with the professor acting as a scribe at the board or the document camera. Challenge Problems 1, 2, and 3 are doable enough that usually at least one group will either have solved the question, or at least will made enough progress that, with some help from classmates or some hints/help from the professor, a solution can quickly be reached building on the group's work. (The professor ideally should have read the solutions in this chapter to those problems, so that even if a solution isn't reached or almost reached by the students on one or two of those problems, the professor can provide a solution at the board or document camera. However, in the ideal case, the solutions of those problems will be heavily student-driven and won't need much, if any, professorial steering.)

Challenge Problem 4 is sufficiently hard that the slides do include both a slide explaining why a certain very natural approach—which is the one students often (quite reasonably) try to make work—cannot possibly work, and thus why the approach that the slides gave to students as a gentle, oblique hint may be the way to go, and then the slides present a solution along those lines.

The difficulty of Challenge Problem 4 has a point. Although this chapter is trying to show students that they *can* do theory research, there is also an obligation not to give an artificial sense that all problems are easily solved. Challenge Problem 4 shows students that some problems can have multiple twists and turns on the way to finding a solution. Ideally, the students won't be put off by this, but rather will appreciate both that solving problems is something they can do, and that in doing so one may well run into obstacles that will take some out-of-the-box thinking to try to get around—obstacles that might take not minutes of thought but rather hours or days or more, as well as working closely with others to share ideas as to what might work.

1.2 Self-Reducibility and SAT

Now that you have read whatever part of Sect. 1.1 applied to you, to get the lay of the land as to what this chapter is trying to provide you, let us discuss the approach that will be our lodestar throughout this chapter.

One of the most central paradigms of computer science is "divide and conquer." Some of the most powerful realizations of that approach occur though the use of what is known as self-reducibility, which is our chapter's central focus.

Loosely put, a set is self-reducible if any membership question regarding the set can be easily resolved by asking (perhaps more than one) membership questions about smaller strings.

That certainly divides, but does it conquer?

The answer varies greatly depending on the setting. Self-reducibility itself, depending on which polynomial-time variant one is looking at, gives upper bounds on a set's complexity. However, those bounds—which in some cases are the complexity classes known as NP and PSPACE—are nowhere near to putting the set into deterministic polynomial time (aka, P).

The magic of self-reducibility, however, comes when one adds another ingredient to one's stew. Often, one can prove that if a set is self-reducible *and has some other property regarding its structure*, then the set *is* feasible, i.e., is in deterministic polynomial time.

This tutorial will ask the reader to—and help the reader to—discover for him- or herself the famous proofs of three such magic cases (due to Selman, Berman, and Fortune), and then of a fourth case that is about the "counting" analogue of what was described in the previous paragraph.

Beyond that, I hope you'll keep the tool/technique of self-reducibility in mind for the rest of your year, decade, and lifetime—and on each new challenge will spend at least a few moments asking, "Can self-reducibility play a helpful role in my study of this problem?" And with luck, sooner or later, the answer may be, "Yes! Goodness... what a surprise!"

Throughout this chapter, our model language (set) will be "SAT," i.e., the question of whether a given Boolean formula, for some way of assigning each of its variables to True or to False, evaluates to True. SAT is a central problem in computer science, and possesses a strong form of self-reducibility. As a quiet bonus, though we won't focus on this in our main traversal of the problems and their solutions, SAT has certain "completeness" properties that make results proven about SAT often yield results for an entire important class of problems known as the "NP-complete" sets; for those interested in that, Sect. 7, "Going Big: Complexity-Class Implications," briefly covers that broader view.

2 Definitions Used Throughout: SAT and Self-Reducibility

The game plan of this chapter, as mentioned above, is this: For each of the four challenge problems (theorems), you will be given definitions and some other background or tools. Then the challenge problem (theorem) will be stated, and you'll be asked to try to solve it, that is, you'll be asked to prove the theorem. After you do, or after you hit a wall so completely that you feel you can't solve the theorem even with additional time, you'll read a proof of the theorem. Each of the four challenge problems has an appendix presenting a proof of the result.

But before we start on the problems, we need to define SAT and discuss its self-reducibility.

Definition 1. SAT *is the set of all satisfiable (propositional) Boolean formulas.*

Example 1. 1. $x \wedge \overline{x} \notin$ SAT, since neither of the two possible assignments to x causes the formula to evaluate to True.
2. $(x_1 \wedge x_2 \wedge \overline{x_3}) \vee (x_4 \wedge \overline{x_4}) \in$ SAT, since that formula evaluates to True under at least one of the eight possible ways that the four variables can each be assigned to be True or False. For example, when we take $x_1 = x_2 = x_4 =$ True and $x_3 =$ False, the formula evaluates to True.

SAT has the following "divide and conquer" property.

Fact 1 (2-disjunctive length-decreasing self-reducibility). *Let $k \geq 1$. Let $F(x_1, x_2, \ldots, x_k)$ be a Boolean formula (without loss of generality, assume that each of the variables actually occurs in the formula). Then*

$$F(x_1, x_2, \ldots, x_k) \in \text{SAT} \iff$$
$$\left(F(\text{True}, x_2, \ldots, x_k) \in \text{SAT} \vee F(\text{False}, x_2, \ldots, x_k) \in \text{SAT}\right).$$

The above fact says that *SAT is self-reducible* (in particular, in the lingo, it says that SAT is 2-disjunctive length-decreasing self-reducible).

Note: We won't at all focus in this chapter on details of the encoding of formulas and other objects. That indeed is an issue if one wants to do an utterly detailed discussion/proof. But the issue is not a particularly interesting one, and certainly can be put to the side during a first traversal, such as that which this chapter is inviting you to make.

A Bit of History. In this chapter, we typically won't focus much on references. It is best to immerse oneself in the challenges, without getting overwhelmed with a lot of detailed historical context. However, so that those who are interested in history can have some sense of the history, and so that those who invented the concepts and proved the theorems are property credited, we will for most sections have an "A Bit of History" paragraph that extremely briefly gives literature citations and sometimes a bit of history and context. As to the present section, self-reducibility dates back to the 1970s, and in particular is due to the work of Schnorr [34] and Meyer and Paterson [33].

3   Challenge Problem 1: Is SAT Even *Semi*-feasible?

Pretty much no one believes that SAT has a polynomial-time decision algorithm, i.e., that SAT $\in$ P [8]. This section asks you to show that it even is unlikely that SAT has a polynomial-time *semi-decision* algorithm—a polynomial-time algorithm that, given any two formulas, always outputs one of them and does so in such a way that if at least one of the input formulas is satisfiable then the formula that is output is satisfiable.

3.1   Needed Definitions

A set L is said to be feasible (in the sense of belonging to P) if there is a polynomial-time algorithm that decides membership in L.

A set is said to be semi-feasible (aka P-selective) if there is a polynomial-time algorithm that semi-decides membership, i.e., that given any two strings, outputs one that is "more likely" to be in the set (to be formally cleaner, since the probabilities are all 0 and 1 and can tie, what is really meant is "no less likely" to be in the set). The following definition makes this formal. (Here and elsewhere, Σ will denote our (finite) alphabet and Σ^* will denote the set of finite strings over that alphabet.)

Definition 2. *A set L is P-selective if there exists a polynomial-time function, $f : \Sigma^* \times \Sigma^* \to \Sigma^*$ such that,*

$$(\forall a, b \in \Sigma^*)[f(a,b) \in \{a,b\} \wedge (\{a,b\} \cap L \neq \emptyset \implies f(a,b) \in L)].$$

It is known that some P-selective sets can be very hard. Some even have the property known as being "undecidable." Despite that, our first challenge problem is to prove that SAT cannot be P-selective unless SAT is outright easy computationally, i.e., SAT $\in$ P. Since it is close to an article of faith in computer science that SAT $\notin$ P, showing that some hypothesis implies that SAT $\in$ P is considered, with the full weight of modern computer science's current understanding and intuition, to be extremely strong evidence that the hypothesis is unlikely to be true. (In the lingo, the hypothesis is implying that P = NP. Although it is possible that P = NP is true, basically no one believes that it is [8]. However, the issue is the most important open issue in applied mathematics, and there is currently a \$1,000,000 prize for whoever resolves the issue [5].)

A Bit of History. Inspired by an analogue in computability theory, P-selectivity was defined by Selman in a seminal series of papers [35–38], which included a proof of our first challenge theorem. The fact, alluded to above, that P-selective sets can be very hard is due to Alan L. Selman's above work and to the work of the researcher in whose memory this chapter is written, Ker-I Ko [24]. In that same paper, Ko also did very important early work showing that P-selective sets are unlikely to have what is known as "small circuits." For those particularly interested in the P-selective sets, they and their nondeterministic cousins are the subject of a book, *Theory of Semi-feasible Algorithms* [21].

3.2   Can SAT Be P-Selective?

Challenge Problem 1. *(Prove that) if* SAT *is* P*-selective, then* SAT $\in$ P.

Keep in mind that what you should be trying to do is this. You may assume that SAT is P-selective. So you may act as if you have in hand a polynomial-time computable function, f, that in the sense of Definition 2 shows that SAT is P-selective. And your task is to give a polynomial-time algorithm for SAT, i.e., a program that in time polynomial in the number of bits in its input determines whether the input string belongs to SAT. (Your algorithm surely will be making calls to f—possibly quite a few calls.)

So that you have them easily at hand while working on this, here are some of the key definitions and tools that you might want to draw on while trying to prove this theorem.

SAT. SAT is the set of all satisfiable (propositional) Boolean formulas.

Self-Reducibility. Let $k \geq 1$. Let $F(x_1, x_2, \ldots, x_k)$ be a Boolean formula (without loss of generality assume that each of the variables occurs in the formula). Then $F(x_1, x_2, \ldots, x_k) \in$ SAT $\iff$ ($F(\text{True}, x_2, \ldots, x_k) \in$ SAT $\lor$ $F(\text{False}, x_2, \ldots, x_k) \in$ SAT).

P-Selectivity. A set L is P-selective if there exists a polynomial-time function, $f : \Sigma^* \times \Sigma^* \to \Sigma^*$ such that, $(\forall a, b \in \Sigma^*)[f(a,b) \in \{a,b\} \land (\{a,b\} \cap L \neq \emptyset \implies f(a,b) \in L)]$.

My suggestion to you would be to work on proving this theorem until either you find a proof, or you've put in at least 20 minutes of thought, are stuck, and don't think that more time will be helpful.

When you've reached one or the other of those states, please go on to Appendix A to read a proof of this theorem. Having that proof will help you check whether your proof (if you found one) is correct, and if you did not find a proof, will show you a proof. Knowing the answer to this challenge problem before going on to the other three challenge problems is important, since an aspect of this problem's solution will show up in the solutions to the other challenge problems.

I've put the solutions in separate appendix sections so that you can avoid accidentally seeing them before you wish to. But please do (unless you are completely certain that your solution to the first problem is correct) read the solution for the first problem before moving on to the second problem. If you did not find a proof for this first challenge problem, don't feel bad; everyone has days when we see things and days when we don't. On the other hand, if you did find a proof of this first challenge theorem, wonderful, and if you felt that it was easy, well, the four problems get steadily harder, until by the fourth problem almost anyone would have to work very, very hard and be a bit lucky to find a solution.

4   Challenge Problem 2: Low Information Content and SAT, Part 1: Can SAT Reduce to a Tally Set?

Can SAT have low information content? To answer that, one needs to formalize what notion of low information content one wishes to study. There are many such notions, but a particularly important one is whether a given set can "many-one polynomial-time reduce" to a tally set (a set over a 1-letter alphabet).

A Bit of History. Our second challenge theorem was stated and proved by Berman [2]. Berman's paper started a remarkably long and productive line of work, which we will discuss in more detail in the "A Bit of History" note accompanying the third challenge problem. That same note will provide pointers to surveys of that line of work, for those interested in additional reading.

4.1   Needed Definitions

ϵ will denote the empty string.

Definition 3. *A set T is a tally set if $T \subseteq \{\epsilon, 0, 00, 000, \dots\}$.*

Definition 4. *We say that $A \leq_m^p B$ (A many-one polynomial-time reduces to B) if there is a polynomial-time computable function g such that*

$$(\forall x \in \Sigma^*)[x \in A \iff g(x) \in B].$$

Informally, this says that B is so powerful that each membership query to A can be efficiently transformed into a membership query to B that gets the same answer as would the question regarding membership in A.

4.2   Can SAT Reduce to a Tally Set?

Challenge Problem 2. *(Prove that) if there exists a tally set T such that* SAT $\leq_m^p T$, *then* SAT $\in$ P.

Keep in mind that what you should be trying to do is this. You may assume that there exists a tally set T such that SAT $\leq_m^p T$. You may not assume that you have a polynomial-time algorithm for T; you are assuming that T exists, but for all we know, T might well be very hard. On the other hand, you *may* assume that you have in hand a polynomial-time computable function g that reduces from SAT to T in the sense of Definition 4. (After all, that reduction is (if it exists) a finite-sized program.) Your task here is to give a polynomial-time algorithm for SAT, i.e., a program that in time polynomial in the number of bits in its input determines whether the input string belongs to SAT. (Your algorithm surely will be making calls to g—possibly quite a few calls.)

So that you have them easily at hand while working on this, here are some of the key definitions and tools that you might want to draw on while trying to prove this theorem.

SAT. SAT is the set of all satisfiable (propositional) Boolean formulas.

Self-Reducibility. Let $k \geq 1$. Let $F(x_1, x_2, \ldots, x_k)$ be a Boolean formula (without loss of generality assume that each of the x_i actually occurs in the formula). Then $F(x_1, x_2, \ldots, x_k) \in \text{SAT} \iff (F(\text{True}, x_2, \ldots, x_k) \in \text{SAT} \vee F(\text{False}, x_2, \ldots, x_k) \in \text{SAT})$.

Tally Sets. A set T is a tally set if $T \subseteq \{\epsilon, 0, 00, 000, \ldots\}$.

Many-One Reductions. We say that $A \leq_m^p B$ if there is a polynomial-time computable function g such that $(\forall x \in \Sigma^*)[x \in A \iff g(x) \in B]$.

My suggestion to you would be to work on proving this theorem until either you find a proof, or you've put in at least 30 minutes of thought, are stuck, and don't think that more time will be helpful.

When you've reached one or the other of those states, please go on to Appendix B to read a proof of this theorem. The solution to the third challenge problem is an extension of this problem's solution, so knowing the answer to this challenge problem before going on to the third challenge problem is important.

5   Challenge Problem 3: Low Information Content and SAT, Part 2: Can $\overline{\text{SAT}}$ Reduce to a Sparse Set?

This problem challenges you to show that even a class of sets that is far broader than the tally sets, namely, the so-called sparse sets, cannot be reduced to from $\overline{\text{SAT}}$ unless SAT $\in$ P.

A Bit of History. This third challenge problem was stated and proved by Fortune [7]. It was another step in what was a long line of advances—employing more and more creative and sometimes difficult proofs—that eventually led to the understanding that, unless SAT $\in$ P, no sparse set can be hard for SAT even with respect to extremely flexible types of reductions. The most famous result within this line is known as Mahaney's Theorem: If there is a sparse set S such that SAT $\leq_m^p S$, then SAT $\in$ P [30]. There are many surveys of the just-mentioned line of work, e.g., [11,31,32,41]. The currently strongest result in that line is due to Glaßer [9] (see the survey/treatment of that in [10], and see also the results of Arvind et al. [1]).

5.1   Needed Definitions

Let $\|S\|$ denote the cardinality of set S, e.g., $\|\{\epsilon, 0, 0, 0, 00\}\| = 3$.

For any set L, let $\overline{L}$ denote the complement of L.

Let $|x|$ denote the length string x, e.g., $|\text{moon}| = 4$.

Definition 5. *A set S is sparse if there exists a polynomial q such that, for each natural number $n \in \{0, 1, 2, \ldots\}$, it holds that*

$$\|\{x \mid x \in S \wedge |x| \leq n\}\| \leq q(n).$$

Informally put, the sparse sets are the sets whose number of strings up to a given length is at most polynomial. $\{0,1\}^*$ is, for example, not a sparse set, since up to length n it has $2^{n+1} - 1$ strings. But all tally sets are sparse, indeed all via the bounding polynomial $q(n) = n + 1$.

5.2   Can $\overline{\text{SAT}}$ Reduce to a Sparse Set?

Challenge Problem 3. *(Prove that) if there exists a sparse set S such that* $\overline{\text{SAT}} \leq_m^p S$, *then* $\text{SAT} \in \text{P}$.

Keep in mind that what you should be trying to do is this. You may assume that there exists a sparse set S such that $\overline{\text{SAT}} \leq_m^p S$. You may not assume that you have a polynomial-time algorithm for S; you are assuming that S exists, but for all we know, S might well be very hard. On the other hand, you *may* assume that you have in hand a polynomial-time computable function g that reduces from $\overline{\text{SAT}}$ to S in the sense of Definition 4. (After all, that reduction is—if it exists—a finite-sized program.) And you may assume that you have in hand a polynomial that upper-bounds the sparseness of S in the sense of Definition 5. (After all, one of the countably infinite list of simple polynomials $n^k + k$—for $k = 1, 2, 3, \ldots$—will provide such an upper bound, if any such polynomial upper bound exists.) Your task here is to give a polynomial-time algorithm for SAT, i.e., a program that in time polynomial in the number of bits in its input determines whether the input string belongs to SAT. (Your algorithm surely will be making calls to g—possibly quite a few calls.)

One might wonder why I just said that you should build a polynomial-time algorithm for SAT, given that the theorem speaks of $\overline{\text{SAT}}$. However, since it is clear that $\text{SAT} \in \text{P} \iff \overline{\text{SAT}} \in \text{P}$ (namely, given a polynomial-time algorithm for SAT, if we simply reverse the answer on each input, then we now have a polynomial-time algorithm for $\overline{\text{SAT}}$), it is legal to focus on SAT—and most people find doing so more natural and intuitive.

Do be careful here. Solving this challenge problem may take an "aha!... insight" moment. Knowing the solution to Challenge Problem 2 will be a help here, but even with that knowledge in hand one hits an obstacle. And then the challenge is to find a way around that obstacle.

So that you have them easily at hand while working on this, here are some of the key definitions and tools that you might want to draw on while trying to prove this theorem.

SAT and $\overline{\text{SAT}}$. SAT is the set of all satisfiable (propositional) Boolean formulas. $\overline{\text{SAT}}$ denotes the complement of SAT.

Self-Reducibility. Let $k \geq 1$. Let $F(x_1, x_2, \ldots, x_k)$ be a Boolean formula (without loss of generality assume that each of the x_i actually occurs in the formula). Then $F(x_1, x_2, \ldots, x_k) \in \text{SAT} \iff (F(\text{True}, x_2, \ldots, x_k) \in \text{SAT} \vee F(\text{False}, x_2, \ldots, x_k) \in \text{SAT})$.

Sparse Sets. A set S is sparse if there exists a polynomial q such that, for each natural number $n \in \{0, 1, 2, \ldots\}$, it holds that $\|\{x \mid x \in S \wedge |x| \leq n\}\| \leq q(n)$.

Many-One Reductions. We say that $A \leq_m^p B$ if there is a polynomial-time computable function g such that $(\forall x \in \Sigma^*)[x \in A \iff g(x) \in B]$.

My suggestion to you would be to work on proving this theorem until either you find a proof, or you've put in at least 40 minutes of thought, are stuck, and don't think that more time will be helpful.

When you've reached one or the other of those states, please go on to Appendix C to read a proof of this theorem.

6   Challenge Problem 4: Is #SAT as Hard to (Enumeratively) Approximate as It Is to Solve Exactly?

This final challenge is harder than the three previous ones. To solve it, you'll have to have multiple insights—as to what approach to use, what building blocks to use, and how to use them.

The problem is sufficiently hard that the solution is structured to give you, if you did not solve the problem already, a second bite at the apple! That is, the solution—after discussing why the problem can be hard to solve—gives a very big hint, and then invites you to re-try to problem with that hint in hand.

A Bit of History. The function #SAT, the counting version of SAT, will play a central role in this challenge problem. #SAT was introduced and studied by Valiant [39,40]. This final challenge problem, its proof (including the lemma given in the solution I give here and the proof of that lemma), and the notion of enumerators and enumerative approximation are due to Cai and Hemachandra [3]. The challenge problem is a weaker version of the main result of that paper, which proves the result for not just 2-enumerators but even for sublinear-enumerators; later work showed that the result even holds for all polynomial-time computable enumerators [4].

6.1   Needed Definitions

$|F|$ will denote the length of (the encoding of) formula F.

#SAT is the function that given as input a Boolean formula $F(x_1, x_2, \ldots, x_k)$—without loss of generality assume that each of the variables occurs in F—outputs the number of satisfying assignments the formula has (i.e., of the 2^k possible assignments of the variables to True/False, the number of those under which F evaluates to True; so the output will be a natural number in the interval $[0, 2^k]$). For example, #SAT$(x_1 \vee x_2) = 3$ and #SAT$(x_1 \wedge \overline{x_1}) = 0$.

Definition 6. *We say that #SAT has a polynomial-time 2-enumerator (aka, is polynomial-time 2-enumerably approximable) if there is a polynomial-time computable function h such that on each input x,*

1. *$h(x)$ outputs a list of two (perhaps identical) natural numbers, and*
2. *#SAT(x) appears in the list output by $h(x)$.*

So a 2-enumerator h outputs a list of (at most) two candidate values for the value of #SAT on the given input, and the actual output is always somewhere in that list. This notion generalizes in the natural way to other list cardinalities, e.g., 1492-enumerators and, for each $k \in \{1, 2, 3, \dots\}$, $\max(1, |F|^k)$-enumerators.

6.2 Food for Thought

You'll certainly want to use some analogue of the key self-reducibility observation, except now respun by you to be about the number of solutions of a formula and how it relates to or is determined by the number of solutions of its two "child" formulas.

But doing that is just the first step your quest. So... please play around with ideas and approaches. Don't be afraid to be bold and ambitious. For example, you might say "Hmmmm, if we could do/build XYZ (where perhaps XYZ might be some particular insight about combining formulas), that would be a powerful tool in solving this, and I suspect we can do/build XYZ." And then you might want to work both on building XYZ and on showing in detail how, if you did have tool XYZ in hand, you could use it to show the theorem.

6.3 Is #SAT as Hard to (Enumeratively) Approximate as It Is to Solve Exactly?

Challenge Problem 4 (Cai and Hemachandra). *(Prove that) if #SAT has a polynomial-time 2-enumerator, then there is a polynomial-time algorithm for #SAT.*

Keep in mind that what you should be trying to do is this. You may assume that you have in hand a polynomial-time 2-enumerator for #SAT. Your task here is to give a polynomial-time algorithm for #SAT, i.e., a program that in time polynomial in the number of bits in its input determines the number of satisfying assignments of the (formula encoded by the) input string. (Your algorithm surely will be making calls to the 2-enumerator—possibly quite a few calls.)

Do be careful here. Proving this may take about three "aha!... insight" moments; Sect. 6.2 gave slight hints regarding two of those.

So that you have them easily at hand while working on this, here are some of the key definitions and tools that you might want to draw on while trying to prove this theorem.

#SAT. #SAT is the function that given as input a Boolean formula $F(x_1, x_2, \dots, x_k)$—without loss of generality assume that each of the variables occurs in F—outputs the number of satisfying assignments the formula has (i.e., of the 2^k possible assignments of the variables to True/False, the number of those under which F evaluates to True; so the output will be a natural number in the interval $[0, 2^k]$). For example, $\#\mathrm{SAT}(x_1 \vee x_2) = 3$ and $\#\mathrm{SAT}(x_1 \wedge \overline{x_1}) = 0$.

Enumerative Approximation. We say that #SAT has a polynomial-time 2-enumerator (aka, is polynomial-time 2-enumerably approximable) if there is a polynomial-time computable function h such that on each input x, (a) $h(x)$ outputs a list of two (perhaps identical) natural numbers, and (b) #SAT(x) appears in the list output by $h(x)$.

My suggestion to you would be to work on proving this theorem until either you find a proof, or you've put in at least 30–60 minutes of thought, are stuck, and don't think that more time will be helpful.

When you've reached one or the other of those states, please go on to Appendix D, where you will find first a discussion of what the most tempting dead end here is, why it is a dead end, and a tool that will help you avoid the dead end. And then you'll be urged to attack the problem again with that extra tool in hand.

7   Going Big: Complexity-Class Implications

During all four of our challenge problems, we focused just on the concrete problem, SAT, in its language version or in its counting analogue, #SAT.

However, the challenge results in fact apply to broader classes of problems. Although we (mostly) won't prove those broader results in this chapter, this section will briefly note some of those (and the reader may well be able to in most cases easily fill in the proofs). The original papers, cited in the "A Bit of History" notes, are an excellent source to go to for more coverage. None of the claims below, of course, are due to the present tutorial paper, but rather they are generally right from the original papers. Also often of use for a gentler treatment than the original papers is the textbook, *The Complexity Theory Companion* [18], in which coverage related to our four problems can be found in, respectively, Chapters 1, 1 [sic], 3, and (using a different technique and focusing on a concrete but different target problem) 6.

Let us define the complexity class NP by NP $= \{L \mid L \leq_m^p \text{SAT}\}$. NP more commonly is defined as the class of sets accepted by nondeterministic polynomial-time Turing machines; but that definition in fact yields the same class of sets as the alternate definition just given, and would require a detailed discussion of what Turing machines are.

Recall that $\overline{L}$ denotes the complement of L. Let us define the complexity class coNP by coNP $= \{L \mid \overline{L} \in \text{NP}\}$.

A set H is said to be hard for a class $\mathcal{C}$ if for each set $L \in \mathcal{C}$ it holds that $L \leq_m^p H$. If in addition $H \in \mathcal{C}$, then we say that H is $\mathcal{C}$-complete. It is well known—although it takes quite a bit of work to show and showing this was one of the most important steps in the history of computer science—that SAT is NP-complete [6,23,29].

The following theorem follows easily from our first challenge theorem, basically because if some NP-hard set is P-selective, that causes SAT to be P-selective. (Why? Our P-selector function for SAT will simply polynomial-time reduce each of its two inputs to the NP-hard set, will run that set's P-selector function on those two strings, and then will select as the more likely to belong to SAT whichever input string corresponded to the selected string, and for definiteness will choose its first argument in the degenerate case where both its arguments map to the same string.)

Theorem 1. *If there exists an* NP-*hard,* P-*selective set, then* P = NP.

The converse of the above theorem also holds, since if P = NP then SAT and indeed all of NP is P-selective, since P sets unconditionally are P-selective.

The following theorem follows easily from our second challenge theorem.

Theorem 2. *If there exists an* NP-*hard tally set, then* P = NP.

The converse of the above theorem also holds.

The following theorem follows easily from our third challenge theorem.

Theorem 3. *If there exists a* coNP-*hard sparse set, then* P = NP.

The converse of the above theorem also holds.

To state the complexity-class analogue of the fourth challenge problem takes a bit more background, since the result is about function classes rather than language classes.

There is a complexity class, which we will not define here, defined by Valiant and known as #P [39], that is the set of functions that count the numbers of accepting paths of what are known as nondeterministic polynomial-time Turing machines.

Metric reductions give a reduction notion that applies to the case of functions rather than languages, and are defined as follows. A function $f : \Sigma^* \to \{0, 1, 2, \ldots\}$ is said to polynomial-time metric reduce to a function $g : \Sigma^* \to \{0, 1, 2, \ldots\}$ if there exist two polynomial-time computable functions, φ and ψ, such that $(\forall x \in \Sigma^*)[f(x) = \psi(x, g(\varphi(x)))]$ [28]. (We are assuming that our output natural numbers are naturally coded in binary.) We say a function f is hard for #P with respect to polynomial-time metric reductions if for every $f' \in$ #P it holds that f' polynomial-time metric reduces to f; if in addition $f \in$ #P, we say that f is #P-complete with respect to polynomial-time metric reductions.

With that groundwork in hand, we can now state the analogue, for counting classes, of our fourth challenge theorem. Since we have not defined #P here, we'll state the theorem both in terms of #SAT and in terms of #P (the two statements below in fact turn out to be equivalent).

Theorem 4. *1. If there exists a function f such that* #SAT *polynomial-time metric reduces to f and f has a 2-enumerator, then there is a polynomial-time algorithm for* #SAT.
2. If there exists a function that is #P-*hard with respect to polynomial-time metric reductions and has a 2-enumerator, then there is a polynomial-time algorithm for* #SAT.

The converse of each of the theorem parts also holds. The above theorem parts (and their converses) even hold if one asks not about 2-enumerators but rather about polynomial-time enumerators that have no limit on the number of elements in their output lists (aside from the polynomial limit that is implicit from the fact that the enumerators have only polynomial time to write their lists).

8   Conclusions

In conclusion, self-reducibility provides a powerful tool with applications across a broad range of settings.

Myself, I have found self-reducibility and its generalizations to be useful in understanding topics ranging from election manipulation [12] to backbones of and backdoors to Boolean formulas [16,17] to the complexity of sparse sets [20], space-efficient language recognition [19], logspace computation [15], and approximation [14,22].

My guess and hope is that perhaps you too may find self-reducibility useful in your future work. That is, please, if it is not already there, consider adding this tool to *your* personal research toolkit: When you face a problem, think (if only for a moment) whether the problem happens to be one where the concept of self-reducibility will help you gain insight. Who knows? One of these years, you might be happily surprised in finding that your answer to such a question is "Yes!"

Acknowledgments. I am grateful to the students and faculty at the computer science departments of RWTH Aachen University, Heinrich Heine University Düsseldorf, and the University of Rochester. I "test drove" this chapter at each of those schools in the form of a lecture or lecture series. Particular thanks go to Peter Rossmanith, Jörg Rothe, and Muthu Venkitasubramaniam, who invited me to speak, and to Gerhard Woeginger regarding the counterexample in Appendix D. My warm appreciation to Ding-Zhu Du, Bin Liu, and Jie Wang for inviting me to contribute to this project that they have organized in memory of the wonderful Ker-I Ko, whose work contributed so richly to the beautiful, ever-growing tapestry that is complexity theory.

Appendices

A   Solution to Challenge Problem 1

Before we start on the proof, let us put up a figure that shows the flavor of a structure that we will use to help us understand and exploit SAT's self-reducibility. The structure is known as the self-reducibility tree of a formula. At the root of this tree sits the formula. At the next level as the root's children, we have the formula with its first variable assigned to True and to False. At the level below that, we have the two formulas from the second level, except with each of *their* first variables (i.e., the second variable of the original formula) assigned to both True and False. Figure 1 shows the self-reducibility tree of a two-variable formula.

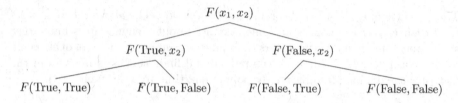

Fig. 1. The self-reducibility tree (completely unpruned) of a two-variable formula, represented generically.

Self-reducibility tells us that, for each node N in such a self-reducibility tree (except the leaves, since they have no children), N is satisfiable if and only if at least one of its two children is satisfiable. Inductively, the formula at the root of the tree is satisfiable if and only if each level of the self-reducibility tree has at least one satisfiable node. And, also, the formula at the root of the tree is satisfiable if and only if every level of the self-reducibility tree has at least one satisfiable node.

How helpful is this tree? Well, we certainly don't want to solve SAT by checking every leaf of the self-reducibility tree. On formulas with k variables, that would take time at least 2^k—basically a brute-force exponential-time algorithm. Yuck! That isn't surprising though. After all, the tree is really just listing all assignments to the formula.

But the magic here, which we will exploit, is that the "self-reducibility" relationship between nodes and their children as to satisfiability will, at least with certain extra assumptions such as about P-selectivity, allow us to *not* explore the whole tree. Rather, we'll be able to prune away, quickly, all but a polynomially large subtree. In fact, though on its surface this chapter is about four questions from complexity theory, it really is about tree-pruning—a topic more commonly associated with algorithms than with complexity. To us, though, that is not a problem but an advantage. As we mentioned earlier, complexity is largely about building algorithms, and that helps make complexity far more inviting and intuitive than most people realize.

That being said, let us move on to giving a proof of the first challenge problem. Namely, in this section we sketch a proof of the result:

If SAT is P-selective, then SAT $\in$ P.

So assume that SAT is P-selective, via (in the sense of Definition 2) polynomial-time computable function f. Let us give a polynomial-time algorithm for SAT. Suppose the input to our algorithm is the formula $F(x_1, x_2, \ldots, x_k)$. (If the input is not a syntactically legal formula we immediately reject, and if the input is a formula that has zero variables, e.g., True $\wedge$ True $\wedge$ False, we simply evaluate it and accept if and only if it evaluates to True.) Let us focus on F and F's two children in the self-reducibility tree, as shown in Fig. 2.

Now, run f on F's two children. That is, compute, in polynomial time, $f(F(\text{True}, x_2, \ldots, x_k), F(\text{False}, x_2, \ldots, x_k))$. Due to the properties of P-selectivity and self-reducibility, note that the output of that application of f is

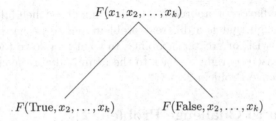

Fig. 2. F and F's two children.

a formula/node that has the property that the original formula is satisfiable if and only if that child-node is satisfiable.

In particular, if $f(F(\text{True}, x_2, \ldots, x_k), F(\text{False}, x_2, \ldots, x_k)) = F(\text{True}, x_2, \ldots, x_k)$ then we know that $F(x_1, x_2, \ldots, x_k)$ is satisfiable if and only if $F(\text{True}, x_2, \ldots, x_k)$ is satisfiable. And if $f(F(\text{True}, x_2, \ldots, x_k), F(\text{False}, x_2, \ldots, x_k)) \neq F(\text{True}, x_2, \ldots, x_k)$ then we know that $F(x_1, x_2, \ldots, x_k)$ is satisfiable if and only if $F(\text{False}, x_2, \ldots, x_k)$ is satisfiable.

Either way, we have in time polynomial in the input's size eliminated the need to pay attention to one of the two child nodes, and now may focus just on the other one.

Repeat the above process on the child that, as per the above, was selected by the selector function. Now, "split" that formula by assigning x_2 both possible ways. That will create two children, and then analogously to what was done above, use the selector function to decide which of those two children is the more promising branch to follow.

Repeat this until we have assigned all variables. We now have a fully assigned formula, but due to how we got to it, we know that it evaluates to True if and only if the original formula is satisfiable. So if that fully assigned formula evaluates to True, then we state that the original formula is satisfiable (and indeed, our path down the self-reducibility tree has outright put into our hands a satisfying assignment). And, more interestingly, if the fully assigned formula evaluates to False, then we state that the original formula is not satisfiable. We are correct in stating that, because at each iterative stage we know that if the formula we start that stage focused on is satisfiable, then the child the selector function chooses for us will also be satisfiable.

The process above is an at most polynomial number of at most polynomial-time "descend one level having made a linkage" stages, and so overall itself runs in polynomial time. Thus we have given a polynomial-time algorithm for SAT, under the hypothesis that SAT is P-selective. This completes the proof sketch.

Our algorithm was mostly focused on tree pruning. Though F induces a giant binary tree as to doing variable assignments one variable at a time in all possible ways, thanks to the guidance of the selector function, we walked just a single path through that tree.

Keeping this flavor of approach in mind might be helpful on Challenge Problem 2, although that is a different problem and so perhaps you'll have to bring some new twist, or greater flexibility, to what you do to tackle that.

And now, please pop right on back to the main body of the chapter, to read and tackle Challenge Problem 2!

B   Solution to Challenge Problem 2

In this section we sketch a proof of the result:

If there exists a tally set T such that SAT $\leq_m^p T$, then SAT $\in$ P.

So assume that there exists a tally set T such that SAT $\leq_m^p T$. Let g be the polynomial-time computable function performing that reduction, in the sense of Definition 4. (Keep in mind that we may *not* assume that $T \in$ P. We have no argument line in hand that would tell us that that happens to be true.) Let us give a polynomial-time algorithm for SAT.

Suppose the input to our algorithm is the formula $F(x_1, x_2, \ldots, x_k)$. (If the input is not a syntactically legal formula we immediately reject, and if the input is a formula that has zero variables we simply evaluate it and accept if and only if it evaluates to True.)

Let us focus first on F. Compute, in polynomial time, $g(F(x_1, x_2, \ldots, x_k))$. If $g(F(x_1, x_2, \ldots, x_k)) \notin \{\epsilon, 0, 00, \ldots\}$, then clearly $F(x_1, x_2, \ldots, x_k) \notin$ SAT, since we know that (a) $T \subseteq \{\epsilon, 0, 00, \ldots\}$ and (b) $F(x_1, x_2, \ldots, x_k) \in$ SAT $\iff$ $g(F(x_1, x_2, \ldots, x_k)) \in T$. So in that case, we output that $F(x_1, x_2, \ldots, x_k) \notin$ SAT. Otherwise, we descend to the next level of the "self-reducibility tree" as follows.

We consider the nodes (i.e., in this case, formulas) $F(\text{True}, x_2, \ldots, x_k)$ and $F(\text{False}, x_2, \ldots, x_k)$. Compute $g(F(\text{True}, x_2, \ldots, x_k))$ and $g(F(\text{False}, x_2, \ldots, x_k))$. If either of our two nodes in question does not, under the action just computed of g, map to a string in $\{\epsilon, 0, 00, \ldots\}$, then that node certainly is not a satisfiable formula, and we can henceforward mentally ignore it and the entire tree (created by assigning more of its variables) rooted at it. This is one key type of pruning that we will use: eliminating from consideration nodes that map to "nontally" strings.

But there is a second type of pruning that we will use: If it happens to be the case that $g(F(\text{True}, x_2, \ldots, x_k)) \in \{\epsilon, 0, 00, \ldots\}$ and $g(F(\text{True}, x_2, \ldots, x_k)) = g(F(\text{False}, x_2, \ldots, x_k))$, then at this point it may not be clear to us whether $F(\text{True}, x_2, \ldots, x_k)$ is or is not satisfiable. However, what is clear is that

$$F(\text{True}, x_2, \ldots, x_k) \in \text{SAT} \iff F(\text{False}, x_2, \ldots, x_k) \in \text{SAT}.$$

How do we know this? Since g reduces SAT to T, we know that

$$g(F(\text{True}, x_2, \ldots, x_k)) \in T \iff F(\text{True}, x_2, \ldots, x_k) \in \text{SAT}$$

and

$$g(F(\text{False}, x_2, \ldots, x_k)) \in T \iff F(\text{False}, x_2, \ldots, x_k) \in \text{SAT}.$$

By those observations, the fact that $g(F(\text{True}, x_2, \ldots, x_k)) = g(F(\text{False}, x_2, \ldots, x_k))$, and the transitivity of "$\Longleftrightarrow$", we indeed have that $F(\text{True}, x_2, \ldots, x_k) \in \text{SAT} \iff F(\text{False}, x_2, \ldots, x_k) \in \text{SAT}$. But since that says that either both or neither of these nodes is a formula belonging to SAT, there is no need at all for us to further explore more than one of them, since they stand or fall together as to membership in SAT. So if we have $g(F(\text{True}, x_2, \ldots, x_k)) = g(F(\text{False}, x_2, \ldots, x_k))$, we can mentally dismiss $F(\text{False}, x_2, \ldots, x_k)$—and of course also the entire subtree rooted at it—from all further consideration.

After doing the two types of pruning just mentioned, we will have either one or two nodes left at the level of the tree—the level one down from the root—that we are considering. (If we have zero nodes left, we have pruned away all possible paths and can safely reject). Also, if $k = 1$, then we can simply check whether at least one node that has not been pruned away evaluates to True, and if so we accept and if not we reject.

But what we have outlined can iteratively be carried out in a way that drives us right down through the tree, one level at a time. At each level, we take all nodes (i.e., formulas; we will speak interchangeably of the node and the formula that it is representing) that have not yet been eliminated from consideration, and for each, take the next unassigned variable and make two child formulas, one with that variable assigned to True and one with that variable assigned to False. So if at a given level after pruning we end up with j formulas, we in this process start the next level with $2j$ formulas, each with one fewer variable. Then for those $2j$ formulas we do the following: For each of them, if g applied to that formula outputs a string that is not a member of $\{\epsilon, 0, 00, \ldots\}$, then eliminate that node from all further consideration. After all, the node clearly is not a satisfiable formula. Also, for all nodes among the $2j$ such that the string z that g maps them to belongs to $\{\epsilon, 0, 00, \ldots\}$ and z is mapped to by g by at least one other of the $2j$ nodes, for each such cluster of nodes that map to the same string z (of the form $\{\epsilon, 0, 00, \ldots\}$) eliminate all but one of the nodes from consideration. After all, by the argument given above, either all of that cluster are satisfiable or none of them are, so we can eliminate all but one from consideration, since eliminating all the others still leaves one that is satisfiable, if in fact the nodes in the cluster are satisfiable.

Continue this process until (it internally terminates with a decision, or) we reach a level where all variables are assigned. If there were j nodes at the level above that after pruning, then at this no-variables-left-to-assign level we have at most $2j$ formulas. The construction is such that $F(x_1, x_2, \ldots, x_k) \in \text{SAT}$ if and only if at least one of these at most $2j$ variable-free formulas belongs to SAT, i.e., evaluates to True. But we can easily check that in time polynomial in $2j \times |F(x_1, x_2, \ldots, x_k)|$.

Is the proof done? Not yet. If j can be huge, we're dead, as we might have just sketched an exponential-time algorithm. But fortunately, and this was the key insight in Piotr Berman's paper that proved this result, as we go down the tree, level by level, the tree *never* grows too wide. In particular, it is at most polynomially wide!

How can we know this? The insight that Berman (and with luck, also you!) had is that there are not many "tally" strings that can be reached by the reduction g on the inputs that it will be run on in our construction on a given input. And that fact will ensure us that after we do our two kinds of pruning, we have at most polynomially many strings left at the now-pruned level.

Let us be more concrete about this, since it is not just the heart of this problem's solution, but also might well (*hint!, hint!*) be useful when tacking the third challenge problem.

In particular, we know that g is polynomial-time computable. So there certainly is some natural number k such that, for each natural number n, the function g runs in time at most $n^k + k$ on all inputs of length n. Let $m = |F(x_1, x_2, \ldots, x_k)|$. Note that, at least if the encoding scheme is reasonable and we if needed do reasonable, obvious simplifications (e.g., True $\wedge y \equiv y$, True $\vee y \equiv$ True, $\neg$True $\equiv$ False, and $\neg$False $\equiv$ True), then each formula in the tree is of length less than or equal to m. Crucially, g applied to strings of length less than or equal to m can never output any string of length greater than $m^k + k$. And so there are at most $m^k + k + 1$ strings (the "$+1$" is because the empty string is one of the strings that can be reached) in $\{\epsilon, 0, 00, \ldots\}$ that can be mapped to by any of the nodes that are part of our proof's self-reducibility tree when the input is $F(x_1, x_2, \ldots, x_k)$. So at each level of our tree-pruning, we eliminate all nodes that map to strings that do not belong to $\{\epsilon, 0, 00, \ldots\}$, and since we leave at most one node mapping to each string that is mapped to in $\{\epsilon, 0, 00, \ldots\}$, and as we just argued that there are at most $m^k + k + 1$ of those, at the end of pruning a given level, at most $m^k + k + 1$ nodes are still under consideration. But m is the length of our problem's input, so each level, after pruning, finishes with at most at most $m^k + k + 1$ nodes, and so the level after it, after we split each of the current level's nodes, will begin with at most $2(m^k + k + 1)$ nodes. And after pruning *that* level, it too ends up with at most $m^k + k + 1$ nodes still in play. The tree indeed remains at most polynomially wide.

Thus when we reach the "no variables left unassigned" level, we come into it with a polynomial-sized set of possible satisfying assignments (namely, a set of at most $m^k + k + 1$ assignments), and we know that the original formula is satisfiable if and only if at least one of these assignments satisfies F.

Thus the entire algorithm is a polynomial number of rounds (one per variable eliminated), each taking polynomial time. So overall it is a polynomial-time algorithm that it is correctly deciding SAT. This completes the proof sketch.

And now, please pop right on back to the main body of the chapter, to read and tackle Challenge Problem 3! While doing so, please keep this proof in mind, since doing so will be useful on Challenge Problem 3... though you also will need to discover a quite cool additional insight—the same one Steve Fortune discovered when he originally proved the theorem that is our Challenge Problem 3.

C   Solution to Challenge Problem 3

In this section we sketch a proof of the result:

If there exists a sparse set S such that $\overline{\text{SAT}} \leq_m^p S$, then SAT $\in$ P.

So assume that there exists a sparse set S such that $\overline{\text{SAT}} \leq_m^p S$. Let g be the polynomial-time computable function performing that reduction, in the sense of Definition 4. (Keep in mind that we may *not* assume that $S \in$ P. We have no argument line in hand that would tell us that that happens to be true.) Let us give a polynomial-time algorithm for SAT.

Suppose the input to our algorithm is the formula $F(x_1, x_2, \ldots, x_k)$. (If the input is not a syntactically legal formula we immediately reject, and if the input is a formula that has zero variables we simply evaluate it and accept if and only if it evaluates to True.)

What we are going to do here is that we are going to mimic the proof that solved Challenge Problem 2. We are going to go level by level down the self-reducibility tree, pruning at each level, and arguing that the tree never gets too wide—at least if we are careful and employ a rather jolting insight that Steve Fortune (and with luck, also you!) had.

Note that of the two types of pruning we used in the Challenge Problem 2 proof, one applies perfectly well here. If two or more nodes on a given level of the tree map under g to the same string, we can eliminate from consideration all but one of them, since either all of them or none of them are satisfiable.

However, the other type of pruning—eliminating all nodes not mapping to a string in $\{\epsilon, 0, 00, \ldots\}$—completely disappears here. Sparse sets don't have too many strings per level, but the strings are not trapped to always being of a specific, well-known form.

Is the one type of pruning that is left to us enough to keep the tree from growing exponentially bushy as we go down it? At first glance, it seems that exponential width growth is very much possible, e.g., imagine the case that every node of the tree maps to a different string than all the others at the node's same level. Then with each level our tree would be doubling in size, and by its base, if we started with k variables, we'd have 2^k nodes at the base level—clearly an exponentially bushy tree.

But Fortune stepped back and realized something lovely. He realized that if the tree ever became too bushy, *then that itself would be an implicit proof that F is satisfiable*! Wow; mind-blowing!

In particular, Fortune used the following beautiful reasoning.

We know g runs in polynomial time. So let the polynomial $r(n)$ bound g's running time on inputs of length n, and without loss of generality, assume that r is nondecreasing. We know that S is sparse, so let the polynomial $q(n)$ bound the number of strings in S up to and including length n, and without loss of generality, assume that q is nondecreasing.

Let $m = |F(x_1, x_2, \ldots, x_k)|$, and as before, note that all the nodes in our proof are of length less than or equal to m.

How many distinct strings in S can be reached by applying g to strings of length at most m? On inputs of length at most m, clearly g maps to strings of length at most $r(m)$. But note that the number of strings in S of length at most $r(m)$ is at most $q(r(m))$.

Now, there are two cases. Suppose that at each level of our tree we have, after pruning, at most $q(r(m))$ nodes left active. Since $q(r(m))$ itself is a polynomial in the input size, m, that means our tree remains at most polynomially bushy (since levels of our tree are never, even right after splitting a level's nodes to create the next level, wider than $2q(r(m))$). Analogously to the argument of Challenge Problem 2's proof, when we reach the "all variables assigned" level, we enter it with a set of at most $2q(r(m))$ no-variables-left formulas such that F is satisfiable if and only if at least one of those formulas evaluates to True. So in that case, we easily do compute in polynomial time whether the given input is satisfiable, analogously to the previous proof.

On the other hand, suppose that on some level, after pruning, we have at least $1+q(r(m))$ nodes. This means that at that level, we had at least $1+q(r(m))$ distinct labels. But there are only $q(r(m))$ distinct strings that g can possibly reach, on our inputs, that belong to S. So at least one of the $1+q(r(m))$ formulas in our surviving nodes maps to a string that does not belong to S. But g was a reduction from $\overline{\text{SAT}}$ to S, so that node that mapped to a string that does not belong to S must itself be a satisfiable formula. Ka-*zam!* That node is satisfiable, and yet that node is simply F with some of its variables fixed. And so F itself certainly is satisfiable. We are done, and so the moment our algorithm finds a level that has $1 + q(r(m))$ distinct labels, our algorithm halts and declares that $F(x_1, x_2, \ldots, x_k)$ is satisfiable.

Note how subtle the action here is. The algorithm is correct in reasoning that, when we have at least $1 + q(r(m))$ distinct labels at a level, at least one of the still-live nodes at that level must be satisfiable, and thus $F(x_1, x_2, \ldots, x_k)$ is satisfiable. However, the algorithm doesn't know a particular one of those at-least-$1 + q(r(m))$-nodes that it can point to as being satisfiable. It merely knows that at least one of them is. And that is enough to allow the algorithm to act correctly. (One can, if one wants, extend the above approach to actually drive onward to the base of the tree; what one does is that at each level, the moment one gets to $1 + q(r(m))$ distinct labels, one stops handling that level, and goes immediately on to the next level, splitting each of those $1 + q(r(m))$ nodes into two at the next level. This works since we know that at least one of the nodes is satisfiable, and so we have ensured that at least node at the next level will be satisfiable.) This completes the proof sketch.

And now, please pop right on back to the main body of the chapter, to read and tackle Challenge Problem 4! There, you'll be working within a related but changed and rather challenging setting: you'll be working in the realms of functions and counting. Buckle up!

D   Solution to Challenge Problem 4

D.1   Why One Natural Approach Is Hopeless

One natural approach would be to run the hypothetical 2-enumerator h on the input formula F and both of F's x_1-assigned subformulas, and to argue that *purely based on the two options that h gives for each of those three, i.e., viewing the formulas for a moment as black boxes* (note: without loss of generality, we may assume that each of the three applications of the 2-enumerator has two distinct outputs; the other cases are even easier), we can either output $\|F\|$ or can identify at least one of the subformulas such that we can show a particular 1-to-1 linkage between which of the two predicted numbers of solutions it has and which of the two predicted numbers of solutions F has. And then we would iteratively walk down the tree, doing that.

But the following example, based on one suggested by Gerhard Woeginger, shows that that is impossible. Suppose h predicts outputs $\{0,1\}$ for F, and h predicts outputs $\{0,1\}$ for the left subformula, and h predicts outputs $\{0,1\}$ for the right subformula. That is, for each, it says "this formula either has zero satisfying assignments or has exactly one satisfying assignment." In this case, note that the values of the root can't be, based solely on the numbers the enumerator output, linked 1-to-1 to those of the left subformula, since 0 solutions for the left subformula can correspond to a root value of 0 $(0 + 0 = 0)$ or to a root value of 1 $(0 + 1 = 1)$. The same clearly also holds for the right subformula.

The three separate number-pairs just don't have enough information to make the desired link! But don't despair: we can make h help us far more powerfully than was done above!

D.2   XYZ Idea/Statement

To get around the obstacle just mentioned, we can try to trick the enumerator into giving us *linked/coordinated* guesses! Let us see how to do that.

What I was thinking of, when I mentioned XYZ in the food-for-thought hint (Sect. 6.2), is the fact that we can efficiently combine two Boolean formulas into a new one such that from the number of satisfying assignments of the new formula we can easily "read off" the number of satisfying assignments of both the original formulas. In fact, it turns out that we can do the combining in such a way that if we concatenate the (appropriately padded as needed) bitstrings capturing the numbers of solutions of the two formulas, we get the (appropriately padded as needed) bitstring capturing the number of solutions of the new "combined" formula. We will, when F is a Boolean formula, use $\|F\|$ to denote the number of satisfying assignments of F.

Lemma 1. *There are polynomial-time computable functions* combiner *and* decoder *such that for any Boolean formulas F and G,* combiner(F, G) *is a Boolean formula and* decoder$(F, G, \|$combiner$(F,G)\|)$ *prints* $\|F\|, \|G\|$.

Proof Sketch. Let $F = F(x_1, \ldots, x_n)$ and $G = G(y_1, \ldots, y_m)$, where $x_1, \ldots, x_n, y_1, \ldots, y_m$ are all distinct. Let z and z' be two new Boolean variables. Then

$$H = (F \wedge z) \vee (\bar{z} \wedge x_1 \wedge \cdots \wedge x_n \wedge G \wedge z')$$

gives the desired combination, since $\|h\| = \|f\|2^{m+1} + \|g\|$ and $\|g\| \leq 2^m$. □

We can easily extend this technique to combine three, four, or even polynomially many formulas.

D.3   Invitation to a Second Bite at the Apple

Now that you have in hand the extra tool that is Lemma 1, this would be a great time, unless you already found a solution to the fourth challenge problem, to try again to solve the problem. My guess is that if you did not already solve the fourth challenge problem, then the ideas you had while trying to solve it will stand you in good stead when you with the combining lemma in hand revisit the problem.

My suggestion to you would be to work again on proving Challenge Problem 4 until either you find a proof, or you've put in at least 15 more minutes of thought, are stuck, and don't think that more time will be helpful.

When you've reached one or the other of those states, please go on to Sect. D.4 to read a proof of the theorem.

D.4   Proof Sketch of the Theorem

Recall that we are trying to prove:

If #SAT is has a polynomial-time 2-enumerator, then there is a polynomial-time algorithm for #SAT.

Here is a quick proof sketch. Start with our input formula, F, whose number of solutions we wish to compute in polynomial time.

If F has no variables, we can simply directly output the right number of solutions, namely, 1 (if F evaluates to True), or 0 (otherwise). Otherwise, self-reduce formula F on its first variable. Using the XYZ trick, twice, combine the original formula and the two subformulas into a single formula, H, whose number of solutions gives the number of solutions of all three. For example, if our three formulas are $F = F(x_1 x_2, x_3, \ldots)$, $F_{left} = F(\text{True}, x_2, x_3, \ldots)$, and $F_{right} = F(\text{False}, x_2, x_3, \ldots)$, our combined formula can be

$$H = \text{combiner}(F, \text{combiner}(F_{left}, F_{right})),$$

and the decoding process is clear from this and Lemma 1 (and its proof). Run the 2-enumerator on H. If either of H's output's two decoded guesses are inconsistent ($a \neq b + c$), then ignore that line and the other one is the truth. If both are consistent and agree on $\|F\|$, then we're also done. Otherwise, the two guesses must each be internally consistent and the two guesses must disagree on $\|F\|$, and so it follows that the two guesses differ in their claims about at least one of $\|F_{left}\|$ and $\|F_{right}\|$. Thus if we know the number of solutions of that one, shorter formula, we know the number of solutions of $\|F\|$.

Repeat the above on *that* formula, and so on, right on down the three, and then (unless the process resolves internally or ripples back up earlier) at the end we have reached a zero-variable formula and for it we by inspection will know how many solutions it has (either 1 or 0), and so using that we can ripple our way all the way back up through the tree, using our linkages between each level and the next, and thus we now have computed $\|F\|$. The entire process is a polynomial number of polynomial-time actions, and so runs in polynomial time overall.

That ends the proof sketch, but let us give an example regarding the key step from the proof sketch, as that will help make clear what is going on.

Which of the Guesses	$\|F(x_1, x_2, x_3, \dots)\|$	$\|F(\text{True}, x_2, x_3, \dots)\|$	$\|F(\text{False}, x_2, x_3, \dots)\|$
First	100	83	17
Second	101	85	16

In this example, note that we can conclude that $\|F\| = 100$ if $\|F(\text{False}, x_2, x_3, \dots)\| = 17$, and $\|F\| = 101$ if $\|F(\text{False}, x_2, x_3, \dots)\| = 16$; and we know that $\|F(\text{False}, x_2, x_3, \dots)\| \in \{16, 17\}$.

So we have in polynomial time completely linked $\|F(x_1, x_2, x_3, \dots)\|$ to the issue of the number of satisfying assignments of the (after simplifying) shorter formula $F(\text{False}, x_2, x_3, \dots)$. This completes our example of the key linking step.

References

1. Arvind, V., Han, Y., Hemachandra, L., Köbler, J., Lozano, A., Mundhenk, M., Ogiwara, M., Schöning, U., Silvestri, R., Thierauf, T.: Reductions to sets of low information content. In: Ambos-Spies, K., Homer, S., Schöning, U. (eds.) Complexity Theory, pp. 1–45. Cambridge University Press, Cambridge (1993)
2. Berman, P.: Relationship between density and deterministic complexity of NP-complete languages. In: Ausiello, G., Böhm, C. (eds.) ICALP 1978. LNCS, vol. 62, pp. 63–71. Springer, Heidelberg (1978). https://doi.org/10.1007/3-540-08860-1_6
3. Cai, J.-Y., Hemachandra, L.: Enumerative counting is hard. Inf. Comput. **82**(1), 34–44 (1989)
4. Cai, J.-Y., Hemachandra, L.: A note on enumerative counting. Inf. Process. Lett. **38**(4), 215–219 (1991)
5. Clay Mathematics Institute: Millennium problems (web page) (2019). https://www.claymath.org/millennium-problems. Accessed 10 July 2019
6. Cook, S.: The complexity of theorem-proving procedures. In: Proceedings of the 3rd ACM Symposium on Theory of Computing, pp. 151–158. ACM Press, May 1971
7. Fortune, S.: A note on sparse complete sets. SIAM J. Comput. **8**(3), 431–433 (1979)
8. Gasarch, W.: The third P =? NP poll. SIGACT News **50**(1), 38–59 (2019)

9. Glaßer, C.: Consequences of the existence of sparse sets hard for NP under a subclass of truth-table reductions. Technical report, TR 245, Institut für Informatik, Universität Würzburg, Würzburg, Germany, January 2000

10. Glaßer, C., Hemaspaandra, L.: A moment of perfect clarity II: consequences of sparse sets hard for NP with respect to weak reductions. SIGACT News **31**(4), 39–51 (2000)

11. Hemachandra, L., Ogiwara, M., Watanabe, O.: How hard are sparse sets? In: Proceedings of the 7th Structure in Complexity Theory Conference, pp. 222–238. IEEE Computer Society Press, June 1992

12. Hemaspaandra, E., Hemaspaandra, L., Menton, C.: Search versus decision for election manipulation problems. In: Proceedings of the 30th Annual Symposium on Theoretical Aspects of Computer Science, vol. 20, pp. 377–388. Leibniz International Proceedings in Informatics (LIPIcs), February/March 2013

13. Hemaspaandra, L.: The power of self-reducibility: selectivity, information, and approximation (2019). File set–providing slides and their source code. http://www.cs.rochester.edu/u/lane/=self-reducibility/. Accessed 10 July 2019

14. Hemaspaandra, L., Hempel, H.: P-immune sets with holes lack self-reducibility properties. Theoret. Comput. Sci. **302**(1–3), 457–466 (2003)

15. Hemaspaandra, L., Jiang, Z.: Logspace reducibility: models and equivalences. Int. J. Found. Comput. Sci. **8**(1), 95–108 (1997)

16. Hemaspaandra, L., Narváez, D.: The opacity of backbones. In: Proceedings of the 31st AAAI Conference on Artificial Intelligence, pp. 3900–3906. AAAI Press, February 2017

17. Hemaspaandra, L.A., Narváez, D.E.: Existence versus exploitation: the opacity of backdoors and backbones under a weak assumption. In: Catania, B., Královič, R., Nawrocki, J., Pighizzini, G. (eds.) SOFSEM 2019. LNCS, vol. 11376, pp. 247–259. Springer, Cham (2019). https://doi.org/10.1007/978-3-030-10801-4_20

18. Hemaspaandra, L., Ogihara, M.: The Complexity Theory Companion. Springer, Heidelberg (2002). https://doi.org/10.1007/978-3-662-04880-1

19. Hemaspaandra, L., Ogihara, M., Toda, S.: Space-efficient recognition of sparse self-reducible languages. Comput. Complex. **4**(3), 262–296 (1994)

20. Hemaspaandra, L., Silvestri, R.: Easily checked generalized self-reducibility. SIAM J. Comput. **24**(4), 840–858 (1995)

21. Hemaspaandra, L., Torenvliet, L.: Theory of Semi-feasible Algorithms. Springer, Heidelberg (2003). https://doi.org/10.1007/978-3-662-05080-4

22. Hemaspaandra, L., Zimand, M.: Strong self-reducibility precludes strong immunity. Math. Syst. Theory **29**(5), 535–548 (1996)

23. Karp, R.: Reducibilities among combinatorial problems. In: Miller, R., Thatcher, J. (eds.) Complexity of Computer Computations, pp. 85–103. Springer, Boston (1972). https://doi.org/10.1007/978-1-4684-2001-2_9

24. Ko, K.: The maximum value problem and NP real numbers. J. Comput. Syst. Sci. **24**(1), 15–35 (1982)

25. Ko, K.: On self-reducibility and weak P-selectivity. J. Comput. Syst. Sci. **26**(2), 209–221 (1983)

26. Ko, K.: On helping by robust oracle machines. Theoret. Comput. Sci. **52**(1–2), 15–36 (1987)

27. Ko, K., Moore, D.: Completeness, approximation, and density. SIAM J. Comput. **10**(4), 787–796 (1981)

28. Krentel, M.: The complexity of optimization problems. J. Comput. Syst. Sci. **36**(3), 490–509 (1988)

29. Levin, L.: Universal sequential search problems. Probl. Inf. Transm. **9**(3), 265–266 (1975)
30. Mahaney, S.: Sparse complete sets for NP: solution of a conjecture of Berman and Hartmanis. J. Comput. Syst. Sci. **25**(2), 130–143 (1982)
31. Mahaney, S.: Sparse sets and reducibilities. In: Book, R. (ed.) Studies in Complexity Theory, pp. 63–118. Wiley, Hoboken (1986)
32. Mahaney, S.: The Isomorphism Conjecture and sparse sets. In: Hartmanis, J. (ed.) Computational Complexity Theory, pp. 18–46. American Mathematical Society (1989). Proceedings of Symposia in Applied Mathematics #38
33. Meyer, A., Paterson, M.: With what frequency are apparently intractable problems difficult? Technical report, MIT/LCS/TM-126, Laboratory for Computer Science, MIT, Cambridge, MA (1979)
34. Schnorr, C.: Optimal algorithms for self-reducible problems. In: Proceedings of the 3rd International Colloquium on Automata, Languages, and Programming, pp. 322–337. Edinburgh University Press, July 1976
35. Selman, A.: P-selective sets, tally languages, and the behavior of polynomial time reducibilities on NP. Math. Syst. Theory **13**(1), 55–65 (1979)
36. Selman, A.: Some observations on NP real numbers and P-selective sets. J. Comput. Syst. Sci. **23**(3), 326–332 (1981)
37. Selman, A.: Analogues of semirecursive sets and effective reducibilities to the study of NP complexity. Inf. Control **52**(1), 36–51 (1982)
38. Selman, A.: Reductions on NP and P-selective sets. Theoret. Comput. Sci. **19**(3), 287–304 (1982). https://doi.org/10.1016/0304-3975(82)90039-1
39. Valiant, L.: The complexity of computing the permanent. Theoret. Comput. Sci. **8**(2), 189–201 (1979)
40. Valiant, L.: The complexity of enumeration and reliability problems. SIAM J. Comput. **8**(3), 410–421 (1979)
41. Young, P.: How reductions to sparse sets collapse the polynomial-time hierarchy: a primer. SIGACT News **23** (1992). Part I (#3, pp. 107–117), Part II (#4, pp. 83–94), and Corrigendum to Part I (#4, p. 94)

Who Asked *Us?*
How the Theory of Computing Answers
Questions about Analysis

Jack H. Lutz[✉] and Neil Lutz

Iowa State University, Ames, IA, USA
{lutz,nlutz}@iastate.edu

Dedicated to the Memory of Ker-I Ko

Abstract. Algorithmic fractal dimensions—constructs of computability theory—have recently been used to answer open questions in *classical* geometric measure theory, questions of mathematical analysis whose statements do not involve computability theory or logic. We survey these developments and the prospects for future such results.

1 Introduction

Ker-I Ko was a pioneer in the computability, and especially the computational complexity, of problems in mathematical analysis. Aside from his visionary work on the complexity theory of functions on the reals, the early part of which is summarized in his well-known 1991 monograph [17], he did groundbreaking work on computability and complexity aspects of fractal geometry and other topics in geometric measure theory [5–8, 18–22, 48].

This chapter surveys recent developments in which algorithmic fractal dimensions, which are constructs of the theory of computing, have been used to answer open questions in *classical* fractal geometry, questions of mathematical analysis whose statements do not involve the theory of computing.

The results surveyed here concern the classical Hausdorff and packing dimensions of sets in Euclidean spaces $\mathbb{R}^n$. These fractal dimensions are duals of each other that were developed in 1918 and the early 1980s, respectively [15, 45, 46]. They assign every set $E \subseteq \mathbb{R}^n$ a *Hausorff dimension* $\dim_H(E)$ and a *packing dimension* $\dim_P(E)$, which are real numbers satisfying $0 \le \dim_H(E) \le \dim_P(E) \le n$ [12, 14]. These dimensions are both 0 if E consists of a single point, 1 if E is a smooth curve, 2 if E is a smooth surface, etc., but, for any two real numbers α and β satisfying $0 \le \alpha \le \beta \le n$, there are $2^{\mathfrak{c}}$ many sets $E \subseteq \mathbb{R}^n$ such that $\dim_H(E) = \alpha$ and $\dim_P(E) = \beta$, where $\mathfrak{c} = 2^{\aleph_0}$ is the cardinality of the continuum. So-called "fractals" (a term with no accepted formal definition)

J.H. Lutz—Research supported in part by National Science Foundation grants 1545028 and 1900716.

D.-Z. Du and J. Wang (Eds.): Ko Festschrift, LNCS 12000, pp. 48–56, 2020.
https://doi.org/10.1007/978-3-030-41672-0_4

are typically sets $E \subseteq \mathbb{R}^n$ with non-integral Hausdorff and packing dimensions. (Note: Hausdorff and packing dimensions are well-defined in arbitrary metric spaces, but this generality is not needed in the present survey.)

In contrast with the above classical fractal dimensions, the algorithmic fractal dimensions developed in [1,24] and defined in Sect. 2 below use computability theory to assign each *individual point* x in a Euclidean space $\mathbb{R}^n$ a *dimension* $\dim(x)$ and a *strong dimension* $\mathrm{Dim}(x)$ satisfying $0 \leq \dim(x) \leq \mathrm{Dim}(x) \leq n$. Intuitively, $\dim(x)$ and $\mathrm{Dim}(x)$ are the lower and upper densities of the algorithmic information in x. Computable points x (and many other points) satisfy $\dim(x) = \mathrm{Dim}(x) = 0$. In contrast, points x that are algorithmically random in the sense of Martin-Löf [33] (and many other points) satisfy $\dim(x) = \mathrm{Dim}(x) = n$. In general, for any two real numbers α and β satisfying $0 \leq \alpha \leq \beta \leq n$, the set of points $x \in \mathbb{R}^n$ such that $\dim(x) = \alpha$ and $\mathrm{Dim}(x) = \beta$ has the cardinality $\mathfrak{c}$ of the continuum.

The algorithmic fractal dimensions $\dim(x)$ and $\mathrm{Dim}(x)$ were known from their inceptions to be closely related to—and in fact Σ_1^0 versions of—their respective classical forerunners $\dim_H(E)$ and $\dim_P(E)$ [1,24]. However, it was only recently [26] that the *point-to-set principles* discussed in Sect. 3 below were proven, giving complete characterizations of $\dim_H(E)$ and $\dim_P(E)$ in terms of oracle relativizations of $\dim(x)$ and $\mathrm{Dim}(x)$, respectively.

The point-to-set principles are so named because they enable one to infer a bound—especially a difficult lower bound—on the classical fractal dimensions of a set $E \subseteq \mathbb{R}^n$ from a bound on the relativized algorithmic dimension of a single, judiciously chosen point $x \in E$. The power of this point-to-set reasoning has quickly become apparent. Sections 4 through 7 below survey recent research in which this method has been used to prove new theorems in classical fractal geometry. Several of these theorems answered well-known open questions in the field, completely classical questions whose statements do not involve computability or logic. Section 8 discusses the prospects for future such results.

2   Algorithmic Information and Algorithmic Dimensions

The *Kolmogorov complexity*, or *algorithmic information content*, of a string $x \in \{0,1\}^*$ is

$$K(x) = \min\left\{|\pi| \mid \pi \in \{0,1\}^* \text{ and } U(\pi) = x\right\},$$

where U is a fixed universal prefix Turing machine, and π is the length of a binary "program π for x." Extensive discussions of the history and intuition behind this notion, including its essential invariance with respect to the choice of the universal Turing machine U, may be found in any of the standard texts [11, 23,40,41]. By routine encoding we extend this notion to let x range over various countable sets, so that $K(x)$ is well defined when x is an element of $\mathbb{N}$, $\mathbb{Q}$, $\mathbb{Q}^n$, etc.

We "lift" Kolmogorov complexity to Euclidean space in two steps. We first define the Kolmogorov complexity of a set $E \subseteq \mathbb{R}^n$ to be

$$K(E) = \min\{K(q) \mid q \in \mathbb{Q}^n \cap E\},$$

i.e., the amount of information to specify *some* rational point in E. (A similar notion was used for a different purpose in [42].) Note that

$$E \subseteq F \implies K(E) \geq K(F).$$

We then define the *Kolmogorov complexity* of a point $x \in \mathbb{R}^n$ at a *precision* $r \in \mathbb{N}$ to be

$$K_r(x) = K\big(B_{2^{-r}}(x)\big),$$

where $B_\varepsilon(x)$ is the open ball of radius ε about x. That is, $K_r(x)$ is the number of bits required to specify *some* rational point q whose Euclidean distance from x is less than 2^{-r}.

The *(algorithmic) dimension* of a point $x \in \mathbb{R}^n$ is

$$\dim(x) = \liminf_{r \to \infty} \frac{K_r(x)}{r}, \tag{2.1}$$

and the *strong (algorithmic) dimension* of a point $x \in \mathbb{R}^n$ is

$$\mathrm{Dim}(x) = \limsup_{r \to \infty} \frac{K_r(x)}{r}. \tag{2.2}$$

(The adjectives "constructive" and "effective" are sometimes used in place of "algorithmic" here.) We should note that the identities (2.1) and (2.2) were originally *theorems* proven in [27] (following a key breakthrough in [37]) characterizing the algorithmic dimensions $\dim(x)$ and $\mathrm{Dim}(x)$ that had first been developed using algorithmic betting strategies called gales [1,24]. The characterizations (2.1) and (2.2) support the intuition that $\dim(x)$ and $\mathrm{Dim}(x)$ are the lower and upper asymptotic densities of algorithmic information in the point $x \in \mathbb{R}^n$.

By giving the underlying universal prefix Turing machine oracle access to a set $A \subseteq \mathbb{N}$, the quantities in this section can all be defined *relative* to A. We denote these relativized complexities and dimensions by $K^A(x)$, $K_r^A(x)$, $\dim^A(x)$, etc. When A encodes a point $y \in \mathbb{R}^n$, we may instead write $K^y(x)$, $K_r^y(x)$, $\dim^y(x)$, etc. The following easily verified result is frequently useful.

Theorem 1 (chain rule for algorithmic dimensions). *For all $x \in \mathbb{R}^m$ and $y \in \mathbb{R}^n$,*

$$\dim^y(x) + \dim(y) \leq \dim(x, y)$$
$$\leq \mathrm{Dim}^y(x) + \dim(y)$$
$$\leq \mathrm{Dim}(x, y)$$
$$\leq \mathrm{Dim}^y(x) + \mathrm{Dim}(y).$$

3   Point-to-Set Principles

One of the oldest and most beautiful theorems of computable analysis says that a function $f : \mathbb{R} \to \mathbb{R}$ is continuous if and only if there is an oracle $A \subseteq \mathbb{N}$

relative to which f is computable [39, 43]. That is, relativization allows us to characterize continuity—a completely classical notion—in terms of computability. The following two recent theorems are very much in the spirit of this old theorem.

Theorem 2 (point-to-set principle for Hausdorff dimension [26]). *For every set $E \subseteq \mathbb{R}^n$,*

$$\dim_H(E) = \min_{A \subseteq \mathbb{N}} \sup_{x \in E} \dim^A(x). \tag{3.1}$$

Theorem 3 (point-to-set principle for packing dimension [26]). *For every set $E \subseteq \mathbb{R}^n$,*

$$\dim_P(E) = \min_{A \subseteq \mathbb{N}} \sup_{x \in E} \text{Dim}^A(x). \tag{3.2}$$

For purposes of this survey, readers unfamiliar with Hausdorff and packing dimensions may use Theorems 2 and 3 as their *definitions*, but it should be kept in mind that these characterizations are theorems that were proven a century after Hausdorff developed his beautiful dimension.

Two remarks on the point-to-set principles are in order here. First, as the principles state, the minima on the right-hand sides of (3.1) and (3.2) are actually achieved. In other words, if we define a *Hausdorff oracle* for a set $E \subseteq \mathbb{R}^n$ to be an oracle $A \subseteq \mathbb{N}$ such that

$$\dim_H(E) = \sup_{x \in E} \dim^A(x), \tag{3.3}$$

and we similarly define a *packing oracle* for a set $E \subseteq \mathbb{R}^n$ to be an oracle $A \subseteq \mathbb{N}$ such that

$$\dim_P(E) = \sup_{x \in E} \text{Dim}^A(x), \tag{3.4}$$

then the point-to-set principles are assertions that *every* set $E \subseteq \mathbb{R}^n$ has Hausdorff and packing oracles. It is easy to show that, if A is a Hausdorff oracle for a set $E \subseteq \mathbb{R}^n$, and if A is Turing reducible to a set $B \subseteq \mathbb{N}$, then B is also a Hausdorff oracle for E, and similarly for packing oracles. This is useful, because it often enables one to combine Hausdorff or packing oracles with other oracles in a proof.

The second remark on the point-to-set principles concerns their use. Some of the most challenging problems in fractal geometry and dynamical systems involve finding lower bounds on the fractal dimensions of various sets. The point-to-set principles allow us to infer lower bounds on the fractal dimensions of sets from lower bounds on the corresponding relativized algorithmic fractal dimensions of judiciously chosen *individual points* in those sets. For example, to prove, for a given set $E \subseteq \mathbb{R}^n$, that $\dim_H(E) \geq \alpha$, it suffices to show that, for every Hausdorff oracle A for E and every $\varepsilon > 0$, there is a point $x \in E$ such that $\dim^A(x) > \alpha - \varepsilon$. In some applications, the ε here is not even needed, because one can readily show that there is a point $x \in E$ such that $\dim^A(x) \geq \alpha$. Most of the rest of this survey is devoted to illustrating the power of this point-to-set reasoning about fractal dimensions.

4   Fractal Products

Marstrand's product formula [14,32] states that for all sets $E, F \subseteq \mathbb{R}^n$,

$$\dim_{\mathrm{H}}(E) \leq \dim_{\mathrm{H}}(E \times F) - \dim_{\mathrm{H}}(F).$$

The proof of this fact for Borel sets is simple [14], but Marstrand's original proof of the general result is more difficult [36]. Using the point-to-set principle for Hausdorff dimension, the general result is an almost trivial consequence of the chain rule, Theorem 1 [28]. Tricot [46] proved related inequalities about packing dimension, including the fact that for all $E, F \subseteq \mathbb{R}^n$,

$$\dim_{\mathrm{P}}(E) \geq \dim_{\mathrm{H}}(E \times F) - \dim_{\mathrm{H}}(F).$$

Xiao [47] showed that for every Borel set $E \subseteq \mathbb{R}^n$ and $\varepsilon > 0$, there exists a Borel set $F \subseteq \mathbb{R}^n$ such that

$$\dim_{\mathrm{P}}(E) \leq \dim_{\mathrm{H}}(E \times F) - \dim_{\mathrm{H}}(F) + \varepsilon. \tag{4.1}$$

Bishop and Peres [4] independently showed that for Borel (or analytic) E there exists a compact F satisfying (4.1); they also later commented that that it would be straightforward to modify their construction to achieve $\varepsilon = 0$.

Using the point-to-set principles, N. Lutz proved for arbitrary sets E that $\varepsilon = 0$ can be achieved in (4.1), albeit not necessarily by a compact or Borel set F.

Theorem 4 ([29]). *For every set $E \subseteq \mathbb{R}^n$,*

$$\dim_{\mathrm{P}}(E) = \max_{F \subseteq \mathbb{R}^n} \left(\dim_{\mathrm{H}}(E \times F) - \dim_{\mathrm{H}}(F) \right).$$

The particular set F constructed in the proof of this theorem is the set of all points $x \in \mathbb{R}^n$ with $\dim^A(x) \leq n - \dim_{\mathrm{P}}(E)$, for a carefully chosen oracle A.

5   Fractal Intersections

Given a parameter $x \in \mathbb{R}$ and a set $E \subseteq \mathbb{R}^2$ with $\dim_{\mathrm{H}}(E) \geq 1$, what can we say about the Hausdorff dimension of the vertical *slice* $E_x = \{y : (x, y) \in E\}$? Without further information, we can only give the trivial upper bound,

$$\dim_{\mathrm{H}}(E_x) \leq 1. \tag{5.1}$$

For instance, equality holds in (5.1) whenever $\{x\} \times [0,1] \subseteq E$. It would be more informative, then, to ask about the Hausdorff dimension of a *random* vertical slice of E. The Marstrand slicing theorem tells us that if E is a Borel set, then for Lebesgue almost every $x \in E$,

$$\dim_{\mathrm{H}}(E_x) \leq \dim_{\mathrm{H}}(E_x) - 1.$$

Several more general results giving upper bounds on the Hausdorff dimension of the intersections of random transformations of restricted classes of sets have been proven, including theorems by Mattila [34–36] and Kahane [16]; in particular, Falconer [14] showed that when $E, F \subseteq \mathbb{R}^n$ are Borel sets,

$$\dim_H(E \cap (F + z)) \leq \max\{0, \dim_H(E \times F) - n\} \qquad (5.2)$$

holds for Lebesgue almost every $z \in \mathbb{R}$. Using the point-to-set principle, N. Lutz showed that this inequality holds even when the Borel assumption is removed.

Theorem 5 ([28]). *For all $E, F \subseteq \mathbb{R}^n$, and for Lebesgue almost every $z \in \mathbb{R}^n$,*

$$\dim_H(E \cap (F + z)) \leq \max\{0, \dim_H(E \times F) - n\}.$$

6   Kakeya Sets and Generalized Furstenberg Sets

A *Kakeya set* in $\mathbb{R}^n$ is a set that contains unit-length line segments in all directions. That is, a set $E \subseteq \mathbb{R}^n$ such that for every direction $a \in S^{n-1}$ (the $(n-1)$-dimensional unit sphere in $\mathbb{R}^n$), there exists $b \in \mathbb{R}^n$ with $\{ax + b \mid x \in [0,1]\} \subseteq E$.

Besicovitch [2,3] proved that Kakeya sets in $\mathbb{R}^n$ can have measure 0, and Davies [9] proved that Kakeya sets in $\mathbb{R}^2$ must have Hausdorff dimension 2.

Lutz and Lutz gave computability theoretic proofs of both of these facts. They showed that the former corresponds to the existence of lines in all directions that contain no random points [25], and that the latter corresponds to the fact that for any random pair $(a, x) \in \mathbb{R}^2$, $\dim(x, ax + b) = 2$ holds for all $b \in \mathbb{R}$ [26].

A set $E \subseteq \mathbb{R}^2$ is an (α, β)-*generalized Furstenberg set*, for parameters $\alpha, \beta \in [0,1]$, if E contains α-dimensional subsets of lines in all of a β-dimensional set of directions. That is, E is an (α, β)-generalized Furstenberg set if there is a set $J \subseteq S^1$ such that $\dim_H(H) = \beta$ and, for every direction $a \in J$, there exist $b \in \mathbb{R}^2$ and $F_a \subseteq \mathbb{R}$ with $\dim_H(F_a) = \alpha$ and $\{ax + b \mid x \in S_a\} \subseteq E$.

It is known that (α, β)-generalized Furstenberg sets of Hausdorff dimension $\alpha + \frac{\alpha+\beta}{2}$ exist. Molter and Rela [38] gave a lower bound on the Hausdorff dimension of such sets:

$$\dim_H(E) \geq \alpha + \max\left\{\frac{\beta}{2}, \alpha + \beta - 1\right\}. \qquad (6.1)$$

Stull [44] gave a new computability theoretic proof of (6.1), based on the point-to-principle. N. Lutz and Stull used the point-to-set principle to give a bound that improves on (6.1) whenever $\alpha, \beta < 1$ and $\beta < 2\alpha$.

Theorem 6 ([30]). *For all $\alpha, \beta \in (0,1]$ and every set $E \in F_{\alpha\beta}$,*

$$\dim_H(E) \geq \alpha + \min\{\beta, \alpha\}.$$

7   Fractal Projections

In recent decades, Marstrand's projection theorem has become one of the most central results in fractal geometry [13]. It says that almost all orthogonal projections of a Borel set onto a line have the maximum possible dimension. More formally, letting proj_a denote orthogonal projection onto a line in direction a, Marstrand's projection theorem states that for all Borel $E \subseteq \mathbb{R}^2$ and Lebesgue almost every $a \in S^1$,

$$\dim_H(\text{proj}_a E) = \min\{1, \dim_H(E)\}. \qquad (7.1)$$

Given Theorems 4 and 5, it is natural to hope that the point-to-set principle for Hausdorff dimension might allow us to remove the Borel assumption here as well. But Davies [10], assuming the continuum hypothesis, constructed a non-Borel set E for which (7.1) does not hold. Nevertheless, N. Lutz and Stull used the point-to-set principles to prove the following.

Theorem 7 ([31]). *Let $E \subseteq \mathbb{R}^2$ be any set such that $\dim_H(E) = \dim_P(E)$. Then for Lebesgue almost every $a \in S^1$,*

$$\dim_H(\text{proj}_a E) = \min\{1, \dim_H(E)\}.$$

Theorem 8 ([31]). *Let $E \subseteq \mathbb{R}^2$ be any set. Then for Lebesgue almost every $a \in S^1$,*

$$\dim_P(\text{proj}_a E) \geq \min\{1, \dim_H(E)\}.$$

8   Conclusion

As the preceding four sections show, the point-to-set principles have enabled the theory of computing to make significant advances in classical fractal geometry in a very short time. There is every indication that more such advances are on the near horizon. But a scientist with Ker-I Ko's vision would already be asking about more distant horizons. What other areas of classical mathematical analysis can be advanced by analogous methods? Are there intrinsic limits of such methods? We look forward to seeing the answers to these questions take shape.

References

1. Athreya, K.B., Hitchcock, J.M., Lutz, J.H., Mayordomo, E.: Effective strong dimension in algorithmic information and computational complexity. SIAM J. Comput. **37**(3), 671–705 (2007)
2. Besicovitch, A.S.: Sur deux questions d'intégrabilité des fonctions. J. de la Soci?t? de physique et de mathematique de l'Universite de Perm **2**, 105–123 (1919)
3. Besicovitch, A.S.: On Kakeya's problem and a similar one. Math. Z. **27**, 312–320 (1928)

4. Bishop, C.J., Peres, Y.: Packing dimension and Cartesian products. Trans. Am. Math. Soc. **348**, 4433–4445 (1996)
5. Chou, A.W., Ko, K.: Computational complexity of two-dimensional regions. SIAM J. Comput. **24**(5), 923–947 (1995)
6. Chou, A.W., Ko, K.: On the complexity of finding paths in a two-dimensional domain I: shortest paths. Math. Log. Q. **50**(6), 551–572 (2004)
7. Chou, A.W., Ko, K.: The computational complexity of distance functions of two-dimensional domains. Theor. Comput. Sci. **337**(1–3), 360–369 (2005)
8. Chou, A.W., Ko, K.: On the complexity of finding paths in a two-dimensional domain II: piecewise straight-line paths. Electr. Notes Theor. Comput. Sci. **120**, 45–57 (2005)
9. Davies, R.O.: Some remarks on the Kakeya problem. In: Proceedings of the Cambridge Philosophical Society, vol. 69, pp. 417–421 (1971)
10. Davies, R.O.: Two counterexamples concerning Hausdorff dimensions of projections. Colloq. Math. **42**, 53–58 (1979)
11. Downey, R., Hirschfeldt, D.: Algorithmic Randomness and Complexity. Springer, New York (2010). https://doi.org/10.1007/978-0-387-68441-3
12. Edgar, G.: Measure, Topology, and Fractal Geometry, 2nd edn. Springer, New York (2008). https://doi.org/10.1007/978-0-387-74749-1
13. Falconer, K., Fraser, J., Jin, X.: Sixty years of fractal projections. In: Bandt, C., Falconer, K., Zähle, M. (eds.) Fractal Geometry and Stochastics V. PP, vol. 70, pp. 3–25. Springer, Cham (2015). https://doi.org/10.1007/978-3-319-18660-3_1
14. Falconer, K.J.: Fractal Geometry: Mathematical Foundations and Applications, 3rd edn. Wiley, Hoboken (2014)
15. Hausdorff, F.: Dimension und äusseres Mass. Math. Ann. **79**, 157–179 (1918)
16. Kahane, J.P.: Sur la dimension des intersections. In: Barroso, J.A. (ed.) Aspects of Mathematics and Its Applications, pp. 419–430. Elsevier (1986). N.-Holl. Math. Libr. **34**
17. Ko, K.: Complexity Theory of Real Functions. Birkhäuser, Boston (1991)
18. Ko, K.: A polynomial-time computable curve whose interior has a nonrecursive measure. Theor. Comput. Sci. **145**(1&2), 241–270 (1995)
19. Ko, K.: On the computability of fractal dimensions and Hausdorff measure. Ann. Pure Appl. Logic **93**(1–3), 195–216 (1998)
20. Ko, K.: On the complexity of computing the Hausdorff distance. J. Complex. **29**(3–4), 248–262 (2013)
21. Ko, K., Weihrauch, K.: On the measure of two-dimensional regions with polynomial-time computable boundaries. In: Proceedings of the Eleveth Annual IEEE Conference on Computational Complexity, Philadelphia, Pennsylvania, USA, 24–27 May 1996, pp. 150–159 (1996)
22. Ko, K., Yu, F.: Jordan curves with polynomial inverse moduli of continuity. Theor. Comput. Sci. **381**(1–3), 148–161 (2007)
23. Li, M., Vitányi, P.M.: An Introduction to Kolmogorov Complexity and Its Applications, 3rd edn. Springer, New York (2008). https://doi.org/10.1007/978-0-387-49820-1
24. Lutz, J.H.: The dimensions of individual strings and sequences. Inf. Comput. **187**(1), 49–79 (2003)
25. Lutz, J.H., Lutz, N.: Lines missing every random point. Computability **4**(2), 85–102 (2015)
26. Lutz, J.H., Lutz, N.: Algorithmic information, plane Kakeya sets, and conditional dimension. ACM Trans. Comput. Theory **10**(2), 7:1–7:22 (2018)

27. Lutz, J.H., Mayordomo, E.: Dimensions of points in self-similar fractals. SIAM J. Comput. **38**(3), 1080–1112 (2008)
28. Lutz, N.: Fractal intersections and products via algorithmic dimension. In: 42nd Proceedings of the International Symposium on Mathematical Foundations of Computer Science, Aalborg, Denmark, 21–25 August 2017 (2017)
29. Lutz, N.: Fractal intersections and products via algorithmic dimension (extended version) (2019). https://arxiv.org/abs/1612.01659
30. Lutz, N., Stull, D.M.: Bounding the dimension of points on a line. Information and Computation (to appear)
31. Lutz, N., Stull, D.M.: Projection theorems using effective dimension. In: 43rd International Symposium on Mathematical Foundations of Computer Science, MFCS 2018, Liverpool, UK, 27–31 August 2018, pp. 71:1–71:15 (2018)
32. Marstrand, J.M.: Some fundamental geometrical properties of plane sets of fractional dimensions. Proc. Lond. Math. Soc. **4**(3), 257–302 (1954)
33. Martin-Löf, P.: The definition of random sequences. Inf. Control **9**(6), 602–619 (1966)
34. Mattila, P.: Hausdorff dimension and capacities of intersections of sets in n-space. Acta Math. **152**, 77–105 (1984)
35. Mattila, P.: On the Hausdorff dimension and capacities of intersections. Mathematika **32**, 213–217 (1985)
36. Mattila, P.: Geometry of Sets and Measures in Euclidean Spaces: Fractals and Rectifiability. Cambridge University Press, Cambridge (1995)
37. Mayordomo, E.: A Kolmogorov complexity characterization of constructive Hausdorff dimension. Inf. Process. Lett. **84**(1), 1–3 (2002)
38. Molter, U., Rela, E.: Furstenberg sets for a fractal set of directions. Proc. Am. Math. Soc. **140**, 2753–2765 (2012)
39. Moschovakis, Y.N.: Descriptive Set Theory. North-Holland Publishing, Amsterdam (1980)
40. Nies, A.: Computability and Randomness. Oxford University Press Inc., New York (2009)
41. Shen, A., Uspensky, V.A., Vereshchagin, N.: Kolmogorov Complexity and Algorithmic Randomness. AMS, Boston (2017)
42. Shen, A., Vereshchagin, N.K.: Logical operations and Kolmogorov complexity. Theoret. Comput. Sci. **271**(1–2), 125–129 (2002)
43. Soare, R.I.: Turing oracle machines, online computing, and three displacements in computability theory. Ann. Pure Appl. Log. **160**, 368–399 (2009)
44. Stull, D.M.: Results on the dimension spectra of planar lines. In: 43rd International Symposium on Mathematical Foundations of Computer Science, MFCS 2018, Liverpool, UK, 27–31 August 2018, pp. 79:1–79:15 (2018)
45. Sullivan, D.: Entropy, Hausdorff measures old and new, and limit sets of geometrically finite Kleinian groups. Acta Math. **153**(1), 259–277 (1984)
46. Tricot, C.: Two definitions of fractional dimension. Math. Proc. Camb. Philos. Soc. **91**(1), 57–74 (1982)
47. Xiao, Y.: Packing dimension, Hausdorff dimension and Cartesian product sets. Math. Proc. Camb. Philos. Soc. **120**(3), 535–546 (1996)
48. Yu, F., Chou, A.W., Ko, K.: On the complexity of finding circumscribed rectangles and squares for a two-dimensional domain. J. Complex. **22**(6), 803–817 (2006)

Promise Problems on Probability Distributions

Jan-Hendrik Lorenz and Uwe Schöning$^{(\boxtimes)}$

Ulm University, Institute of Theoretical Computer Science, 89081 Ulm, Germany
{jan-hendrik.lorenz,uwe.schoening}@uni-ulm.de

Abstract. We consider probability distributions which are associated with the running time of probabilistic algorithms, given for algorithmic processing in symbolic form. The considered decision (also counting) problems deal with the question whether a complete restart of the underlying probabilistic algorithm after some number of steps t gives an advantage. Since deciding whether a given symbolic formula indeed represents a probability distribution (either as probability mass function or as cumulative distribution function) is itself a difficult problem to decide, we discuss the issue in terms of promise problems.

Keywords: Promise problems · Probability distributions · Restart strategies

1 Introduction

The concept of a promise problem was initiated by Even, Selman and Yacobi [1] and was especially popularized by Selman [9]. It also has applications in cryptographic complexity (see the extensive survey by Goldreich [2]).

A promise problem is given by a pair of languages (decision problems) (Q, R) and can be represented in the following way:

Input: x
Promise: $Q(x)$
Question: $R(x)$

Intuitively, the issue is: what is the complexity of deciding property R, given that the input x has property Q. Vaguely, this has some similarity with the definition of conditional probability. Notice that it might be harder to decide Q than it is to decide R. Also there can be a difference between deciding R as such and deciding R under the precondition that Q already holds.

We will concentrate on the following definitions: A promise problem (Q, R) is solvable in polynomial-time if there is a polynomial-time Turing machine M such that for all inputs x,

$$Q(x) \rightarrow (M(x) = \text{"yes"} \leftrightarrow R(x)).$$

© Springer Nature Switzerland AG 2020
D.-Z. Du and J. Wang (Eds.): Ko Festschrift, LNCS 12000, pp. 57–66, 2020.
https://doi.org/10.1007/978-3-030-41672-0_5

Equivalently, if L is the language accepted by machine M, then it holds:

$$Q \cap R \subseteq L \text{ and } Q \cap \overline{R} \subseteq \overline{L}.$$

On the other hand, a promise problem (Q, R) is NP-hard, if every language L which satisfies the property $Q \cap R \subseteq L$ and $Q \cap \overline{R} \subseteq \overline{L}$ is NP-hard in the usual sense. That is, to show that a promise problem (Q, R) is NP-hard, it is sufficient to construct a polynomial-time computable function f such that for all x, if $x \in \text{SAT}$ (a well-known NP-complete problem), then $f(x)$ is in $Q \cap R$, and if $x \in \overline{\text{SAT}}$, then $f(x) \in Q \cap \overline{R}$.

We apply the promise-problem framework to a setting which is concerned with probability distributions.

Probability distributions are considered in statistics, mostly in terms of estimating or verifying their parameters. In algorithmics, probability distributions are often used to describe the runtime behavior of probabilistic algorithms. Typically, the distributions are used to characterize the mean runtime, the tail behavior, or other properties which are unambiguously described by the probability distributions. It is possible to consider complexity issues when some probability distribution is given in an appropriate form, see [6]. In particular, the question whether restarts are advantageous is addressed. Restarting is a paradigm used in some algorithms. After a fixed number of steps t, a stochastic search process is reset and reinitialized with a new random seed. Potentially, restarts can dramatically speed up the expected run time of algorithms (e.g. [3, 8]). However, Lorenz [6] showed that deciding whether an algorithm benefits from restarts is NP-hard.

Usually it is assumed that the probability distribution is known and given as a formula in symbolic form. However, the underlying decision problem is only reasonable if the formula indeed describes a probability distribution. For other formulas, the behavior is ill-defined. For instance, let f be a function which is suspected (but not proven) to be a probability function. If f is used in the model to decide whether restarts are beneficial, then the answer is conditional on f being a distribution. Formally, this means that the problems described in [6] are in fact *promise problems*.

2 Preliminaries

For the remainder of this paper it is assumed that the probability distribution is known either as a cumulative distribution function or as a probability mass function in symbolic form (as described below). For each such function it is only required that it is well defined on a bounded interval $I = \{0, \ldots, a\}$ with $0 < a < \infty$. Here, a is typically exponentially large, say $a = 2^n - 1$, so that the binary representation of a (or $i \le a$) takes n bits.

Definition 1. *Let $f : \{0, \ldots, a\} \mapsto \mathbb{Q}$ be a function and let X be some integer-valued random variable. The function f is a **cumulative distribution function** (cdf) of X if and only if for all $i \in \{0, \ldots, a\}$,*

$$f(i) = \Pr(X \le i). \tag{1}$$

Equivalently, f is a cdf if and only if

$$f(a) \le 1 \text{ and } \forall t \in \{1, \ldots, a\} : f(t) - f(t-1) \ge 0.$$

Definition 2. *Let* $f : \{0, \ldots, a\} \mapsto \mathbb{Q}$ *be a function and let* X *be some integer-valued random variable. The function* f *is the* **probability mass function** *(pmf) of* X *if and only if for all* $i \in \{0, \ldots, a\}$

$$f(i) = \Pr(X = i). \tag{2}$$

Equivalently, f is a pmf if and only if $F(t) = \sum_{i=0}^{t} f(i)$ *is a cdf.*

In this work, each function $F : \{0,1\}^n \mapsto \mathbb{Q}$ uses a binary encoded input and F is given in symbolic form, e.g. as a straight-line program calculating on the binary representation of input $i \le a$. In each line of such a program, either a Boolean operation is evaluated, or an arithmetical operation is performed. In each intermediate step a number of bits are provided as are necessary to represent the intermediate result. In such a form, $F(i)$ can be evaluated in some fixed polynomial-time (relative to the size of the straight line program F and $|i|$).

We often argue about the effect of restarts on the expected runtime of an algorithm. The fixed-cutoff strategy is a theoretically optimal restart strategy. For the remainder of this work, whenever we refer to restarts, we implicitly mean a fixed-cutoff strategy.

Definition 3. ([7]). *Let* $\mathcal{A}(x)$ *be an algorithm* $\mathcal{A}$ *on input* x. *Let* t *be a positive integer. A modified algorithm* $\mathcal{A}_t$ *is obtained from* $\mathcal{A}$ *by introducing a step counter to* $\mathcal{A}$, *and each time the counter has reached the value* t, $\mathcal{A}(x)$ *is reset with a new random seed and the counter set to zero again. If at any point an instantiation of* $\mathcal{A}(x)$ *finds a solution, then* $\mathcal{A}_t(x)$ *stops and returns this solution, otherwise the computation continues. The integer* t *is called* **restart time**. *This algorithmic approach is called* **fixed-cutoff strategy**.

Let X and X_t be discrete random variables corresponding to the running times of $\mathcal{A}$ and $\mathcal{A}_t$. Let the distribution of $\mathcal{A}$ be described in terms of cdf F. Luby et al. [7] calculated the expected value $E[X_t]$ as follows:

$$E[X_t] = \frac{1}{F(t)} \left(t - \sum_{x < t} F(x) \right) \le \frac{t}{F(t)}. \tag{3}$$

For the rest of this work, the quotient $t/F(t)$ is called the *upper bound* (of the expected value with restarts).

3 On the Hardness of Probability Distributions

Many algorithmic questions are, in fact, promise problems. For example, consider the well-known 3-SAT problem. In this case, the promise is that the input consists of a Boolean formula in conjunctive normal form (CNF) with each clause having three literals. Such a promise is taken as normal and not mentioned explicitly since in this case it is easily verified in polynomial-time. On the other hand, consider the 1SAT problem (e.g. discussed in [5]). The promise here is that the input CNF formula has either zero or exactly one satisfying assignment, and the question is whether the formula is satisfiable. However, deciding whether a given CNF formula has either zeo or exactly one satisfying assignment is at least as hard as SAT itself.

The goal of this section is evaluating the hardness of several non-trivial promise problems related to probability destributions (being non-trivial meaning that the promise itself is equivalent to some complexity theoretic statement). Consider the following problem introduced in [6]. The definition is rewritten as a promise problem.

RESTARTPMF
Input: A formula f in symbolic form and an integer k.
Promise: The formula f fulfills the definition of a pmf
              with respect to the interval $\{0, 1, \ldots, k\}$.
Question: Is there a restart time t such that $\frac{t}{F(t)} < k$

$$where \quad F(t) = \sum_{i=0}^{t} f(i)?$$

It is known that RESTARTPMF is NP-hard [6]. However, the promise to this problem is that f is a pmf. We will show that this promise is non-trivial. First, we define the promise explicitly as a decision problem and restrict ourselves to functions of the type $c(i)/M$ where c is a function mapping to the integers, and M is an integer.

PMF
Input: An integer k, a function $c : \{0, \ldots, k\} \mapsto \mathbb{Z}$ in symbolic form,
              and an integer M.
Question: Is $\frac{c(i)}{M}$ a probability mass function on $\{1, \ldots, k\}$?

We will show that this problem is $P(\#P)$-complete, where $\#P$ is the class of functions $f : \{0,1\}^* \to \mathbb{N}$ corresponding to the number of accepting computations which a nondeterministic, polynomial-time Turing machine on input $x \in \{0,1\}^*$ can achieve. Toda [11] showed that $P(\#P)$, i.e. the class of decision problems which can be solved in polynomial-time on oracle Turing machines with a (functional) oracle from $\#P$, is a powerful class that includes the entire polynomial-time hierarchy.

Theorem 1. PMF *is $P(\#P)$-complete with respect to polynomial-time Turing reductions.*

Proof. Let $c : \{0, \ldots, k\} \mapsto \mathbb{Z}$ be an arbitrary function and let M be any integer. In the following, $f(i)$ denotes $\frac{c(i)}{M}$. According to Definition 2, f is a probability mass function on $\{1, \ldots, k\}$ if and only if $F(x) = \sum_{i=1}^{x} f(i)$ is a cdf. The function F is a cdf if F is monotone increasing and upper bounded by 1. F being monotone increasing means that for all i, $f(i)$ is non-negative. This can be expressed by a CoNP predicate. The function F can be represented as a $\#P$ function. A nondeterministic Turing machine can guess $i \leq x$, calculate $f(i)$ and afterwards produce $f(i)$ many accepting computation paths. Therefore, both conditions can be verified in $P(\#P)$. It remains to be shown that PMF is $P(\#P)$-hard. Toda [10] showed that the following problem is $P(\#P)$-complete:

LEXICAL K-TH SAT
Input: Boolean CNF formula G, integer k
Question: Is $x_n = 1$ in the k-th satisfying assignment of G?

We show that LEXICAL K-TH SAT can be reduced to PMF with a polynomial-time Turing reduction. Let G be an arbitrary CNF formula and k an arbitrary integer. The CNF formula G can be evaluated at any assignment, and integers (their binary representation) can be interpreted as assignments. Let α_t be the assignment associated with t, and let G_{α_t} be the result of evaluating G on assignment α_t. Define:

$$p_G(t) = \frac{G_{\alpha_{t-1}}}{k}. \tag{4}$$

From the construction, we know that $p_G(t) \geq 0$ for all t. Then, $\mathrm{PMF}(j, G, k)$ yields YES if and only if there are at most k satisfying assignments in G in the interval $\{0, \ldots, j\}$. The minimal j with $\mathrm{PMF}(j, G, k) = \mathrm{YES}$ can be found by using binary search. This j corresponds to the k-th satisfying assignment of G. The binary encoding of j can be used to answer whether $x_n = 1$ in the k-th satisfying assignment. Therefore, LEXICAL K-THE SAT can be reduced to PMF.   □

RESTARTPMF requires a pmf as input. The same kind of problem which takes a cdf as input can be defined. This raises the question how hard it is to recognize a cdf. In the following, we show that this problem is significantly easier than its pmf version.

CDF
Input: A function $F : \mathbb{N} \mapsto \mathbb{Q}$ in symbolic form and an integer k.
Question: Is F a cumulative distribution function on $\{1, \ldots, k\}$?

Theorem 2. CDF *is CoNP-complete.*

Proof. An arbitrary function f is a cumulative distribution function on $\{1, \ldots, k\}$ if and only if f is monotone and is bounded by 1 (compare Definition 1). Both conditions can be verified by a CoNP-machine. Thus, CDF is in CoNP.

We show CoNP-hardness by reducing UNSAT to CDF. Let G be an arbitrary SAT formula in CNF and define the following formula:

$$F_G(t) = \begin{cases} \frac{1}{2^n+1}, & t = 1 \\ \frac{t}{2^n+1}(1 - G_{\alpha_{t-2}}), & t \in \{2, \ldots, 2^n + 1\} \\ 1, & t > 2^n + 1 \end{cases} \tag{5}$$

If G is unsatisfiable, then F_G is the cdf of a uniform distribution on $\{1, \ldots, 2^n + 1\}$. Consider the case when G is satisfiable and let x be the index of the first satisfying assignment of G. Then, by definition $F_G(x+2) = 0$ and $F(x+1) = \frac{x+1}{2^n+1}$, i.e., $F_G(x+2) - F_G(x+1) < 0$ and F_G is not a cdf. Therefore, F_G is a cdf if and only if G is unsatisfiable. This completes the proof. $\qquad\square$

Lorenz [6] examined the cdf version of RESTARTPMF, called RESTARTCDF. As before, it should be considered as a promise problem. RESTARTCDF in fact is NP-complete [6].

4 Approximating the Restart Time

In RESTARTCDF the question is whether there is a restart time t such that $t/F(t) < k$ for some fixed k. Accordingly, the corresponding optimization problem is finding a restart time t which minimizes $t/F(t)$. Since RESTARTCDF is NP-complete, it is unlikely that there is an efficient algorithm which finds the optimal restart time. Nevertheless, this raises the question whether it is possible to approximate the optimal restart time in polynomial-time. Lorenz [6] showed that the pmf version RESTARTPMF does not admit an efficient approximation algorithm unless $P = NP$. On the other hand, we show that for the cdf version there is a fully polynomial-time approximation scheme (FPTAS).

Consider the following algorithm. It uses as promise that F is a cdf.

function APPROXRESTARTTIME(F, ε, k)
  $x := minx := 1$
  $minvalue := x/F(x)$
  while $x \leq k$ **do**
    $x := (1 + \varepsilon) \cdot x$
    if $x/F(x) < minvalue$ **then**
      $minx := x$
      $minvalue := x/F(x)$
    end if
  end while
  return $minx$
end function

Theorem 3. *For every $\varepsilon > 0$ APPROXRESTARTTIME(F, ε, k) returns a restart time x with $\dfrac{x}{F(x)} \leq (1+\varepsilon) \dfrac{t^*}{F(t^*)}$ where $t^* = \arg\min\limits_{1 \leq t \leq k} \dfrac{t}{F(t)}$.*

Proof. Let x be a number with $t^* \leq x \leq (1+\varepsilon)t^*$. Since F is monotone, we have $F(x) \geq F(t^*)$. Then,

$$\frac{x}{F(x)} \leq (1+\varepsilon)\frac{t^*}{F(x)} \leq (1+\varepsilon)\frac{t^*}{F(t^*)}. \qquad (6)$$

Sometime during the loop $t^* \leq x \leq (1+\varepsilon)t^*$ is fulfilled. Therefore, the output of the algorithm is a restart time x with $\frac{x}{F(x)} \leq (1+\varepsilon)\frac{t^*}{F(t^*)}$. $\qquad\square$

Corollary 1. APPROXRESTARTTIME *is an FPTAS which terminates in* $\lceil \frac{\ln k}{\ln (1+\varepsilon)} \rceil$ *loop cycles.*

Proof. Let l be the smallest integer with $(1+\varepsilon)^l \geq k$. Simple arithmetics yield $l \geq \frac{\ln k}{\ln (1+\varepsilon)}$. By definition, l is $\lceil \frac{\ln k}{\ln (1+\varepsilon)} \rceil$. After l loop cycles x takes the value $(1+\varepsilon)^l \geq k$ and the algorithm terminates.

We use the well-known inequality $\frac{2x}{2+x} \leq \ln (1+x)$ to bound the required number of loop cycles from above:

$$\left\lceil \frac{\ln k}{\ln (1+\varepsilon)} \right\rceil \leq \left\lceil \left(\frac{1}{\varepsilon} + \frac{1}{2} \right) \ln k \right\rceil. \tag{7}$$

Thus, the runtime of APPROXRESTARTTIME is in $O\left(\frac{1}{\varepsilon} \ln k \right)$ which is linear in both $\frac{1}{\varepsilon}$ and the size of k. Therefore, APPROXRESTARTTIME is an FPTAS. □

In other words, APPROXRESTARTTIME finds a good approximation for the restart time on the interval $\{1, \ldots, k\}$ with respect to the $t/F(t)$. However, $t/F(t)$ is just an upper bound for the expected runtime $E[X_t]$ according to Eq. 3. When choosing a restart time, it is more appropriate to choose a restart time which has certain guarantees with respect to the true optimal expected value $E[X_{t^*}]$, where $t^* = \arg\inf_t E[X_t]$ is an optimal restart time.

In fact, we show that it is possible to use a slight modification of APPROX-RESTARTTIME to obtain a $4 + \varepsilon$ approximation algorithm for restart times with respect to the expected value. To this end, we use a property first stated by Luby et al [7].

Lemma 1. *Let X be any positive random variable and let F be its associated cumulative distribution function. Then,*

$$\inf_t E[X_t] \leq \inf_x \frac{x}{F(x)} \leq 4 \inf_t E[X_t] \tag{8}$$

holds.

A notable property of the upper bound $t/F(t)$ is that it always approaches its minimum for a finite value of t. Therefore, for an appropriate choice of k APPROXRESTARTTIME(F, ε, k) yields a restart time t with

$$\frac{t}{F(t)} \leq (1+\varepsilon) \inf_x \frac{x}{F(x)} \leq 4(1+\varepsilon) \inf_t E[X_t]. \tag{9}$$

Particularly, if a suitable interval $\{1, \ldots, k\}$ is known in advance, then the algorithm is an efficient approximation algorithm. However, the a priori knowledge of a suitable candidate k is not reasonable for many probability distributions.

The question then becomes whether another condition can replace the knowledge of k. In the following, we answer this question in the affirmative. First, consider the role of k in APPROXRESTARTTIME, there k is only used as the condition $x \leq k$ for the while loop. We replace this condition by $x \leq minvalue$.

Consider the following slightly modified algorithm.

function APPROXRESTARTTIMEEXACT(F, ε)
    $x := minx := 1$
    $minvalue := x/F(x)$
    while $x \leq minvalue$ **do**
        $x := (1 + \varepsilon) \cdot x$
        if $x/F(x) < minvalue$ **then**
            $minx := x$
            $minvalue := x/F(x)$
        end if
    end while
    return $minx$
end function

Corollary 2. *Let y be an optimal restart time with respect to $t/F(t)$ and let t^* be an optimal restart time with respect to $E[X_{t^*}]$. For every $\varepsilon > 0$ APPROX-RESTARTTIMEEXACT returns a restart time t with $E[X_t] \leq (4 + \varepsilon)E[X_{t^*}]$. The algorithm terminates after $O(\frac{q(|y|)+\log_2 y}{\log_2 (1+\varepsilon)})$ iterations where $|y|$ is the length of y.*

Proof. First, it is shown that APPROXRESTARTTIMEEXACT terminates after a finite number of iterations. Consider the case when $x > minvalue$. Then,

$$\frac{x}{F(x)} > \frac{minvalue}{F(x)} \geq minvalue$$

because $F(x) \leq 1$ for all x. In other words, all $x > minvalue$ are worse restart times than $minx$ and therefore the modified algorithm eventually terminates.

It remains to be shown how many iterations the modified algorithm needs until it terminates. As usual, we assume that F can be evaluated in polynomial-time. To be more precise, let $y = \arg\min_t \frac{t}{F(t)}$ be an optimal restart time. Since F can be evaluated in polynomial-time, the length of the binary representation $|F(y)|$ is bounded by some polynomial $q(|y|)$. Therefore, $F(y)$ is at least $2^{-q(|y|)}$ and we conclude

$$\frac{y}{F(y)} \leq y2^{q(|y|)}. \tag{10}$$

After some number of iterations, $minvalue$ is at most $(1 + \varepsilon)y2^{q(|y|)}$. Let $l + 1$ be the number of iterations until x is at least as big as $minvalue$, i.e., l is the number of iterations after which the algorithm terminates.

$$x = (1 + \varepsilon)^{l+1} \geq (1 + \varepsilon)y2^{q(|y|)} \tag{11}$$

$$\Leftrightarrow (1 + \varepsilon)^l \geq y2^{q(|y|)} \tag{12}$$

$$\Leftrightarrow l \geq \frac{q(|y|) + \log_2 y}{\log_2 (1 + \varepsilon)} \tag{13}$$

By definition, l is the smallest integer for which inequality 13 holds. Therefore, the algorithm terminates after $O(\frac{q(|y|)+\log_2 y}{\log_2 (1+\varepsilon)})$ iterations which is polynomial in both the size of y and $\frac{1}{\varepsilon}$.

The approximation factor is a consequence of Theorem 3 and Lemma 1. This completes the proof.   □

5   Conclusion and Outlook

We discussed several problems corresponding to probability distributions. Two models were studied: One assumes that the probability mass function (pmf) is given and the other uses the cumulative distribution function (cdf). Both models require that pmf / cdf is provided in a symbolic form.

We note that for the problems studied in this work the cdf versions are considerably simpler than its corresponding pmf versions. Specifically, we studied the complexity of verifying whether a given function is a pmf/cdf. Theorem 1 shows that the pmf version of this problem is $P(\#P)$-complete. On the other hand, the cdf version of this problem is CoNP-complete (Theorem 2). For the case when the cdf is known, an FPTAS for the optimal restart time is analyzed in Theorem 3. Unless $P = NP$, Lorenz [6] showed that there is no polynomial-time approximation algorithm for the pmf case.

This raises the question whether the pmf versions of a problem is always harder than its cdf counterpart. This question can be answered in the negative. Lorenz [6] found that the calculation of the mean and all other moments is #P-complete for both the pmf and the cdf version. However, the cdf version of every problem is always at most as hard as its pmf version. The reason is that for a cdf F the corresponding pmf is given by $F(t) - F(t-1)$. Therefore, there is always a polynomial-time reduction from the cdf problem to the corresponding pmf problem. This argument only holds for discrete probability distributions. Thus, for continuous distributions there might be properties for which the probability density function version of a problem is significantly easier than the cdf version.

More generally, it would be interesting to study the continuous generalizations of the problems presented here. This will need such framework as studied by Ko [4]. A starting point could be the approximation algorithms. The promise that the cdf can be evaluated in polynomial-time also means that the binary representation of each value is finite and its length is bounded by a polynomial. For real-valued functions, this property does not generally hold. Yet, not all bits have to be evaluated for deciding the condition $x/F(x) < minvalue$. This could be achieved by a lazy data structure. Nonetheless, a more careful approach and analysis are necessary.

There are also other loose ends to this work. So far, the computational complexity of probability distributions has only been considered in the context of restarts. Other questions could be parameter estimations and distinguishing between several types of distributions.

References

1. Even, S., Selman, A.L., Yacobi, Y.: The complexity of promise problems with applications to public-key cryptography. Inf. Control **61**(2), 159–173 (1984)

2. Goldreich, O.: On promise problems: a survey. In: Goldreich, O., Rosenberg, A.L., Selman, A.L. (eds.) Theoretical Computer Science. LNCS, vol. 3895, pp. 254–290. Springer, Heidelberg (2006). https://doi.org/10.1007/11685654_12
3. Gomes, C.P., Selman, B., Kautz, H.: Boosting combinatorial search through randomization. In: National Conference on Artificial Intelligence, pp. 431–437. AAAI Press (1998)
4. Ko, K.: Complexity Theory of Real Functions. Birkhäuser, Boston (1991)
5. Köbler, J., Schöning, U., Torán, J.: The Graph Isomorphism Problem - Its Structural Complexity. Birkhäuser, Boston (1993)
6. Lorenz, J.-H.: On the complexity of restarting. In: van Bevern, R., Kucherov, G. (eds.) CSR 2019. LNCS, vol. 11532, pp. 250–261. Springer, Cham (2019). https://doi.org/10.1007/978-3-030-19955-5_22
7. Luby, M., Sinclair, A., Zuckerman, D.: Optimal speed-up of las vegas algorithms. Inf. Process. Lett. **47**(4), 173–180 (1993)
8. Schöning, U.: A probabilistic algorithm for k-SAT and constraint satisfaction problems. In: 40th Annual Symposium on Foundations of Computer Science, pp. 410–414. IEEE (1999)
9. Selman, A.L.: Promise problems complete for complexity classes. Inf. Comput. **78**(2), 87–98 (1988)
10. Toda, S.: The complexity of finding medians. In: Proceedings 31th Annual Symposium on Foundations of Computer Science, pp. 778–787. IEEE (1990)
11. Toda, S.: PP is as hard as the polynomial-time hierarchy. SIAM J. Comput. **20**, 865–877 (1991)

On Nonadaptive Reductions to the Set of Random Strings and Its Dense Subsets

Shuichi Hirahara[1] and Osamu Watanabe[2(✉)]

[1] National Institute of Informatics, Tokyo, Japan
s_hirahara@nii.ac.jp
[2] Tokyo Institute of Technology, Tokyo, Japan
watanabe@c.titech.ac.jp

Abstract. We explain our recent results [21] on the computational power of an arbitrary distinguisher for (not necessarily computable) hitting set generators. This work is motivated by the desire of showing the limits of black-box reductions to some distributional NP problem. We show that a black-box nonadaptive randomized reduction to any distinguisher for (not only polynomial-time but also) exponential-time computable hitting set generators can be simulated in $\mathsf{AM} \cap \mathsf{coAM}$; we also show an upper bound of $\mathsf{S}_2^{\mathsf{NP}}$ even if there is no computational bound on a hitting set generator. These results provide additional evidence that the recent worst-case to average-case reductions within NP shown by Hirahara (2018, FOCS) are inherently non-black-box. (We omit all detailed arguments and proofs, which can be found in [21].)

Dedication to Ker-I from Osamu

I, Osamu Watanabe, (with my co-author, Shuichi Hirahara) dedicate this article to my senior colleague and good friend Ker-I Ko. I met Ker-I in 1985 when I visited University of California, Santa Barbara (UCSB) for participating in a small work shop organized by Ron, Professor Ronald V. Book. We then met again when I was a Key Fan visiting professor at Department of Mathematics, UCSB from 1987 to 1988. He was visiting Ron around that time. We discussed a lot on various things almost every day with me sitting in his office for many hours. I still recall him saying "Osamu, you know what?", which was usually followed by an interesting episode of famous researchers, politicians, among other things. This period was very important for me to develop my career as a computer scientist, in particular, in theoretical computer science. Certainly, I learnt a lot from Ker-I. I am also proud of having the following sentence in the acknowledgement of his paper [25]:

> The author would like to thank Ronald Book and Osamu Watanabe. Without their *help*, this work would never be finished *in polynomial time*.

During that time, we discussed a lot on the structure of complexity classes such as reducibilities, relativiations, sparse sets, approximability, etc. For example, we spent a lot of time trying to improve Mahaney's theorem: For any NP-complete set L, if L is polynomial-time many-one reducible to a sparse set, then L

© Springer Nature Switzerland AG 2020
D.-Z. Du and J. Wang (Eds.): Ko Festschrift, LNCS 12000, pp. 67–79, 2020.
https://doi.org/10.1007/978-3-030-41672-0_6

is indeed in P; that is, it is polynomial-time computable. Since then, the complexity theory has been developed (not so rapidly but) steadily. Several important notions have been introduced, and many powerful computational/mathematical tools have been developed for analyzing computability of various types. In this article, we are glad to explain our result that is much stronger (in several aspects emphasized below with underlined comments) than Mahaney's theorem. One of the results stated in Theorem 1 here can be interpreted as follows: For any set L (for which no complexity class assumption is needed) if L is randomized polynomial-time nonadaptively and "robustly" reducible (which is much more general than the one considered in Mahaney's theorem) to a relatively small density set (that could be much larger than sparse sets), then L is indeed in $\mathsf{S}_2^{\mathsf{NP}}$. Another interesting and exciting point of our results is that it is motivated from a question in a quite different context, the average-case vs. the worst-case complexity in NP, which was also one of the topics that I discussed with Ker-I with no idea at all of how to attack it at that time. Hope Ker-I would like these results and the following explanation.

1   Introduction

We explain our recent investigation on what can be reduced to the set of random strings, and its dense subset, which is related to several lines of research of complexity theory – including average-case complexity and black-box reductions, hitting set generators, the Minimum Circuit Size Problem, and the computational power of the set of random strings.

The underlying theme that unifies these research lines is Kolmogorov complexity. *Kolmogorov complexity* enables us to quantify how a finite string looks "random" in terms of compressibility. For a string $x \in \{0,1\}^*$, its Kolmogorov complexity is the length of the shortest program d such that running d will print x. More specifically, we fix an arbitrary universal Turing machine U, and the Kolmogorov complexity of x is defined as $\mathrm{K}_U(x) := \min\{\,|d| \mid U(d) = x\,\}$. A string x is called *random* (with threshold s) if $\mathrm{K}_U(x) \geq s$, i.e., x cannot be compressed into a short program. While Kolmogorov complexity is not computable, by either imposing a time constraint on U or taking another "decoder" U, we are led to several important concepts of complexity theory mentioned above. Below, we review these concepts through the lens of Kolmogorov complexity.

An important motivation for this work is the case when a decoder U is defined as a circuit interpreter G^{int}: Let G^{int} denote the function that takes a description of a Boolean circuit C, and outputs the truth table of the function computed by C. Here a *truth table* of a function $f\colon \{0,1\}^n \to \{0,1\}$ is the string of length 2^n that can be obtained by concatenating $f(x)$ for every input $x \in \{0,1\}^n$, and we often identify a function with its truth table. Taking $U = G^{\text{int}}$, the Kolmogorov complexity $\mathrm{K}_{G^{\text{int}}}(f)$ is approximately equal to the minimum circuit size for computing f. Therefore, a circuit lower bound question can be seen as a question of finding a random string f with respect to $\mathrm{K}_{G^{\text{int}}}$. For example, one of the central open questions in complexity theory, $\mathsf{E} \not\subset \mathsf{SIZE}(2^{\epsilon n})$ for some constant $\epsilon > 0$, can be equivalently rephrased as the question whether there exists a polynomial-time algorithm that, on input 1^N, finds a "random" string f of length N such that $K_{G^{\text{int}}}(f) = N^{\Omega(1)}$ for infinitely many N. The problem of computing $K_{G^{\text{int}}}(f)$ on input f is called the Minimum Circuit Size Problem (MCSP) [24], which is intensively studied recently.

A dense subset of random strings (with respect to $K_{G^{\text{int}}}$) is also one of the important concepts in complexity theory, which was called a natural property by Razborov and Rudich [30]. In their influential work, Razborov and Rudich introduced the notion of natural proof, and explained the limits of current proof techniques for showing circuit lower bounds. A *natural property* $R \subset \{0,1\}^*$ is a polynomial-time computable $1/\text{poly}(\ell)$-dense subset of random strings with respect to $K_{G^{\text{int}}}$. Here, a set is called γ-*dense* if $\Pr_{x \in_R \{0,1\}^\ell}[x \in R] \geq \gamma(\ell)$ for every $\ell \in \mathbb{N}$. It is known that a natural property is equivalent to an errorless average-case algorithm for MCSP [19].

More generally, a dense subset of random strings with respect to K_G can be seen as an adversary for a hitting set generator G. We consider a family of functions $G = \{G_\ell : \{0,1\}^{s(\ell)} \to \{0,1\}^\ell\}_{\ell \in \mathbb{N}}$. A *hitting set generator* (HSG) is the notion that is used to derandomize one-sided-error randomized algorithms. For a set $R \subset \{0,1\}^*$, we say that G is a hitting set generator (with parameter γ) for R if $\Pr_{r \in_R \{0,1\}^\ell}[r \in R] \geq \gamma(\ell)$ implies $R \cap \text{Im}(G_\ell) \neq \varnothing$, for every $\ell \in \mathbb{N}$. Conversely, R is said to γ-*avoid* G if G is not a hitting set generator for R, that is, (1) $\Pr_{r \in_R \{0,1\}^\ell}[r \in R] \geq \gamma(\ell)$ for all $\ell \in \mathbb{N}$ (i.e., R is γ-dense), and (2) $R \cap \text{Im}(G_\ell) = \varnothing$ (i.e., R does not intersect with the image $\text{Im}(G_\ell)$ of G_ℓ). Since $\text{Im}(G_\ell)$ contains all the non-random strings with respect to K_{G_ℓ}, this definition means that R is a γ-dense subset of random strings with respect to K_G.

Next, we proceed to reviewing each research line. We start with average-case complexity and black-box reductions.

2 Reducing from the Worst-Case to the Average-Case: Limits of Black-Box Reductions

The security of modern cryptography is based on average-case hardness of some computational problems in NP. It is, however, a challenging question to find a problem in NP that is hard with respect to a random input generated efficiently. The fundamental question of average-case complexity is to find a problem in NP whose average-case hardness is based on the worst-case complexity of an NP-complete problem.

A line of work was devoted to understanding why resolving this question is so difficult. Given our limited understanding of unconditional lower bounds, the most prevailing proof technique in complexity theory for showing intractability of a problem is by means of reductions. Moreover, almost all reduction techniques are *black-box* in the sense that, given two computational problems A and B, a reduction R solves A given any oracle (i.e., a black-box algorithm) solving B. The technique of reductions led to the discovery of a large number of NP-complete problems computationally equivalent to each other—in the *worst-case* sense. On the other hand, it turned out that the power of black-box reductions is limited for the purpose of showing intractability of average-case problems based on worst-case problems.

Building on the work of Feigenbaum and Fortnow [11], Bogdanov and Trevisan [9] showed that if a worst-case problem L is reducible to some average-case problem in NP via a nonadaptive black-box randomized polynomial-time reduction, then L must be in NP/poly $\cap$ coNP/poly. This in particular shows that the hardness of any average-case problem in NP cannot be based on the worst-case hardness of an NP-complete problem via such a reduction technique (unless the polynomial-time hierarchy collapses [34]). Akavia, Goldreich, Goldwasser and Moshkovitz [1,2] showed that, in the special case of a nonadaptive reduction to the task of inverting a one-way function, the upper bound of [9] can be improved to AM $\cap$ coAM, thereby removing the advice "/poly". Bogdanov and Brzuska [8] showed that even a general (i.e. adaptive) reduction to the task of inverting a size-verifiable one-way function cannot be used for any problem outside AM $\cap$ coAM. Applebaum, Barak, and Xiao [7] studied black-box reductions to PAC learning, and observed that the technique of [1] can be applied to (some restricted type of) a black-box reduction to the task of inverting an auxiliary-input one-way function.

3   A Motivation for Investigating Non-black-box Reductions Further

It was very recent that the first worst-case to average-case reductions from worst-case problems conjectured to be outside coNP to some average-case problems in NP were found: Hirahara [18] showed that approximation versions of the minimum time-bounded Kolmogorov complexity problem (MINKT [26]) and MCSP admit worst-case to average-case reductions. These problems ask, given a string x and a threshold s, whether x can be compressed by certain types of algorithms of size s. For example, MCSP asks whether x can be compressed as a truth table of a circuit of size at most s. For a constant $\epsilon > 0$, its approximation version $\mathrm{Gap}_\epsilon\mathrm{MCSP}$ is the problem of approximating the minimum circuit size for a function $f \colon \{0,1\}^n \to \{0,1\}$ (represented as its truth table) within a factor of $2^{(1-\epsilon)n}$. Specifically, the YES instances of $\mathrm{Gap}_\epsilon\mathrm{MCSP}$ consists of (f,s) such that $\mathsf{size}(f) \leq s$, and the NO instances of $\mathrm{Gap}_\epsilon\mathrm{MCSP}$ consists of (f,s) such that $\mathsf{size}(f) > 2^{(1-\epsilon)n}s$. MCSP can be defined as $\mathrm{Gap}_1\mathrm{MCSP}$. It is easy to see that MCSP $\in$ NP and MINKT $\in$ NP, but these are important examples of problems for which there is currently neither a proof of NP-completeness nor evidence against NP-completeness. Allender and Das [4] showed that MCSP is SZK-hard, but this hardness result is unlikely to be improved to NP-hardness using "oracle-independent" reduction techniques: Hirahara and Watanabe [20] showed that a one-query randomized polynomial-time reduction to MCSP^A for every oracle A can be simulated in AM$\cap$coAM. Nonetheless, MCSP and MINKT are (indirectly) conjectured to be outside coNP/poly by Rudich [31] based on some assumptions of average-case complexity: He conjectured that there exists a (certain type of) hitting set generator secure even against nondeterministic polynomial-size circuits. We also mention that the approximation version of MINKT is harder than Random 3SAT, which is conjectured by Ryan O'Donnell (cf. [19]) to not be solvable by coNP algorithms.

The work of Hirahara motivates us to study black-box reductions further. We ask whether the technique used in [18] is inherently non-black-box or not. As mentioned above, there are several results and techniques developed in order to simulate black-box reductions by AM∩coAM algorithms. Why can't we combine these techniques with the (seemingly non-black-box) reductions of [18] in order to prove $Gap_\epsilon MCSP \in coAM$ and refute Rudich's conjecture? Note that refuting Rudich's conjecture would significantly change our common belief about average-case complexity and the power of nondeterministic algorithms. We emphasize that while the proof of [18] seems to yield only non-black-box reductions, it does not necessarily mean that there is no alternative proof that yields a black-box reduction.

In order to address the question, we aim at improving our understanding of the limits of black-box reductions. We summarize a landscape around average-case complexity in Fig. 1.

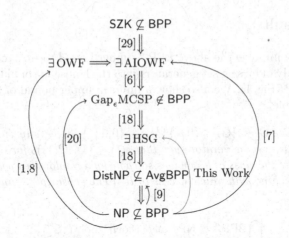

Fig. 1. Average-case complexity and limits of black-box reductions. "$A \to B$" means that there is no black-box (or oracle-independent) reduction technique showing "$A \Rightarrow B$" under reasonable complexity theoretic assumptions. The security of all cryptographic primitives is with respect to an almost-everywhere polynomial-time randomized adversary.

A couple of remarks about implications written in Fig. 1 are in order: First, the implication from the existence of an auxiliary-input one-way function (AIOWF) to $Gap_\epsilon MCSP \notin BPP$ was implicitly proved in [3] and explicitly in [6], based on [13,17,30]. The implication from $SZK \not\subseteq BPP$ to the existence of an auxiliary-input one-way function is due to Ostrovsky [29] (see also [33]). Second, building on [10, 19], it was shown in [18, Theorem VI.5] that $\mathrm{Gap}_\epsilon\mathsf{MCSP} \notin \mathsf{BPP}$ implies the nonexistence of natural properties, which yields a hitting set generator $G^{\mathrm{int}} = \{G_{2^n} : \{0,1\}^{\tilde{O}(2^{\epsilon' n})} \to \{0,1\}^{2^n}\}_{n\in\mathbb{N}}$ defined as a "circuit interpreter": a function that takes a description of a circuit of size $2^{\epsilon' n}$ and outputs its truth table (cf. [18, Definition V.3]). The existence of a hitting set generator naturally induces a hard problem in DistNP with respect to AvgBPP algorithms (cf. [18, Lemma VI.4]). Therefore, the reduction of [18] can be regarded as a non-black-box (in fact, nonadaptive) reduction to a distinguisher for the hitting set generator G^{int}.

We thus continue the study of the limits of black-box reductions to a distinguisher for a hitting set generator, initiated by Gutfreund and Vadhan [15]. Motivated by the question on whether derandomization is possible under uniform assumptions (cf. [32]), they investigated what can be reduced to any oracle avoiding a hitting set generator in a black-box way.[1] They showed that any polynomial-time randomized nonadaptive black-box reductions to any oracle avoiding an exponential-time computable hitting set generator G can be simulated in $\mathsf{BPP}^{\mathsf{NP}}$, which is a trivial upper bound when G is polynomial-time computable.

4   Our Results

We significantly improve the above $\mathsf{BPP}^{\mathsf{NP}}$ upper bound to $\mathsf{AM} \cap \mathsf{coAM}$, thereby putting the study of hitting set generators into the landscape of black-box reductions within NP (Fig. 1). We also show a uniform upper bound of $\mathsf{S}_2^{\mathsf{NP}}$ even if G is not computable.

Theorem 1. *Let $G = \{G_\ell : \{0,1\}^{s(\ell)} \to \{0,1\}^\ell\}_{\ell\in\mathbb{N}}$ be any (not necessarily computable) hitting set generator such that $s(\ell) \leq (1 - \Omega(1))\ell$ for all large $\ell \in \mathbb{N}$. Let $\mathsf{BPP}_\|^R$ denote the class of languages solvable by a randomized polynomial-time nonadaptive machine with oracle access to R. (The subscript $\|$ stands for parallel queries.) Then,*

$$\bigcap_R \mathsf{BPP}_\|^R \subset \mathsf{NP}/\mathsf{poly} \cap \mathsf{coNP}/\mathsf{poly} \cap \mathsf{S}_2^{\mathsf{NP}},$$

where the intersection is taken over all oracles R that $(1 - 1/\mathrm{poly}(\ell))$-avoid G. Moreover, if G_ℓ is computable in $2^{O(\ell)}$, then we also have

$$\bigcap_R \mathsf{BPP}_\|^R \subset \mathsf{AM} \cap \mathsf{coAM}.$$

---

[1] As a *black-box* reduction to any distinguisher for G, it is required in [15] that there exists a *single* machine that computes a reduction to every oracle avoiding G. On the other hand, as stated in Theorem 1, we allow reductions to depend on oracles, which makes our results stronger.

Compared to the line of work showing limits of black-box reductions within NP, a surprising aspect of Theorem 1 is that it generalizes to any function G that may not be computable. Indeed, almost all the previous results [1,7,9,11] crucially exploit the fact that a verifier can check the correctness of a certificate for an NP problem; thus a dishonest prover can cheat the verifier only for one direction, by not providing a certificate for a YES instance. In our situation, a verifier cannot compute G and thus cannot prevent dishonest provers from cheating in this way. At a high level, our technical contributions are to overcome this difficulty by combining the ideas of Gutfreund and Vadhan [15] with the techniques developed in [9,11].

Moreover, we present a new S_2^p-type algorithm for simulating reductions to an oracle R avoiding G. Indeed, at the core of Theorem 1 is the following two types of algorithms simulating reductions: One is an S_2^p algorithm that simulates any query $q \stackrel{?}{\in} R$ of length at most $\Theta(\log n)$, and the other is an AM∩coAM algorithm that simulates any query $q \stackrel{?}{\in} R$ of length at least $\Theta(\log n)$. In particular, when G is exponential-time computable, the S_2^p algorithm can be replaced with a polynomial-time algorithm and obtain the AM ∩ coAM upper bound.

We remark that Theorem 1 improves all the previous results mentioned before in some sense. Compared to [9], our results show that the advice "/poly" is not required in order to simulate black-box reductions to any oracle avoiding an exponential-time computable hitting set generator. Compared to [1,7], our results "conceptually" improve their results because the existence of one-way functions imply the existence of hitting set generators; on the other hand, since the implication goes through the *adaptive* reduction (from the task of inverting a one-way function to a distinguisher for a PRG) of [17], technically speaking, our results are incomparable with their results.[2]Similarly, our results conceptually improve the result of [20], but these are technically incomparable, mainly because the implication goes through the non-black-box reduction of [18].

5   Why Are the Reductions of [18] Non-black-box?

Based on Theorem 1, we now argue that the reductions of [18] are inherently non-black-box in a certain formal sense, without relying on any unproven assumptions: The reason is that the idea of [18] can be applied to not only time-bounded Kolmogorov complexity but also any other types of Kolmogorov complexity, including resource-unbounded Kolmogorov complexity. Therefore, if this generalized reduction could be made black-box, then (as outlined below) by Theorem 1

---

[2] We emphasize that we are concerned the nonadaptivity of reductions used in the security proof of pseudorandom generators. Several simplified constructions of pseudorandom generators G^f from one-way functions f (e.g., [16,23]) are nonadaptive in the sense that G^f can be efficiently computed with nonadaptive oracle access to f; however, the security reductions of these constructions are adaptive because of the use of Holenstein's uniform hardcore lemma [22]. Similarly, the reduction of [17, Lemma 6.5] is adaptive. (We note that, in the special case when the degeneracy of a one-way function is efficiently computable, the reduction of [17] is nonadaptive.).

we would obtain a finite algorithm S_2^{NP} that approximates resource-unbounded Kolmogorov complexity, which is a contradiction, *unconditionally*.

To give one specific example, we briefly outline how the reductions of [18] can be generalized to the case of Levin's Kt-complexity [27]: Fix any efficient universal Turing machine U, and the Kt-complexity of a string x is defined as

$$\text{Kt}(x) := \min\{|d| + \log t \mid U(d) \text{ outputs } x \text{ within } t \text{ steps }\}.$$

We define a hitting set generator $G = \{G_\ell : \{0,1\}^{\ell/2} \to \{0,1\}^\ell\}_{\ell \in \mathbb{N}}$ as $G_\ell(d,t) := U(d)$ for $(d,t) \in \{0,1\}^{\ell/2}$ when $|U(d)| = \ell$ and $U(d)$ halts within t steps, which is computable in exponential time. Note that $\text{Im}(G)$ contains all strings with low Kt-complexity. Given an efficient algorithm D that γ-avoids G, we can approximate $\text{Kt}(x)$ by the following algorithm: Fix any input x. Take any list-decodable code Enc, and let $\text{NW}^{\text{Enc}(x)}(z)$ denote the Nisan-Wigderson generator [28] instantiated with $\text{Enc}(x)$ as the truth table of a hard function, where z is a seed of the generator. Then check whether the distinguishing probability $|\mathbb{E}_{z,w}[D(\text{NW}^{\text{Enc}(x)}(z)) - D(w)]|$ is large or small by sampling, whose outcome tells us whether $\text{Kt}(x)$ is small or large, respectively. Indeed, if the distinguishing probability is large, then by using the security proof of the Nisan-Wigderson generator, we obtain a short description (with oracle access to D) for x. Conversely, if $\text{Kt}(x)$ is small, then since D γ-avoids G, the distinguishing probability is at least γ. Now, if we could make this analysis work for any oracle that γ-avoids G, then by Theorem 1 we would put a problem of approximating $\text{Kt}(x)$ in AM, which is not possible unless EXP = PH. (Note that the minimization problem of Kt is EXP-complete under NP reductions [3].)

6   Our Techniques

We outline our proof strategy for Theorem 1 below. Suppose that we have some reduction from L to any oracle R that avoids a hitting set generator G. Let $\mathcal{Q}$ denote the query distribution that a reduction makes. We focus on the case when the length of each query is larger than $\Theta(\log n)$, and explain the ideas of the AM $\cap$ coAM simulation algorithms.

As a warm-up, consider the case when the support $\text{supp}(\mathcal{Q})$ of $\mathcal{Q}$ is small (i.e., $|\text{supp}(\mathcal{Q}) \cap \{0,1\}^\ell| \ll 2^\ell$ for any length $\ell \in \mathbb{N}$). In this case, we can define an oracle R_1 so that $R_1 := \{0,1\}^* \setminus \text{supp}(\mathcal{Q}) \setminus \text{Im}(G)$; this is a dense subset and avoids the hitting set generator G. Therefore, we can simulate the reduction by simply answering all the queries by saying "No"; hence such a reduction can be simulated in BPP.

In general, we cannot hope that $\text{supp}(\mathcal{Q})$ is small enough. To generalize the observation above, let us recall the notion of α-heaviness [9]: We say that a query q is α-*heavy* (with respect to $\mathcal{Q}$) if the query q is α times more likely to be sampled under $\mathcal{Q}$ than the uniform distribution on $\{0,1\}^{|q|}$; that is, $\Pr_{w \sim \mathcal{Q}}[w = q] \geq \alpha 2^{-|q|}$. Now we define our new oracle $R_2 := \{0,1\}^* \setminus \{q \in \{0,1\}^* \mid q : \alpha\text{-heavy} \setminus \text{Im}(G)\}$, which can be again shown to avoid G because the fraction of α-heavy queries is at most $1/\alpha$ ($\ll 1$).

The problem now is that it is difficult to simulate the new oracle R_2; it appears that, given a query q, we need to test whether $q \overset{?}{\in} \mathrm{Im}(G)$, which is not possible in AM $\cap$ coAM. However, it turns out that we do not need to test it, as we explain next: Observe that the size of $\mathrm{Im}(G)$ is very small; it is at most $2^{s(\ell)}$ $\left(\ll 2^{\ell} \right)$. Thus, the probability that a query q is in $\mathrm{Im}(G)$ and q is not α-heavy (i.e., q is rarely queried) is at most $\alpha \cdot 2^{s(\ell) - \ell}$, where ℓ is the length of q. As a consequence, the reduction cannot "distinguish" the oracle R_2 and a new oracle $R_3 := \{0,1\}^* \setminus \{ q \in \{0,1\}^* \mid q \colon \alpha\text{-heavy} \}$; hence we can simulate the reduction if, given a query q, we are able to decide whether $q \overset{?}{\in} R_3$ in AM$\cap$coAM.

This task, however, still appears to be difficult for AM$\cap$coAM; indeed, at this point, Gutfreund and Vadhan [15] used the fact that the approximate counting is possible in BPP^{NP}, and thereby simulated the oracle R_3 by BPP^{NP}.

Our main technical contribution is to develop a way of simulating the reduction to R_3. First, note that the lower bound protocol of Goldwasser and Sipser [14] enables us to give an AM certificate for α-heaviness; we can check, given a query q, whether q is $\alpha(1 + \epsilon)$-heavy or α-light for any small error parameter $\epsilon > 0$. Thus, we have an AM protocol for $\{0,1\}^* \setminus R_3$ for every query q (except for $\alpha(1 \pm \epsilon)$-heavy and light queries).

If, in addition, we had an AM protocol for R_3, then we would be done; unfortunately, it does not seem possible in general. The upper bound protocol of Fortnow [12] does a similar task, but the protocol can be applied only for a limited purpose: we need to keep the randomness used to generate a query $q \sim \mathcal{Q}$ from being revealed to the prover. When the number of queries of the reduction is limited to 1, we may use the upper bound protocol in order to give an AM certificate for R_3; on the other hand, if the reduction makes two queries $(q_1, q_2) \sim \mathcal{Q}$, we cannot simultaneously provide AM certificates of the upper bound protocol for *both* of q_1 and q_2, because the fact that q_1 and q_2 are sampled *together* may reveal some information about the private randomness. To summarize, the upper bound protocol works only for the *marginal* distribution of each query, but does not work for the *joint* distribution of several queries.

That is, what we can obtain by using the upper bound protocol is information about *each* query. For example, the heavy-sample protocol of Bogdanov and Trevisan [9] (which combines the lower and upper bound protocol and sampling) estimates, in AM $\cap$ coAM, the probability that a query q sampled from $\mathcal{Q}$ is α-heavy.

Our idea is to overcome the difficulty above by generalizing the Feigenbaum-Fortnow protocol [11]. Feigenbaum and Fortnow developed an AM$\cap$coAM protocol that simulates a nonadaptive reduction to an NP oracle R, given as advice the probability that a query is a positive instance of R. We generalize the protocol in the case when the oracle $\{0,1\}^* \setminus R_3$ is solvable by AM on average (which can be done by the lower bound protocol [14]), and given as advice the probability that a query q is in $\{0,1\}^* \setminus R_3$ (which can be estimated by the heavy-sample protocol [9]):

Theorem 2 (Generalized Feigenbaum-Fortnow Protocol; Informal).
Suppose that M is a randomized polynomial-time nonadaptive reduction to oracle R whose queries are distributed according to $\mathcal{Q}$, and that R is solvable by AM *on average (that is, there exists an* AM *protocol Π_R such that, with probability $1 - 1/\mathrm{poly}(n)$ over the choice of $q \sim \mathcal{Q}$, the protocol Π_R computes R on input q). Then, there exists an* AM $\cap$ coAM *protocol Π_M such that, given a probability $p^* \approx \mathrm{Pr}_{q \sim \mathcal{Q}}[q \in R]$ as advice, the protocol Π_M simulates the reduction M with probability at least $1 - 1/\mathrm{poly}(n)$.*

On the Case of Adaptive Reductions. We mention that Theorem 1 cannot be extended to the case of *adaptive* reductions. Indeed, Trevisan and Vadhan [32] constructed an exponential-time computable pseudorandom generator based on the intractability of some PSPACE-complete problem, and its security reduction is black-box in the sense of Theorem 1 and adaptive. If Theorem 1 could be extended to the case of adaptive reductions, we would obtain PSPACE = AM, which is unlikely to be true.

7   Some Evidence for the Tightness of Our Upper Bounds

Theorem 1 leads us to the natural question whether the upper bound is tight. We present evidence that our two types of simulation algorithms are nearly tight.

First consider the AM $\cap$ coAM-type simulation algorithms. In [21] we observe that the SZK-hardness of MCSP [4] also holds for an average-case version of MCSP:

Theorem 3. *Let $\epsilon > 0$ be any constant, and R be any oracle $\frac{1}{2}$-avoiding $G^{\mathrm{int}} = \{G_n^{\mathrm{int}} : \{0,1\}^{n^\epsilon} \to \{0,1\}^n\}_{n \in \mathbb{N}}$. Then,* SZK $\subset$ BPP^R.

The reduction of Theorem 3 is adaptive because of the use of [17]. We conjecture that SZK $\subset \bigcap_R$ BPP$_\parallel^R$, which implies that the AM $\cap$ coAM upper bound of Theorem 1 cannot be significantly improved.

Next consider our S_2^p-type simulation algorithm. This is in fact completely tight in a certain setting. Let G be a universal Turing machine. We consider an exponential-time analogue of Theorem 1 when the reduction can make only short queries. Specifically, for an oracle R, denote by EXP$^{R^{\leq \mathrm{poly}}}$ the class of languages that can be computed by a $2^{n^{O(1)}}$-time algorithm that can query $q \overset{?}{\in} R$ of length $\leq n^{O(1)}$, on inputs of length n. Then by an exponential-time analogue of Theorem 1 (more specifically, by using the S_2^p-type simulation algorithm), we can show the following upper bound on the computational power of EXP$^{R^{\leq \mathrm{poly}}}$ where R is an arbitrary dense subset of Kolmogorov-random strings, i.e., R is a set avoiding the outputs of a universal Turing machine U on short inputs. (We note that all the queries of polynomial length can be asked by an exponential-time reduction, and thus the adaptivity does not matter here.)

Theorem 4. *Fix any universal Turing machine* U. *Then we have*

$$\bigcap_{R:\ \frac{1}{2}\text{-avoids } U} \mathsf{EXP}^{R^{\leq poly}} \subseteq \bigcap_{R:\ \frac{1}{2}\text{-avoids}U} \mathsf{BPEXP}^{R^{\leq poly}} \subseteq \mathsf{S}_2^{exp}.$$

Here $R^{\leq poly}$ *means that the length of queries is restricted to be at most a polynomial in the input length. We also have* $\mathsf{EXP}^{\mathsf{NP}} \subset \bigcap_R \mathsf{S}_2^R \subset \mathsf{S}_2^{exp}$.

Previously, Allender, Friedman and Gasarch [5] showed that black-box BPP reductions to any avoiding oracle can be simulated in EXPSPACE. Theorem 4 significantly improves their upper bound to S_2^{exp}. What is interesting here is that we can also show [21] the same lower bound, that is,

$$\bigcap_{R:\ \frac{1}{2}\text{-avoids}U} \mathsf{EXP}^{R^{\leq poly}} \supseteq \mathsf{S}_2^{exp}$$

Thus, a complexity class, i.e., the exponential-time analogue of S_2^p, is exactly characterized by using Kolmogorov-random strings. The above lower bound also shows the tightness of the exponential-time analogue of the S_2^p-type simulation algorithm.

References

1. Akavia, A., Goldreich, O., Goldwasser, S., Moshkovitz, D.: On basing one-way functions on NP-hardness. In: Proceedings of the Symposium on Theory of Computing (STOC), pp. 701–710 (2006)
2. Akavia, A., Goldreich, O., Goldwasser, S., Moshkovitz, D.: Erratum for: on basing one-way functions on NP-hardness. In: Proceedings of the Symposium on Theory of Computing (STOC), pp. 795–796 (2010)
3. Allender, E., Buhrman, H., Koucký, M., van Melkebeek, D., Ronneburger, D.: Power from random strings. SIAM J. Comput. **35**(6), 1467–1493 (2006)
4. Allender, E., Das, B.: Zero knowledge and circuit minimization. Inf. Comput. **256**, 2–8 (2017)
5. Allender, E., Friedman, L., Gasarch, W.I.: Limits on the computational power of random strings. Inf. Comput. **222**, 80–92 (2013)
6. Allender, E., Hirahara, S.: New insights on the (non-) hardness of circuit minimization and related problems. In: Proceedings of the International Symposium on Mathematical Foundations of Computer Science (MFCS), pp. 54:1–54:14 (2017)
7. Applebaum, B., Barak, B., Xiao, D.: On basing lower-bounds for learning on worst-case assumptions. In: Proceedings of the Symposium on Foundations of Computer Science (FOCS), pp. 211–220 (2008)
8. Bogdanov, A., Brzuska, C.: On basing size-verifiable one-way functions on NP-hardness. In: Dodis, Y., Nielsen, J.B. (eds.) TCC 2015. LNCS, vol. 9014, pp. 1–6. Springer, Heidelberg (2015). https://doi.org/10.1007/978-3-662-46494-6_1
9. Bogdanov, A., Trevisan, L.: On worst-case to average-case reductions for NP problems. SIAM J. Comput. **36**(4), 1119–1159 (2006)
10. Carmosino, M.L., Impagliazzo, R., Kabanets, V., Kolokolova, A.: Learning algorithms from natural proofs. In: Proceedings of the Conference on Computational Complexity (CCC), pp. 10:1–10:24 (2016)

11. Feigenbaum, J., Fortnow, L.: Random-self-reducibility of complete sets. SIAM J. Comput. **22**(5), 994–1005 (1993)
12. Fortnow, L.: The complexity of perfect zero-knowledge. Adv. Comput. Res. **5**, 327–343 (1989)
13. Goldreich, O., Goldwasser, S., Micali, S.: How to construct random functions. J. ACM **33**(4), 792–807 (1986)
14. Goldwasser, S., Sipser, M.: Private coins versus public coins in interactive proof systems. In: Proceedings of the Symposium on Theory of Computing (STOC), pp. 59–68 (1986)
15. Gutfreund, D., Vadhan, S.: Limitations of hardness vs. randomness under uniform reductions. In: Goel, A., Jansen, K., Rolim, J.D.P., Rubinfeld, R. (eds.) APPROX/RANDOM -2008. LNCS, vol. 5171, pp. 469–482. Springer, Heidelberg (2008). https://doi.org/10.1007/978-3-540-85363-3_37
16. Haitner, I., Reingold, O., Vadhan, S.P.: Efficiency improvements in constructing pseudorandom generators from one-way functions. SIAM J. Comput. **42**(3), 1405–1430 (2013)
17. Håstad, J., Impagliazzo, R., Levin, L.A., Luby, M.: A pseudorandom generator from any one-way function. SIAM J. Comput. **28**(4), 1364–1396 (1999)
18. Hirahara, S.: Non-black-box worst-case to average-case reductions within NP. In: Proceedings of the Symposium on Foundations of Computer Science (FOCS), pp. 247–258 (2018)
19. Hirahara, S., Santhanam, R.: On the average-case complexity of MCSP and its variants. In: Proceedings of the Computational Complexity Conference (CCC), pp. 7:1–7:20 (2017)
20. Hirahara, S., Watanabe, O.: Limits of minimum circuit size problem as oracle. In: Proceedings of the Conference on Computational Complexity (CCC), pp. 18:1–18:20 (2016)
21. Hirahara, S., Watanabe, O.: On nonadaptive reductions to the set of random strings and its dense subsets. In: Electronic Colloquium on Computational Complexity (ECCC), vol. 26, p. 25 (2019)
22. Holenstein, T.: Key agreement from weak bit agreement. In: Proceedings of the Symposium on Theory of Computing (STOC), pp. 664–673 (2005)
23. Holenstein, T.: Pseudorandom generators from one-way functions: a simple construction for any hardness. In: Halevi, S., Rabin, T. (eds.) TCC 2006. LNCS, vol. 3876, pp. 443–461. Springer, Heidelberg (2006). https://doi.org/10.1007/11681878_23
24. Kabanets, V., Cai, J.: Circuit minimization problem. In: Proceedings of the Symposium on Theory of Computing (STOC), pp. 73–79 (2000)
25. Ko, K.: On helping by robust oracle machines. Theor. Comput. Sci. **52**, 15–36 (1987)
26. Ko, K.: On the complexity of learning minimum time-bounded turing machines. SIAM J. Comput. **20**(5), 962–986 (1991)
27. Levin, L.A.: Randomness conservation inequalities; information and independence in mathematical theories. Inf. Control **61**(1), 15–37 (1984)
28. Nisan, N., Wigderson, A.: Hardness vs Randomness. J. Comput. Syst. Sci. **49**(2), 149–167 (1994)
29. Ostrovsky, R.: One-way functions, hard on average problems, and statistical zero-knowledge proofs. In: Proceedings of the Structure in Complexity Theory Conference, pp. 133–138 (1991)
30. Razborov, A.A., Rudich, S.: Natural proofs. J. Comput. Syst. Sci. **55**(1), 24–35 (1997)

31. Rudich, S.: Super-bits, demi-bits, and *NP/qpoly*-natural proofs. In: Rolim, J. (ed.) RANDOM 1997. LNCS, vol. 1269, pp. 85–93. Springer, Heidelberg (1997). https://doi.org/10.1007/3-540-63248-4_8
32. Trevisan, L., Vadhan, S.P.: Pseudorandomness and average-case complexity via uniform reductions. Comput. Complex. **16**(4), 331–364 (2007)
33. Vadhan, S.P.: An unconditional study of computational zero knowledge. SIAM J. Comput. **36**(4), 1160–1214 (2006)
34. Yap, C.: Some consequences of non-uniform conditions on uniform classes. Theor. Comput. Sci. **26**, 287–300 (1983)

Computability of the Solutions to Navier-Stokes Equations via Effective Approximation

Shu-Ming Sun[1], Ning Zhong[2]($\boxtimes$), and Martin Ziegler[3]

[1] Department of Mathematics, Virginia Tech, Blacksburg, VA 24061, USA
sun@math.vt.edu
[2] Department of Mathematical Sciences, University of Cincinnati, Cincinnati, OH 45221, USA
zhongn@ucmail.uc.edu
[3] School of Computing, KAIST, 291 Daehak-ro, Daejeon 34141, Republic of Korea
ziegler@kaist.ac.kr

The paper is dedicated to the memory of Professor Ker-I Ko.

Abstract. As one of the seven open problems in the addendum to their 1989 book *Computability in Analysis and Physics*, Pour-El and Richards proposed "... the recursion theoretic study of particular nonlinear problems of classical importance. Examples are the Navier-Stokes equation, the KdV equation, and the complex of problems associated with Feigenbaum's constant." In this paper, we approach the question of whether the Navier-Stokes Equation admits recursive solutions in the sense of Weihrauch's Type-2 Theory of Effectivity. A natural encoding ("representation") is constructed for the space of divergence-free vector fields on 2-dimensional open square $\Omega = (-1,1)^2$. This representation is shown to render first the mild solution to the Stokes Dirichlet problem and then a strong local solution to the nonlinear inhomogeneous incompressible Navier-Stokes initial value problem uniformly computable. Based on classical approaches, the proofs make use of many subtle and intricate estimates which are developed in the paper for establishing the computability results.

Keywords: Navier-Stokes equations · Computability

1 Introduction

The (physical) Church-Turing Hypothesis [17] postulates that every physical phenomenon or effect can, at least in principle, be simulated by a sufficiently

---

The third author is supported by grant NRF-2017R1E1A1A03071032.

D.-Z. Du and J. Wang (Eds.): Ko Festschrift, LNCS 12000, pp. 80–112, 2020.
https://doi.org/10.1007/978-3-030-41672-0_7

powerful digital computer up to any desired precision. Its validity had been chal-
lenged, though, in the sound setting of Recursive Analysis: with a computable
C^1 initial condition to the Wave Equation leading to an incomputable solution
[11,13]. The controversy was later resolved by demonstrating that, in both phys-
ically [1,30] and mathematically more appropriate Sobolev space settings, the
solution is computable uniformly in the initial data [23]. Recall that functions
f in a Sobolev space are not defined pointwise but by local averages in the L_q
sense[1] (in particular $q = 2$ corresponding to energy) with derivatives under-
stood in the distributional sense. This led to a series of investigations on the
computability of linear and nonlinear partial differential equations [24–26].

The (incompressible) Navier-Stokes Equation

$$\partial_t u - \triangle u + (u \cdot \nabla)u + \nabla P = f, \quad \nabla \cdot u = 0, \quad u(0) = a, \quad u\big|_{\partial\Omega} \equiv 0 \quad (1)$$

describes the motion of a viscous incompressible fluid filling a rigid box $\overline{\Omega}$. The
vector field $u = u(x, t) = (u_1, u_2, \ldots, u_d)$ represents the velocity of the fluid and
$P = P(x, t)$ is the scalar pressure with gradient ∇P; $\triangle$ is the Laplace operator;
$\nabla \cdot u$ denotes componentwise divergence; $u \cdot \nabla$ means, in Cartesian coordinates,
$u_1\partial_{x_1} + u_2\partial_{x_2} + \ldots + u_d\partial_{x_d}$; and the function $a = a(x)$ with $\nabla \cdot a = 0$ provides
the initial velocity and f is a given external force. Equation (1) thus constitutes
a system of $d + 1$ partial differential equations for $d + 1$ functions.

The question of global existence and smoothness of its solutions, even in the
homogeneous case $f \equiv 0$, is one of the seven Millennium Prize Problems posted
by the Clay Mathematics Institute at the beginning of the 21st century. Local
existence has been established, though, in various L_q settings [5]; and unique-
ness of weak solutions in dimension 2, but not in dimension 3 [18, §V.1.5], [2,
§V.1.3.1]. Nevertheless, numerical solution methods have been devised in abun-
dance, often based on pointwise (or even uniform, rather than L_q) approximation
and struggling with nonphysical artifacts [14]. In fact, the very last of seven open
problems listed in the addendum to [12] asks for a "recursion theoretic study
of ... the Navier-Stokes equation". Moreover it has been suggested [16] that
hydrodynamics could in principle be incomputable in the sense of allowing to
simulate universal Turing computation and to thus 'solve' the Halting problem. And indeed recent progress towards (a negative answer to) the Millennium
Problem [20] proceeds by simulating a computational process in the vorticity
dynamics to construct a blowup in finite time for a PDE very similar to (1).

1.1 Overview

Using the sound framework of Recursive Analysis, we assert the computability
of a local strong solution of (1) in the space $L_{2,0}^\sigma(\Omega)$ (see Sect. 2 for definition)
from a given initial condition $a \in L_{2,0}^\sigma(\Omega)$; moreover, the computation is uniform

---

[1] We use $q \in [1, \infty]$ to denote the norm index, P for the pressure field, p for polynomials, $\mathcal{P}$ for a set of trimmed and mollified tuples of the latter, and $\mathbb{P}$ for the Helmholtz
Projection.

in the initial data. We follow a common strategy used in the classical existence proofs [3–5, 18, 21]:

(i) Eliminate the pressure P by applying, to both sides of Eq. (1), the Helmholtz projection $\mathbb{P} : (L_2(\Omega))^2 \to L^\sigma_{2,0}(\Omega)$, thus arriving at the non-linear evolution equation

$$\partial_t u + \mathbb{A}u + \mathbb{B}u = g \quad (t > 0), \qquad u(0) = a \in L^\sigma_{2,0}(\Omega) \qquad (2)$$

where $L_2(\Omega)$ is the set of all square-integrable real-valued functions defined on Ω, $g = \mathbb{P}f$, $u = \mathbb{P}u \in L^\sigma_{2,0}(\Omega)$, $\mathbb{A} = -\mathbb{P}\triangle$ denotes the Stokes operator, and $\mathbb{B}u = \mathbb{P}(u \cdot \nabla)u$ is the nonlinearity.

(ii) Construct a mild solution $v(t)a = e^{-t\mathbb{A}}a$ of the associated homogeneous linear equation

$$\partial_t v + \mathbb{A}v = 0 \quad \text{for } t \geq 0, \qquad v(0) = a \in L^\sigma_{2,0}(\Omega) \qquad (3)$$

(iii) Rewrite (2) using (ii) in an integral form [5, §2]

$$u(t) = e^{-t\mathbb{A}}a + \int_0^t e^{-(t-s)\mathbb{A}}g(s)\,ds - \int_0^t e^{-(t-s)\mathbb{A}}\mathbb{B}u(s)\,ds \quad \text{for } t \geq 0 \ (4)$$

and solve it by a limit/fixed-point argument using the following iteration scheme [5, Eq. (2.1)]:

$$v_0(t) = e^{-t\mathbb{A}}a + \int_0^t e^{-(t-s)\mathbb{A}}g(s)\,ds, \quad v_{n+1}(t) = v_0(t) - \int_0^t e^{-(t-s)\mathbb{A}}\mathbb{B}v_n(s)\,ds \tag{5}$$

(iv) Recover the pressure P from u by solving

$$\nabla P = f - \partial_t u + \triangle u - (u \cdot \nabla)u \qquad (6)$$

To make use of the above strategy for deriving an algorithm to compute the solution of (1), there are several difficulties which need to be dealt with. Firstly, a proper representation is needed for coding the solenoidals. The codes should be not only rich enough to capture the functional characters of these vector fields but also robust enough to retain the coded information under elementary function operations, in particular, integration. Secondly, since the Stokes operator $\mathbb{A} : \text{dom}(\mathbb{A}) \to L^\sigma_{2,0}(\Omega)$ is neither continuous nor its graph dense in $(L^\sigma_{2,0}(\Omega))^2$, there is no convenient way to directly code $\mathbb{A}$ for computing the solution of the linear equation (3). The lack of computer-accessible information on $\mathbb{A}$ makes the computation of the solution $v(t)a = e^{-t\mathbb{A}}a$ of (3) much more intricate. Thirdly, since the nonlinear operator $\mathbb{B}$ in the iteration (5) involves differentiation and multiplication, and a mere $L^\sigma_{2,0}$-code of v_n is not rich enough for carrying out these operations, it follows that there is a need for computationally derive a stronger code for v_n from any given $L^\sigma_{2,0}$-code of a so that $\mathbb{B}v_n$ can be computed. This indicates that the iteration is to move back and forth among different spaces, and thus additional care must be taken in order to keep the computations flowing in and out without any glitches from one space to another. To overcome those difficulties arising in the recursion theoretic study of the Navier-Stokes equation, many estimates - subtle and intricate - are established in addition to the classical estimates.

The paper is organized as follows. Presuming familiarity with Weihrauch's *Type-2 Theory of Effectivity* [22], Sect. 2 recalls the standard representation δ_{L_2} of $L_2(\Omega)$ and introduces a natural representation $\delta_{L_{2,0}^\sigma}$ of $L_{2,0}^\sigma(\Omega)$. Section 3 proves that the Helmholtz projection $\mathbb{P} : \big(L_2(\Omega)\big)^2 \to L_{2,0}^\sigma(\Omega)$ is $\big((\delta_{L_2})^2, \delta_{L_{2,0}^\sigma}\big)$-computable. Section 4 presents the proof that the solution to the linear homogeneous Dirichlet problem (3) is uniformly computable from the initial condition $\boldsymbol{a}$. Section 5 is devoted to show that the solution to the nonlinear Navier-Stokes problem (1) is uniformly computable locally. Subsection 5.1 recalls the Bessel (=fractional Sobolev) space $H_2^s(\Omega) \subseteq L_2(\Omega)$ of s-fold weakly differentiable square-integrable functions on Ω and its associated standard representation $\delta_{H_{2,0}^s}$, $s \geq 0$. For $s > 1$ we assert differentiation $H_2^s(\Omega) \ni w \mapsto \partial_x w \in L_2(\Omega)$ to be $\big(\delta_{H_{2,0}^s}, \delta_{L_2}\big)$-computable and multiplication $H_2^s(\Omega) \times L_2(\Omega) \ni (v, w) \mapsto vw \in L_2(\Omega)$ to be $\big(\delta_{H_{2,0}^s} \times \delta_{L_2}, \delta_{L_2}\big)$-computable. Based on these preparations, Subsect. 5.3 asserts that in the homogeneous case $\boldsymbol{g} \equiv \boldsymbol{0}$, the sequence, generated from the iteration map

$$\mathbb{S} : C\big([0; \infty), L_{2,0}^\sigma(\Omega)\big) \times C\big([0; \infty), L_{2,0}^\sigma(\Omega)\big) \ni (\boldsymbol{v}_0, \boldsymbol{v}_n)$$
$$\mapsto \boldsymbol{v}_{n+1} \in C\big([0; \infty), L_{2,0}^\sigma(\Omega)\big)$$

according to Eq. (5), converges effectively uniformly on some positive (but not necessarily maximal) time interval $[0; T]$ whose length $T = T(\boldsymbol{a}) > 0$ is computable from the initial condition $\boldsymbol{a}$. Subsection 5.4 proves that the iteration map $\mathbb{S}$ is $\big([\rho \to \delta_{L_{2,0}^\sigma}] \times [\rho \to \delta_{L_{2,0}^\sigma}], [\rho \to \delta_{L_{2,0}^\sigma}]\big)$-computable. We conclude in Subsect. 5.5 with the final extensions regarding the inhomogeneity $\boldsymbol{f}$ and pressure P, thus establishing the main result of this work:

Theorem 1. *There exists a $\big(\delta_{L_{2,0}^\sigma} \times [\rho \to \delta_{L_{2,0}^\sigma}], \rho\big)$-computable map T,*

$$T : L_{2,0}^\sigma(\Omega) \times C\big([0; \infty), L_{2,0}^\sigma(\Omega)\big) \to (0; \infty), \quad (\boldsymbol{a}, \boldsymbol{f}) \mapsto T(\boldsymbol{a}, \boldsymbol{f})$$

and a $\big(\delta_{L_{2,0}^\sigma} \times [\rho \to \delta_{L_{2,0}^\sigma}] \times \rho, \delta_{L_{2,0}^\sigma}\big)$-computable partial map $\mathbb{S}$,

$$\mathbb{S} :\subseteq L_{2,0}^\sigma(\Omega) \times C\big([0; \infty), L_{2,0}^\sigma(\Omega)\big) \times [0; \infty) \to L_{2,0}^\sigma(\Omega) \times L_2(\Omega)$$

such that, for every $\boldsymbol{a} \in L_{2,0}^\sigma(\Omega)$ and $\boldsymbol{f} \in C\big([0; \infty), L_{2,0}^\sigma(\Omega)\big)$, the function $(\boldsymbol{u}, P) : [0; T(\boldsymbol{a}, \boldsymbol{f})] \ni t \mapsto \mathbb{S}(\boldsymbol{a}, \boldsymbol{f}, t)$ constitutes a (strong local in time and weak global in space) solution to Eq. (1).

Roughly speaking, a function is computable if it can be approximated by "computer-accessible" functions (such as rational numbers, polynomials with rational coefficients, and so forth) with arbitrary precision, where precision is given as an input; such a sequence of approximations is called an effective approximation. Thus in terms of effective approximations, the theorem states that the solution of Eq. (1) can be effectively approximated locally in the time interval $[0, T(\boldsymbol{a}, \boldsymbol{f})]$, where the time instance $T(\boldsymbol{a}, \boldsymbol{f})$ is effectively approximable.

More precisely, in computable analysis, a map $F : X \to Y$ from a space X with representation δ_X to a space Y with representation δ_Y is said to be (δ_X, δ_Y)-computable if there exists a (Turing) algorithm (or any computer program) that computes a δ_Y-name of $F(x)$ from any given δ_X-name of x. A metric space (X, d), equipped with a partial enumeration $\zeta :\subseteq \mathbb{N} \to X$ of some dense subset, gives rise to a canonical Cauchy representation δ_ζ by encoding each $x \in X$ with a sequence $\bar{s} = (s_0, s_1, s_2, \dots) \in \mathrm{dom}(\zeta)^\omega \subseteq \mathbb{N}^\omega$ such that $d\big(x, \zeta(s_k)\big) \le 2^{-k}$ for all k [22, §8.1]; in other words, $\{\zeta(s_k)\}$ is an effective approximation of x. For example, approximating by (dyadic) rationals thus leads to the standard representation ρ of $\mathbb{R}$; and for a fixed bounded $\Omega \subseteq \mathbb{R}^d$, the standard representation δ_{L_2} of $L_2(\Omega)$ encodes $f \in L_2(\Omega)$ by a sequence $\{p_k : k \in \mathbb{N}\} \subseteq \mathbb{Q}[\mathbb{R}^d]$ of d-variate polynomials with rational coefficients such that $\|f - p_k\|_2 \le 2^{-k}$, where $\|\cdot\|_2 = \|\cdot\|_{L_2}$. Thus if both spaces X and Y admit Cauchy representations, then a function $f : X \to Y$ is computable if there is an algorithm that computes an effective approximation of $f(x)$ on any given effective approximation of x as input. For represented spaces (X, δ_X) and (Y, δ_Y), $\delta_X \times \delta_Y$ denotes the canonical representation of the Cartesian product $X \times Y$. When X and Y are σ-compact metric spaces with respective canonical Cauchy representations δ_X and δ_Y, $[\delta_X \to \delta_Y]$ denotes a canonical representation of the space $C(X, Y)$ of continuous total functions $f : X \to Y$, equipped with the compact-open topology [22, THEOREM 3.2.11+DEFINITION 3.3.13]. The representation $[\delta_X \to \delta_Y]$ supports type conversion in the following sense [22, THEOREM 3.3.15]:

Fact 2. *On the one hand, the evaluation $(f, x) \mapsto f(x)$ is $([\delta_X \to \delta_Y] \times \delta_X , \delta_Y)$-computable. On the other hand, a map $f : X \times Y \to Z$ is $(\delta_X \times \delta_Y , \delta_Z)$-computable iff the map $X \ni x \mapsto \big(y \mapsto f(x, y)\big) \in C(Y, Z)$ is $(\delta_X , [\delta_Y \to \delta_Z])$-computable.*

We mention in passing that all spaces considered in this paper are equipped with a norm. Thus for any space X considered below, a δ_X-name of $f \in X$ is simply an effective approximation of f despite the often cumbersome notations.

2   Representing Divergence-Free L_2 Functions on Ω

Let us call a vector field $\boldsymbol{f}$ satisfying $\nabla \cdot \boldsymbol{f} = 0$ in Ω *divergence-free*. A vector-valued function $\boldsymbol{p}$ is called a polynomial of degree N if each of its components is a polynomial of degree less than or equal to N with respect to each variable and at least one component is a polynomial of degree N. Let $L_{2,0}^\sigma(\Omega)$—or $L_{2,0}^\sigma$ if the context is clear—be the closure in L_2-norm of the set $\{\boldsymbol{u} \in (C_0^\infty(\Omega))^2 : \nabla \cdot \boldsymbol{u} = 0\}$ of all smooth divergence-free functions with support of $\boldsymbol{u}$ and all of its partial derivatives contained in some compact subset of Ω. Let $\mathbb{Q}[\mathbb{R}^2]$ be the set of all polynomials of two real variables with rational coefficients and $\mathbb{Q}_0^\sigma[\mathbb{R}^2]$ the subset of all 2-tuples of such polynomials which are divergence-free in Ω and vanish on $\partial\Omega$. We note that the boundary value of a $L_{2,0}^{\sigma}(\Omega)$-function $\boldsymbol{u}$, $\boldsymbol{u}|_{\partial\Omega}$, is not defined unless $\boldsymbol{u}$ is (weakly) differentiable; if $\boldsymbol{u}$ is (weakly) differentiable, then $\boldsymbol{u}|_{\partial\Omega} = 0$.

Notation 3. *Hereafter we use $\|w\|_2$ for the L_2-norm $\|w\|_{L_2(\Omega)}$ if w is real-valued, or for $\|w\|_{(L_2(\Omega))^2}$ if w is vector-valued (in $\mathbb{R}^2$). We note that $\|\cdot\|_{L_{2,0}^{\sigma}(\Omega)} = \|\cdot\|_{(L_2(\Omega))^2}$. For any subset A of $\mathbb{R}^n$, its closure is denoted as $\overline{A}$.*

Proposition 1. *(a) A polynomial tuple*

$$\boldsymbol{p} = (p_1, p_2) = \Big(\sum_{i,j=0}^{N} a_{i,j}^1 x^i y^j, \sum_{i,j=0}^{N} a_{i,j}^2 x^i y^j \Big)$$

is divergence-free and boundary-free if and only if its coefficients satisfy the following system of linear equations with integer coefficients:

$$(i+1)a_{i+1,j}^1 + (j+1)a_{i,j+1}^2 = 0, \ 0 \leq i,j \leq N-1$$
$$(i+1)a_{i+1,N}^1 = 0, \ \ 0 \leq i \leq N-1 \tag{7}$$
$$(j+1)a_{N,j+1}^2 = 0, \ \ 0 \leq j \leq N-1$$

and for all $0 \leq i,j \leq N$,

$$\sum_{i=0}^{N} a_{i,j}^1 = \sum_{i=0}^{N} a_{i,j}^2 = \sum_{i=0}^{N} (-1)^i a_{i,j}^1 = \sum_{i=0}^{N} (-1)^i a_{i,j}^2 = 0 \tag{8}$$
$$\sum_{j=0}^{N} a_{i,j}^1 = \sum_{j=0}^{N} a_{i,j}^2 = \sum_{j=0}^{N} (-1)^j a_{i,j}^1 = \sum_{j=0}^{N} (-1)^j a_{i,j}^2 = 0 \tag{9}$$

(b) $\mathbb{Q}_0^{\sigma}[\mathbb{R}^2]$ is dense in $L_{2,0}^{\sigma}(\Omega)$ w.r.t. L_2-norm.

The proof of Proposition 1 is deferred to Appendix A.

We may be tempted to use $\mathbb{Q}_0^{\sigma}[\mathbb{R}^2]$ as a set of names for coding/approximating the elements in the space $L_{2,0}^{\sigma}(\Omega)$. However, since the closure of $\mathbb{Q}_0^{\sigma}[\mathbb{R}^2]$ in L_2-norm contains $L_{2,0}^{\sigma}(\Omega)$ as a proper subspace, $\mathbb{Q}_0^{\sigma}[\mathbb{R}^2]$ is "too big" to be used as a set of codes for representing $L_{2,0}^{\sigma}(\Omega)$; one has to "trim" polynomials in $\mathbb{Q}_0^{\sigma}[\mathbb{R}^2]$ so that any convergent sequence of "trimmed" polynomials converges to a limit in $L_{2,0}^{\sigma}(\Omega)$. The trimming process is shown below. For each $k \in \mathbb{N}$ (where $\mathbb{N}$ is the set of all positive integers), let $\Omega_k = (-1+2^{-k}; 1-2^{-k})^2$. And for each $\boldsymbol{p} = (p_1, p_2) \in \mathbb{Q}_0^{\sigma}[\mathbb{R}^2]$, define $\mathbb{T}_k \boldsymbol{p} = (\mathbb{T}_k p_1, \mathbb{T}_k p_2)$, where

$$\mathbb{T}_k p_j(x,y) = \begin{cases} p_j(\frac{x}{1-2^{-k}}, \frac{y}{1-2^{-k}}), & -1+2^{-k} \leq x,y \leq 1-2^{-k} \\ 0, & \text{otherwise} \end{cases} \tag{10}$$

$j = 1, 2$. Then $\mathbb{T}_k p_j$ and $\mathbb{T}_k \boldsymbol{p}$ have the following properties:

(a) $\mathbb{T}_k \boldsymbol{p}$ has compact support $\overline{\varOmega}_k$ contained in $\varOmega$.
(b) $\mathbb{T}_k \boldsymbol{p}$ is a polynomial with rational coefficients defined in $\varOmega_k$.
(c) $\mathbb{T}_k \boldsymbol{p}$ is continuous on $\overline{\varOmega} = [-1, 1]^2$.
(d) $\mathbb{T}_k \boldsymbol{p} = 0$ on $\partial \varOmega_k$, for $\boldsymbol{p}$ vanishes on the boundary of $\varOmega$. Thus $\mathbb{T}_k \boldsymbol{p}$ vanishes in the exterior region of $\varOmega_k$ including its boundary $\partial \varOmega_k$.
e) $\mathbb{T}_k \boldsymbol{p}$ is divergence-free in $\varOmega_k$ following the calculation below: for $(x, y) \in \varOmega_k$, we have $(\frac{x}{1-2^{-k}}, \frac{y}{1-2^{-k}}) \in \varOmega$ and

$$\frac{\partial \mathbb{T}_k p_1}{\partial x}(x, y) + \frac{\partial \mathbb{T}_k p_2}{\partial y}(x, y) = \frac{1}{1-2^{-k}} \frac{\partial p_1}{\partial x'}(x', y') + \frac{1}{1-2^{-k}} \frac{\partial p_2}{\partial y'}(x', y')$$

$$= \frac{1}{1-2^{-k}} \left[\frac{\partial p_1}{\partial x'}(x', y') + \frac{\partial p_2}{\partial y'}(x', y') \right] = 0$$

for $\boldsymbol{p}$ is divergence-free in $\varOmega$, where $x' = \frac{x}{1-2^{-k}}$ and $y' = \frac{y}{1-2^{-k}}$.

It follows from the discussion above that every $\mathbb{T}_k \boldsymbol{p}$ is a divergence-free polynomial of rational coefficients on $\varOmega_k$ that vanishes in $[-1, 1]^2 \setminus \varOmega_k$ and is continuous on $[-1, 1]^2$. However, although the functions $\mathbb{T}_k \boldsymbol{p}$ are continuous on $[-1, 1]^2$ and differentiable in $\varOmega_k$, they can be non-differentiable along the boundary $\partial \varOmega_k \subseteq \varOmega$. To use these functions as names for coding elements in $L_{2,0}^\sigma(\varOmega)$, it is desirable to smoothen them along the boundary $\partial \varOmega_k$ so that they are differentiable in the entire $\varOmega$. A standard technique for smoothing a function is to convolute it with a C^∞ function. We use this technique to modify functions $\mathbb{T}_k \boldsymbol{p}$ so that they become divergence-free and differentiable on the entire region of $\varOmega$. Let

$$\gamma(\boldsymbol{x}) := \begin{cases} \gamma_0 \cdot \exp\left(-\frac{1}{1-\|\boldsymbol{x}\|^2}\right), & \text{if } 1 > \|\boldsymbol{x}\| := \max\{|x_1|, |x_2|\} \\ 0, & \text{otherwise} \end{cases} \tag{11}$$

where γ_0 is a constant such that $\int_{\mathbb{R}^2} \gamma(\boldsymbol{x}) \, d\boldsymbol{x} = 1$ holds. The constant γ_0 is computable, since integration on continuous functions is computable [22, §6.4]. Let $\gamma_k(\boldsymbol{x}) = 2^{2k} \gamma(2^k \boldsymbol{x})$. Then, for all $k \in \mathbb{N}$, γ_k is a C^∞ function having support in the closed square $[-2^{-k}, 2^{-k}]^2$ and $\int_{\mathbb{R}^2} \gamma_k(\boldsymbol{x}) \, d\boldsymbol{x} = 1$. Recall that for differentiable functions $f, g : \mathbb{R}^d \to \mathbb{R}$ with compact support, the convolution $f * g$ is defined as follows:

$$(f * g)(\boldsymbol{x}) = \int_{\mathbb{R}^d} f(\boldsymbol{x} - \boldsymbol{y}) \cdot g(\boldsymbol{y}) \, d\boldsymbol{y} \tag{12}$$

It is easy to see that for $n \geq k+1$ the support of $\gamma_n * \mathbb{T}_k \boldsymbol{p} := (\gamma_n * \mathbb{T}_k p_1, \gamma_n * \mathbb{T}_k p_2)$ is contained in the closed square $[-1 + 2^{-(k+1)}, 1 - 2^{-(k+1)}]^2$. It is also known classically that $\gamma_n * \mathbb{T}_k \boldsymbol{p}$ is a C^∞ function. Since γ_n is a computable function and integration on compact domains is computable, the map $(n, k, \boldsymbol{p}) \mapsto \gamma_n * \mathbb{T}_k \boldsymbol{p}$ is computable. Moreover the following metric is computable:

$$((n, k, \boldsymbol{p}), (n', k', \boldsymbol{p}')) \mapsto \left(\int \left| (\gamma_n * \mathbb{T}_k \boldsymbol{p})(\boldsymbol{x}) - (\gamma_{n'} * \mathbb{T}_{k'} \boldsymbol{p}')(\boldsymbol{x}) \right|^2 d\boldsymbol{x} \right)^{1/2} \tag{13}$$

Lemma 1. *Every function* $\gamma_n * \mathbb{T}_k \boldsymbol{p}$ *is divergence-free in* Ω, *where* $n, k \in \mathbb{N}$, $n \geq k$, *and* $\boldsymbol{p} \in \mathbb{Q}_0^{\sigma}[\mathbb{R}^2]$.

Lemma 2. *The set* $\mathcal{P} = \left\{ \gamma_n * \mathbb{T}_k \boldsymbol{p} : n, k \in \mathbb{N}, n \geq k+1, \boldsymbol{p} \in \mathbb{Q}_0^{\sigma}[\mathbb{R}^2] \right\}$ *is dense in* $L_{2,0}^{\sigma}(\Omega)$.

See Appendices B and C for the proofs.

From Lemmas 1 and 2 it follows that $\mathcal{P}$ is a countable set that is dense in $L_{2,0}^{\sigma}(\Omega)$ (in L_2-norm) and every function in $\mathcal{P}$ is C^{∞}, divergence-free on Ω, and having a compact support contained in Ω; in other words, $\mathcal{P} \subset \{ \boldsymbol{u} \in C_0^{\infty}(\Omega)^2 : \nabla \cdot \boldsymbol{u} = 0 \}$. Thus, $L_{2,0}^{\sigma}(\Omega) =$ the closure of $\mathcal{P}$ in $L_2 -$ norm. This fact indicates that the set $\mathcal{P}$ is qualified to serve as codes for representing $L_{2,0}^{\sigma}(\Omega)$.

Since the function $\phi : \bigcup_{N=0}^{\infty} \mathbb{Q}^{(N+1)^2} \times \mathbb{Q}^{(N+1)^2} \to \{0, 1\}$, where

$$\phi\big((r_{i,j})_{0 \leq i,j \leq N}, (s_{i,j})_{0 \leq i,j \leq N}\big) = \begin{cases} 1, \text{ if } (7), (8), \text{ and } (9) \text{ are satisfied} \\ \quad (\text{with } r_{i,j} = a_{i,j}^1 \text{ and } s_{i,j} = a_{i,j}^2) \\ 0, \text{ otherwise} \end{cases}$$

is computable, there is a total computable function on $\mathbb{N}$ that enumerates $\mathbb{Q}_0^{\sigma}[\mathbb{R}^2]$. Then it follows from the definition of $\mathcal{P}$ that there is a total computable function $\alpha : \mathbb{N} \to \mathcal{P}$ that enumerates $\mathcal{P}$; thus, in view of the computable Eq. (13), $\big(L_{2,0}^{\sigma}(\Omega), (\boldsymbol{u}, \boldsymbol{v}) \mapsto \|\boldsymbol{u} - \boldsymbol{v}\|_2, \mathcal{P}, \alpha\big)$ is a computable metric space. Let $\delta_{L_{2,0}^{\sigma}} : \mathbb{N}^{\omega} \to L_{2,0}^{\sigma}$ be the standard Cauchy representation of $L_{2,0}^{\sigma}$; that is, every function $\boldsymbol{u} \in L_{2,0}^{\sigma}(\Omega)$ is encoded by a sequence $\{\boldsymbol{p}_k : k \in \mathbb{N}\} \subseteq \mathcal{P}$, such that $\|\boldsymbol{u} - \boldsymbol{p}_k\|_2 \leq 2^{-k}$. The sequence $\{\boldsymbol{p}_k\}_{k \in \mathbb{N}}$ is called a $\delta_{L_{2,0}^{\sigma}}$-name of $\boldsymbol{u}$, which is an effective approximation of $\boldsymbol{u}$ (in L_2-norm).

3 Computability of Helmholtz Projection

In this section, we show that the Helmholtz projection $\mathbb{P}$ is computable.

Proposition 2. *The projection* $\mathbb{P} : \big(L_2(\Omega)\big)^2 \to L_{2,0}^{\sigma}(\Omega)$ *is* $\big((\delta_{L_2})^2, \delta_{L_{2,0}^{\sigma}}\big)$-*computable.*

Proof. For simplicity let us set $\Omega = (0, 1)^2$. The proof carries over to $\Omega = (-1, 1)^2$ by a scaling on sine and cosine functions. We begin with two classical facts which are used in the proof:

(i) It follows from [6, p. 40]/[21] that for any $\boldsymbol{u} = (u_1, u_2) \in \big(L_2(\Omega)\big)^2$,

$$\mathbb{P}\boldsymbol{u} = (-\partial_y \varphi, \partial_x \varphi) \tag{14}$$

where the scalar function φ is the solution of the following boundary value problem:

$$\triangle \varphi = \partial_x u_2 - \partial_y u_1 \text{ in } \Omega, \qquad \varphi = 0 \text{ on } \partial\Omega \tag{15}$$

We note that $\mathbb{P}$ is a linear operator.

(ii) Each of $\{\sin(n\pi x)\sin(m\pi y)\}_{n,m\geq 1}$,

$$\{\sin(n\pi x)\cos(m\pi y)\}_{n\geq 1,m\geq 0}, \quad \text{or} \quad \{\cos(n\pi x)\sin(m\pi y)\}_{n\geq 0,m\geq 1}$$

is an orthogonal basis for $L^2(\Omega)$. Thus each $\boldsymbol{u} = (u_1,u_2)$ in $(L_2(\Omega))^2$, u_i, $i = 1$ or 2, can be written in the following form:

$$u_i(x,y) = \sum_{n,m\geq 0} u_{i,n,m}\cos(n\pi x)\sin(m\pi y)$$

$$= \sum_{n,m\geq 0} \tilde{u}_{i,n,m}\sin(n\pi x)\cos(m\pi y)$$

where

$$u_{i,n,m} = \int_0^1\int_0^1 u_i(x,y)\cos(n\pi x)\sin(m\pi y)dxdy, \quad \text{and}$$

$$\tilde{u}_{i,n,m} = \int_0^1\int_0^1 u_i(x,y)\sin(n\pi x)\cos(m\pi y)dxdy$$

with the property that $\|u_i\|_2 = \left(\sum_{n,m\geq 0}|u_{i,n,m}|^2\right)^{1/2} = \left(\sum_{n,m\geq 0}|\tilde{u}_{i,n,m}|^2\right)^{1/2}$. We note that the sequences $\{u_{i,n,m}\}$, $\{\tilde{u}_{i,n,m}\}$, and $\|u_i\|_2$ are computable from $\boldsymbol{u}$; cf. [22].

To prove that the projection is $((\delta_{L_2})^2, \delta_{L_{2,0}^\sigma})$-computable, it suffices to show that there is an algorithm computing, given any accuracy $k \in \mathbb{N}$ and for any $\boldsymbol{u} \in (L^2(\Omega))^2$, a vector function $(p_k,q_k) \in \mathcal{P}$ such that $\|\mathbb{P}\boldsymbol{u} - (p_k,q_k)\|_2 \leq 2^{-k}$. Let us fix k and $\boldsymbol{u} = (u_1,u_2)$. Then a straightforward computation shows that the solution φ of (15) can be explicitly written as

$$\varphi(x,y) = \sum_{n,m=1}^\infty \frac{-nu_{2,n,m} + m\tilde{u}_{1,n,m}}{(n^2+m^2)\pi}\sin(n\pi x)\sin(m\pi y)$$

It then follows that

$$-\partial_y\varphi = \sum_{n,m\geq 1} \frac{mnu_{2,n,m} - m^2\tilde{u}_{1,n,m}}{n^2+m^2}\sin(n\pi x)\cos(m\pi y) \qquad (16)$$

Similarly, we can obtain a formula for $\partial_x\varphi$. Since we have an explicit expression for $(-\partial_y\varphi, \partial_x\varphi)$, a search algorithm is usually a preferred choice for finding a k-approximation (p_k,q_k) of $\mathbb{P}\boldsymbol{u}$ by successively computing the norms

$$\|(-\partial_y\varphi, \partial_x\varphi) - (p,q)\|_2, \qquad (p,q) \in \mathcal{P}.$$

However, since $-\partial_y\varphi$ and $\partial_x\varphi$ are infinite series which involve limit processes, a truncating algorithm is needed so that one can compute approximations of the two limits before a search program can be executed. The truncating algorithm will find some $N(k,\boldsymbol{u}) \in \mathbb{N}$ such that the $N(k,\boldsymbol{u})$-partial sum of $(-\partial_y\varphi, \partial_x\varphi)$ is a

$2^{-(k+1)}$-approximation of the series; in other words, the algorithm chops off the infinite tails of the series within pre-assigned accuracy. The following estimate provides a basis for the desired truncating algorithm:

$$\| - \partial_y \varphi - \sum_{n,m < N} \frac{mnu_{2,n,m} - m^2 \tilde{u}_{1,n,m}}{n^2 + m^2} \sin(n\pi x) \cos(m\pi y) \|_2^2$$

$$= \| \sum_{n,m \geq N} \frac{mnu_{2,n,m} - m^2 \tilde{u}_{1,n,m}}{n^2 + m^2} \sin(n\pi x) \cos(m\pi y) \|_2^2$$

$$= \sum_{n,m \geq N} \left| \frac{mnu_{2,n,m} - m^2 \tilde{u}_{1,n,m}}{n^2 + m^2} \right|^2 \leq 2 \sum_{n,m \geq N} (|u_{2,n,m}|^2 + |\tilde{u}_{1,n,m}|^2)$$

A similar estimate applies to $\partial_x \varphi$. Since

$$\|u_i\|_2^2 = \sum_{n,m \geq 1} |u_{i,n,m}|^2 = \sum_{n,m \geq 1} |\tilde{u}_{i,n,m}|^2, \quad i = 1, 2,$$

is computable, it follows that there is an algorithm computing $N(k, \boldsymbol{u})$ from k and $\boldsymbol{u}$ such that the $N(k, \boldsymbol{u})$-partial sum of $(-\partial_y \varphi, \partial_x \varphi)$ is a $2^{-(k+1)}$-approximation of the series. Now we can search for $(p_k, q_k) \in \mathbb{P}$ that approximates the $N(k, \boldsymbol{u})$-partial sum in L^2-norm within the accuracy $2^{-(k+1)}$ as follows: enumerate $\mathcal{P} = \{\tilde{p}_j\}$, compute the L^2-norm of the difference between the $N(k, \boldsymbol{u})$-partial sum and $\tilde{p}_j$, halt the computation at $\tilde{p}_j$ when the L^2-norm is less that $2^{-(k+1)}$, and then set $(p_k, q_k) = \tilde{p}_j$. We note that each computation halts in finitely many steps. The search will succeed since $\mathbb{P}\boldsymbol{u} = (-\partial_y \varphi, \partial_x \varphi) \in L_{2,0}^\sigma$ and $\mathcal{P}$ is dense in $L_{2,0}^\sigma$. It is then clear that $\|\mathbb{P}\boldsymbol{u} - (p_k, q_k)\|_2 \leq 2^{-k}$.

4   Computability of the Linear Problem

In this section, we show that the solution operator for the linear homogeneous equation (3) is uniformly computable from the initial data. We begin by recalling the Stokes operator and some of its classical properties. Let $\mathbb{A} = -\mathbb{P}\triangle$ be the Stokes operator as defined for instance in [3, §2] or [18, §III.2.1], where $\mathbb{P} : \left(L_2(\Omega)\right)^2 \to L_{2,0}^\sigma$ is the Helmholtz projection. It is known from the classical study that $\mathbb{A}$ is an unbounded but closed positively self-adjoint linear operator whose domain is dense in $L_{2,0}^\sigma$, and thus $-\mathbb{A}$ is the infinitesimal generator of an analytic semigroup; cf. [18, THEOREM III.2.1.1] or [2, §IV.5.2]. In this case, the linear homogeneous equation (3) has the solution $\boldsymbol{u}(t) = e^{-\mathbb{A}t}\boldsymbol{a}$, where $\boldsymbol{u}(0) = \boldsymbol{a}$, $e^{-\mathbb{A}t}$ is the analytic semigroup generated by the infinitesimal generator $-\mathbb{A}$, and $\boldsymbol{u}(t) \in L_{2,0}^\sigma(\Omega)$ for $t \geq 0$. Furthermore, the following lemma shows that the solution $\boldsymbol{u}(t)$ decays in L^2-norm as time t increases.

Lemma 3. *For every* $\boldsymbol{a} \in L_{2,0}^\sigma(\Omega)$ *and* $t \geq 0$,

$$\|\boldsymbol{u}(t)\|_2 = \|e^{-t\mathbb{A}}\boldsymbol{a}\|_2 \leq \|\boldsymbol{a}\|_2 = \|\boldsymbol{u}(0)\|_2 \tag{17}$$

(Recall that $\| \cdot \|_2 = \| \cdot \|_{L_{2,0}^\sigma(\Omega)}$*; see Notation 3.)*

Proof. Classically it is known that for any $a \in L_{2,0}^\sigma(\Omega)$, $u(t) = e^{-t\mathbb{A}}a$ is in the domain of $\mathbb{A}$ for $t > 0$. Thus if $a = u(0)$ itself is in the domain of $\mathbb{A}$, then so is $u(t)$ for $t \geq 0$. Since $\mathbb{A}$ is positively self-adjoint, it follows that $\mathbb{A}^* = \mathbb{A}$ and $\langle \mathbb{A}u(t), u(t) \rangle := \int_\Omega \mathbb{A}u(t)(x) \cdot u(t)(x)\,dx > 0$ for every a in the domain of $\mathbb{A}$ with $a \not\equiv 0$ and $t \geq 0$. Now if we rewrite the Eq. (3) in the form of

$$\langle u_t, u \rangle + \langle \mathbb{A}u, u \rangle = 0$$

or equivalently $\frac{1}{2}\frac{d}{dt}\langle u, u \rangle + \langle \mathbb{A}u, u \rangle = 0$, then $\frac{d}{dt}\langle u, u \rangle \leq 0$ and consequently $\langle u, u \rangle(t) \leq \langle u, u \rangle(0)$; thus (17) holds true for a in the domain of $\mathbb{A}$. Since the domain of $\mathbb{A}$ is dense in $L_{2,0}^\sigma(\Omega)$, it follows that (17) holds true for all $a \in L_{2,0}^\sigma(\Omega)$. □

Proposition 3. *For the linear homogenous equation (3), the solution operator* $S : L_{2,0}^\sigma(\Omega) \to C\big([0;\infty), L_{2,0}^\sigma(\Omega)\big)$, $a \mapsto (t \mapsto e^{-\mathbb{A}t}a)$, *is* $(\delta_{L_{2,0}^\sigma}, [\rho \to \delta_{L_{2,0}^\sigma}])$-*computable.*

By the *First Main Theorem* of Pour-El and Richards [12, §II.3], the unbounded operator $\mathbb{A}$ does not preserve computability. In particular, the naive exponential series $\sum_n (-\mathbb{A}t)^n a/n!$ does not establish Proposition 3.

Convention. For readability we will not notationally distinguish the spaces of vectors u, a and scalar functions u, a in the proof below and the proof of Lemma 6.

Proof. We show how to compute a $\delta_{L_{2,0}^\sigma}$-name of $e^{-t\mathbb{A}}a$ on inputs $t \geq 0$ and $a \in L_{2,0}^\sigma(\Omega)$. Recall that a $\delta_{L_{2,0}^\sigma}$-name of $e^{-t\mathbb{A}}a$ is a sequence $\{q_K\}$, $q_K \in \mathcal{P}$, satisfying $\|e^{-t\mathbb{A}}a - q_K\|_2 \leq 2^{-K}$ for all $K \in \mathbb{N}$. Again, for readability, we assume that $\Omega = (0,1)^2$.

We first consider the case where $a \in \mathcal{P}$ and $t > 0$. The reason for us to start with functions in $\mathcal{P}$ is that these functions have stronger convergence property in the sense that, for any $a \in \mathcal{P}$, if $a = (a^1, a^2)$ is expressed in terms of the orthogonal basis $\{\sin(n\pi x)\sin(m\pi y)\}_{n,m\geq 1}$ for $L_2(\Omega)$: for $i = 1, 2$,

$$a^i = \sum_{n,m\geq 1} a_{n,m}^i \sin(n\pi x)\sin(m\pi y) \tag{18}$$

where $a_{n,m}^i = \int_0^1 \int_0^1 a^i \sin(n\pi x)\sin(m\pi y)\,dx\,dy$, then the following series is convergent

$$\sum_{n,m\geq 1} (1 + n^2 + m^2)|a_{n,m}^i|^2 < \infty \tag{19}$$

The inequality (19) holds true because functions in $\mathcal{P}$ are C^∞. In fact, the series is not only convergent but its sum is also computable (from a) (see, for example, [28]).

Now let $K \in \mathbb{N}$ be any given precision. Since $-\mathbb{A}$ generates an analytic semigroup, it follows from [10, SECTION 2.5] that for $t > 0$,

$$e^{-t\mathbb{A}}a = \frac{1}{2\pi i}\int_\Gamma e^{\lambda t}(\lambda\mathbb{I} + \mathbb{A})^{-1}a\,d\lambda \tag{20}$$

where Γ is the path composed from two rays $re^{i\beta}$ and $re^{-i\beta}$ with $0 < r < \infty$ and $\beta = \frac{3\pi}{5}$. Thus we have an explicit expression for $e^{-t\mathbb{A}}a$, which involves a limit process – an infinite integral – indicating that a search algorithm is applicable for finding a desirable K-approximation provided that a finite approximation of $e^{-t\mathbb{A}}a$ can be computed by some truncating algorithm.

In the following, we construct such a truncating algorithm. We begin by writing the infinite integral in (20) as a sum of three integrals: two are finite and one infinite; the infinite one can be made arbitrarily small. Now for the details. Let l be a positive integer to be determined; let Γ_1 be the path $re^{i\beta}$ with $0 < r \leq l$; Γ_2 the path $re^{-i\beta}$ with $0 < r \leq l$; and $\Gamma_3 = \Gamma \setminus (\Gamma_1 \cup \Gamma_2)$. Since $a \in \mathcal{P}$, it follows that $-\mathbb{A}a = \mathbb{P}\triangle a = \triangle a$, which further implies that

$$(\lambda\mathbb{I} + \mathbb{A})^{-1}a =$$

$$\left(\sum_{n,m\geq 1} \frac{a_{n,m}^1 \sin(n\pi x)\sin(m\pi y)}{\lambda + (n\pi)^2 + (m\pi)^2}, \sum_{n,m\geq 1} \frac{a_{n,m}^2 \sin(n\pi x)\sin(m\pi y)}{\lambda + (n\pi)^2 + (m\pi)^2} \right) \tag{21}$$

Note that for any $\lambda \in \Gamma$, $|\lambda + (n\pi)^2 + (m\pi)^2| \neq 0$. From (20) and (21) we can write $e^{-t\mathbb{A}}a$ as a sum of three terms:

$$e^{-t\mathbb{A}}a = \sum_{j=1}^{3} \frac{1}{2\pi i} \int_{\Gamma_j} \tilde{a}e^{\lambda t} d\lambda$$

$$= \sum_{j=1}^{3} \frac{1}{2\pi i} \sum_{n,m\geq 1} \left[\int_{\Gamma_j} \frac{e^{\lambda t}}{\lambda + (n\pi)^2 + (m\pi)^2} d\lambda \right] a_{n,m}\sin(n\pi x)\sin(m\pi y)$$

$$=: \beta_1 + \beta_2 + \beta_3$$

where $\tilde{a} = (\lambda\mathbb{I} + \mathbb{A})^{-1}a$. The functions β_j, $j = 1, 2, 3$, are in $L_{2,0}^\sigma(\Omega)$ as verified as follows: It follows from $a = (\lambda\mathbb{I} + \mathbb{A})\tilde{a} = (\lambda\mathbb{I} - \mathbb{P}\triangle)\tilde{a}$ and $\mathbb{P}\triangle\tilde{a} = \mathbb{P}(\triangle\tilde{a}) \in L_{2,0}^\sigma(\Omega)$ that $\triangledown(\mathbb{P}\triangle\tilde{a}) = 0$ and

$$0 = \triangledown a = \lambda(\triangledown\tilde{a}) - \triangledown(\mathbb{P}\triangle\tilde{a}) = \lambda(\triangledown\tilde{a}) \tag{22}$$

Since $\lambda \in \Gamma$, it follows that $\lambda \neq 0$; thus $\triangledown\tilde{a} = 0$. This shows that $\tilde{a} \in L_{2,0}^\sigma(\Omega)$. Then it follows from (22) that

$$\triangledown\beta_j = \frac{1}{2\pi i} \int_{\Gamma_j} (\triangledown\tilde{a})e^{\lambda t} d\lambda = 0$$

Hence $\beta_j \in L_{2,0}^\sigma(\Omega)$ for $1 \leq j \leq 3$.

Next we show that β_1 and β_2 can be effectively approximated by finite sums while β_3 tend to zero effectively as $l \to \infty$. We start with β_3. Since $t > 0$ and $\cos\beta = \cos\frac{3\pi}{5} < 0$, it follows that

$$\left| \int_{\Gamma_3} \frac{e^{\lambda t}}{\lambda + (n\pi)^2 + (m\pi)^2} d\lambda \right| \leq 2 \int_{l}^{\infty} \frac{e^{tr\cos\beta}}{r} dr \to 0$$

effectively as $l \to \infty$. Thus there is some $l_K \in \mathbb{N}$, computable from a and t, such that the following estimate is valid for $i = 1, 2$ when we take l to be l_K:

$$
\begin{aligned}
&\|\beta_3^i\|_2 \\
&= \left\| \frac{1}{2\pi i} \sum_{n,m \geq 1} \left[\int_{\Gamma_3} \frac{e^{\lambda t}}{\lambda + (n\pi)^2 + (m\pi)^2} d\lambda \right] a_{n,m}^i \sin(n\pi x) \sin(m\pi y) \right\|_2 \\
&\leq \frac{1}{\pi} \int_{l_K}^{\infty} \frac{e^{tr\cos\beta}}{r} dr \left(\sum_{n,m \geq 1} |a_{n,m}^i|^2 \right)^{1/2} = \frac{1}{\pi} \int_{l_K}^{\infty} \frac{e^{tr\cos\beta}}{r} dr \cdot \|a\|_2 \leq 2^{-(K+7)}
\end{aligned}
$$

where $\beta_3 = (\beta_3^1, \beta_3^2)$. Now let us set $l = l_K$ and estimate β_1. Since $\beta = \frac{3\pi}{5} < \frac{3\pi}{4}$, it follows that $\cos\beta < 0$ and $|\cos\beta| < \sin\beta$. Consequently, for any $\lambda = re^{i\beta}$ on Γ_1, if $r \geq \frac{1}{\sin\beta}$, then $|re^{i\beta} + (n\pi)^2 + (m\pi)^2| \geq r\sin\beta \geq 1$. On the other hand, if $0 < r < \frac{1}{\sin\beta}$, then $r\cos\beta + (n\pi)^2 + (m\pi)^2 \geq \pi^2(n^2 + m^2) - r\sin\beta \geq 2\pi^2 - 1 > 1$, which implies that $|re^{i\beta} + (n\pi)^2 + (m\pi)^2| \geq |r\cos\beta + (n\pi)^2 + (m\pi)^2| > 1$. Thus $|\lambda + (n\pi)^2 + (m\pi)^2| \geq 1$ for every $\lambda \in \Gamma_1$. And so

$$
\begin{aligned}
\left| \int_{\Gamma_1} \frac{e^{\lambda t}}{\lambda + (n\pi)^2 + (m\pi)^2} d\lambda \right| &= \left| \int_0^l \frac{e^{tre^{i\beta}}}{re^{i\beta} + (n\pi)^2 + (m\pi)^2} d(re^{i\beta}) \right| \leq \\
&\leq \int_0^l \frac{|e^{tre^{i\beta}}|}{|re^{i\beta} + (n\pi)^2 + (m\pi)^2|} dr \leq \int_0^l e^{tr\cos\beta} dr \leq \int_0^l e^{tl} dr = le^{tl}
\end{aligned}
$$

This estimate together with (19) implies that there exists a positive integer $k = k(t, a, K)$, computable from $t > 0$, a and K, such that

$$
\frac{1}{1 + 2k^2} \left(\frac{le^{lt}}{2\pi} \right)^2 \left(\sum_{n,m \geq 1} (1 + n^2 + m^2)(|a_{n,m}^1|^2 + |a_{n,m}^2|^2) \right) < 2^{-2(K+7)}
$$

Write $\beta_1(k) = (\beta_1^1(k), \beta_1^2(k))$ with

$$
\beta_1^i(k) = \sum_{1 \leq n,m \leq k} \left(\frac{1}{2\pi i} \int_{\Gamma_1} \frac{e^{\lambda t}}{\lambda + (n\pi)^2 + (m\pi)^2} d\lambda \right) a_{n,m}^i \sin(n\pi x) \sin(m\pi y),
$$

$i = 1, 2$. Then

$$\|\beta_1 - \beta_1(k)\|_2^2$$

$$\leq \sum_{n,m>k} \left| \frac{1}{2\pi i} \int_{\Gamma_1} \frac{e^{\lambda t}}{\lambda + (n\pi)^2 + (m\pi)^2} d\lambda \right|^2 (|a_{n,m}^1|^2 + |a_{n,m}^2|^2)$$

$$\leq \sum_{n,m>k} \frac{1}{1+n^2+m^2} \cdot (1+n^2+m^2) \left(\frac{le^{lt}}{2\pi} \right)^2 (|a_{n,m}^1|^2 + |a_{n,m}^2|^2)$$

$$\leq \frac{1}{1+k^2+k^2} \left(\frac{le^{lt}}{2\pi} \right)^2 \sum_{n,m\geq 1} (1+n^2+m^2)(|a_{n,m}^1|^2 + |a_{n,m}^2|^2)$$

$$< 2^{-2(K+7)}$$

Similarly, if we write $\beta_2(k) = (\beta_2^1(k), \beta_2^2(k))$ with

$$\beta_2^i(k) = \sum_{n,m\leq k} \left(\frac{1}{2\pi i} \int_{\Gamma_2} \frac{e^{\lambda t}}{\lambda + (n\pi)^2 + (m\pi)^2} d\lambda \right) a_{n,m}^i \sin(n\pi x) \sin(m\pi y)$$

then $\|\beta_2 - \beta_2(k)\|_2 \leq 2^{-(K+7)}$. The construction of the truncating algorithm is now complete; the algorithm outputs $\beta_1(k) + \beta_2(k)$ (uniformly) on the inputs $a \in \mathcal{P}$, $t > 0$, and precision K; the output has the property that it is a finite sum involving a finite integral and $\|\beta_1(k) + \beta_2(k) - e^{-t\mathbb{A}}a\|_2 \leq 2^{-(K+4)}$.

Now we are able to search for a desirable approximation in $\mathcal{P}$. Let us list $\mathcal{P} = \{\phi_j : j \in \mathbb{N}\}$ and compute $\|\phi_j - (\beta_1(k) + \beta_2(k))\|_2$. Halt the computation at $j = j(K)$ when

$$\|\phi_j - (\beta_1(k) + \beta_2(k))\|_2 < 2^{-(K+4)}$$

The computation will halt since $\beta_1, \beta_2 \in L_{2,0}^\sigma(\Omega)$, $\|\beta_1 - \beta_1(k)\|_2 \leq 2^{-(K+7)}$, $\|\beta_2 - \beta_2(k)\|_2 \leq 2^{-(K+7)}$, and $\mathcal{P}$ is dense in $L_{2,0}^\sigma(\Omega)$ (in L^2-norm). Set $q_K = \phi_{j(K)}$. Then

$$\|q_K - e^{-t\mathbb{A}}a\|_2$$
$$= \|q_K - (\beta_1 + \beta_2 + \beta_3)\|_2$$
$$\leq \|q_K - (\beta_1(k) + \beta_2(k))\|_2 + \|(\beta_1(k) + \beta_2(k)) - (\beta_1 + \beta_2)\|_2 + \|\beta_3\|_2$$
$$< 2^{-K}$$

Next we consider the case where $a \in L_{2,0}^\sigma(\Omega)$ and $t > 0$. In this case, the input a is presented by (any) one of its $\delta_{L_{2,0}^\sigma}$-names, say $\{a_k\}$, where $a_k \in \mathcal{P}$. It is then clear from the estimate (17) and the discussion above that there is an algorithm that computes a K-approximation $p_K \in \mathcal{P}$ on inputs $t > 0$, a and precision K such that $\|p_K - e^{-t\mathbb{A}}a\|_2 \leq 2^{-K}$.

Finally we consider the case where $t \geq 0$ and $a \in L_{2,0}^\sigma(\Omega)$. Since $e^{-t\mathbb{A}}a = a$ for $t = 0$ and we already derived an algorithm for computing $e^{-t\mathbb{A}}a$ for $t > 0$, it suffices to show that $e^{-t\mathbb{A}}a \to a$ in L^2-norm effectively as $t \to 0$. Let $\{a_k\}$ be a $\delta_{L_{2,0}^\sigma}$-name of a. It follows from Theorem 6.13 of Sect. 2.6 [Paz83] that $\|e^{-t\mathbb{A}}a_k - a_k\| \leq Ct^{1/2}\|\mathbb{A}^{1/2}a_k\|$. Thus

$$\|a - e^{-t\mathbb{A}}a\| \leq \|a - a_k\| + \|a_k - e^{-t\mathbb{A}}a_k\| + \|e^{-t\mathbb{A}}a_k - e^{-t\mathbb{A}}a\|$$

the right-hand side goes to 0 effectively as $t \to 0$.   ⊔

We note that the computation of the approximations q_K of $e^{-t\mathbb{A}}a$ does not require encoding $\mathbb{A}$. Let $W : L_{2,0}^\sigma(\Omega) \times [0, \infty) \to L_{2,0}^\sigma(\Omega)$, $(a,t) \mapsto e^{-t\mathbb{A}}a$. Then it follows from the previous Proposition and Fact 2 that W is computable.

5   Extension to the Nonlinear Problem

We now proceed to the nonlinear problem (2) by solving its integral version (4) via the iteration scheme (5) but first restrict to the homogeneous case $g \equiv 0$:

$$\boldsymbol{u}_0(t) = e^{-t\mathbb{A}}\boldsymbol{a}, \qquad \boldsymbol{u}_{m+1}(t) = \boldsymbol{u}_0(t) - \int_0^t e^{-(t-s)\mathbb{A}}\mathbb{B}\boldsymbol{u}_m(s)\, ds \qquad (23)$$

Classically, it is known that for every initial condition $\boldsymbol{a} \in L_{2,0}^\sigma(\Omega)$ the sequence $\boldsymbol{u}_m = \boldsymbol{u}_m(t)$ converges near $t = 0$ to a unique limit $\boldsymbol{u}$ solving (4) and thus (2). Since there is no explicit formula for the solution $\boldsymbol{u}$, the truncation/search type of algorithms such as those used in the proofs of Propositions 2 and 3 is no longer applicable for the nonlinear case. Instead, we use a method based on the fixed-point argument to establish the computability of $\boldsymbol{u}$. We shall show that the limit of the above sequence $\boldsymbol{u}_m = \boldsymbol{u}_m(t)$ has an effective approximation. The proof consists of two parts: first we study the rate of convergence and show that the sequence converges at a computable rate as $m \to \infty$ for $t \in [0; T]$ with some $T = T_a > 0$, where T_a is computable from $\boldsymbol{a}$; then we show that the sequence – as one entity – can be effectively approximated starting with the given $\boldsymbol{a}$. The precise statements of the two tasks are given in the following two propositions.

Proposition 4. *There is a computable map* $\mathbb{T} : L_{2,0}^\sigma(\Omega) \to (0, \infty)$, $\boldsymbol{a} \mapsto T_a$, *such that the sequence* $\{\boldsymbol{u}_m\}$ *converges effectively in* m *and uniformly for* $t \in [0; T_a]$.

Recall that a sequence $\{x_m\}$ in a metric space (X, d) is effectively convergent if $d(x_m, x_{m+1}) \leq 2^{-m}$. In view of type conversion (Subsect. 1.1), the following proposition asserts (ii):

Proposition 5. *The map* $\mathbb{S} : \mathbb{N} \times L_{2,0}^\sigma(\Omega) \times [0, \infty) \to L_{2,0}^\sigma(\Omega)$, $(m, \boldsymbol{a}, t) \to \boldsymbol{u}_m(t)$ *according to Eq. (23), is* $(\nu \times \delta_{L_{2,0}^\sigma} \times \rho, \delta_{L_{2,0}^\sigma})$-*computable.*

The main difficulties in proving the two propositions are rooted in the nonlinearity of $\mathbb{B}$: the nonlinear operator $\mathbb{B}$ requires greater care in estimating the rate of convergence and demands richer codings for computation. Since information on $\mathbb{B}u_m$ is required in order to compute u_{m+1}, but $\mathbb{B}u_m = \mathbb{P}(u_m \cdot \nabla)u_m$ involves both differentiation and multiplication, it follows that a $\delta_{L_{2,0}^\sigma}$-name of u_m may not contain enough information for computing $\mathbb{B}u_m$. Moreover, since estimates of type $\|\mathbb{A}^\alpha u_m(t)\|_2$, $0 \le \alpha \le 1$, play a key role in proving Propositions 4 and 5, we need to computationally derive a richer code for u_m from a given $\delta_{L_{2,0}^\sigma}$-name of u_m in order to capture the fact that u_m is in the domain of $\mathbb{A}^\alpha$ for $t > 0$.

5.1   Representing and Operating on Space $H_{2,0}^s(\Omega)$

We begin by recalling several definitions and facts. Let $\theta_{n,m}(x,y) := e^{i(nx+my)\pi}$, $n, m \ge 0$. Then, the sequence $\{\theta_{n,m}(x,y)\}_{n,m \ge 0}$ is a computable orthogonal basis of $L_2(\Omega)$. For any $s \ge 0$, $H_2^s(\Omega)$ is the set of all (generalized) functions $w(x,y)$ on Ω satisfying $\sum_{n,m \ge 0}(1 + n^2 + m^2)^s |w_{n,m}|^2 < \infty$, where $w_{n,m} = \int_{-1}^{1}\int_{-1}^{1} w(x,y)\theta_{n,m}(x,y)\,dx\,dy$. $H_2^s(\Omega)$ is a Banach space with a norm $\|w\|_{H_2^s} = (\sum_{n,m \ge 0}(1 + n^2 + m^2)^s |w_{n,m}|^2)^{1/2}$.

Let $D(\mathbb{A}^\alpha)$ be the domain of $\mathbb{A}^\alpha$. Since

$$D(\mathbb{A}) = L_{2,0}^\sigma(\Omega) \bigcap \{u \in (H_2^2(\Omega))^2 : u = 0 \text{ on } \partial\Omega\},$$
$$D(\mathbb{A}^{1/2}) = L_{2,0}^\sigma(\Omega) \bigcap \{u \in (H_2^1(\Omega))^2 : u = 0 \text{ on } \partial\Omega\},$$

and $D(\mathbb{A}^\alpha)$, $0 \le \alpha \le 1$, are the complex interpolation spaces of $L_{2,0}^\sigma(\Omega)$ and $D(\mathbb{A})$, we need to represent the subspace of $H_2^s(\Omega)$ in which the functions vanish on $\partial\Omega$. However, it is usually difficult to design a coding system for such subspaces. Fortunately, for $0 \le s < 3/2$, it is known classically that

$$H_{2,0}^s(\Omega) = \{w \in H_2^s(\Omega) : w = 0 \text{ on } \partial\Omega\} \tag{24}$$

where $H_{2,0}^s(\Omega)$ is the closure in H_2^s-norm of the set of all C^∞-smooth functions defined on compact subsets of Ω. For $H_{2,0}^s(\Omega)$, there is a canonical coding system

$$\mathcal{H} = \{\gamma_n * q : n \in \mathbb{N}, q \in \mathbb{Q}[\mathbb{R}^2]\}$$

(see (11) and (12) for the definitions of γ_n and $\gamma_n * q$). Then every w in $H_{2,0}^s(\Omega)$ can be encoded by a sequence $\{p_k\} \subset \mathcal{H}$ such that $\|p_k - w\|_{H_2^s} \le 2^{-k}$; the sequence $\{p_k\}$, which are mollified polynomials with rational coefficients, is called a $\delta_{H_{2,0}^s}$-name of w. If $w = (w_1, w_2) \in H_{2,0}^s(\Omega) \times H_{2,0}^s(\Omega)$, a $\delta_{H_{2,0}^s}$-name of w is a sequences $\{(p_k, q_k)\}$, $p_k, q_k \in \mathcal{H}$, such that $(\|w_1 - p_k\|_{H_{2,0}^s}^2 + \|w_2 - q_k\|_{H_{2,0}^s}^2)^{1/2} \le 2^{-k}$.

Notation 4. *We use* $\|w\|_{H_2^s}$ *to denote the* H_2^s-norm of w if w is in $H_2^s(\Omega)$ or $H_2^s \times H_2^s$-norm of w if w is in $H_2^s(\Omega) \times H_2^s(\Omega)$. Also for readability we use $[\rho \to \delta_{H_{2,0}^s}]$ to denote the canonical representation of either $C([0;T]; H_{2,0}^s(\Omega))$ or $C([0;T]; H_{2,0}^s(\Omega) \times H_{2,0}^s(\Omega))$.*

Recall that $C([0;T]; H_{2,0}^s(\Omega))$ is the set of all continuous functions from the interval $[0;T]$ to $H_{2,0}^s(\Omega)$. A function $u \in C([0;T]; H_{2,0}^s(\Omega))$ is computable if there is a machine that computes a $\delta_{H_{2,0}^s}$-name of $u(t)$ when given a ρ-name of t as input; and a map $F : X \to C([0;T]; H_{2,0}^s(\Omega))$ from a represented space (X, δ_X) to $C([0;T]; H_{2,0}^s(\Omega))$ is computable if there is a machine that computes a $\delta_{H_{2,0}^s}$-name of $F(x)(t)$ when given a δ_X-name of x and a ρ-name of t. Let X be either $L_2(\Omega)$, $L_{2,0}^\sigma(\Omega)$, $H_{2,0}^s(\Omega)$, or $C([0;T]; H_{2,0}^s(\Omega))$. We remark again that a δ_X-name of $f \in X$ is simply an effective approximation of f because each space X is equipped with a norm.

Lemma 4. *For $s \geq 1$, differentiation $\partial_x, \partial_y : H_{2,0}^s(\Omega) \to L_2(\Omega)$ is $(\delta_{H_{2,0}^s}, \delta_{L_2})$-computable.*

Proof. Let $\{p_k\}$ be a $\delta_{H_{2,0}^s}$-name of $w \in H_{2,0}^s(\Omega)$. Since $\partial_x(\gamma * q) = \gamma * \partial_x q$, the map $p_k \mapsto \partial_x p_k$ is computable; hence a polynomial $\tilde{p}$ in $\mathbb{Q}[\mathbb{R}^2]$ can be computed from p_k such that $\max_{-1 \leq x,y \leq 1} |\partial_x p_k - \tilde{p}_k| < 2^{-k}$. Next let us express w and p_k in the orthogonal basis $\theta_{n,m}$: $w(x,y) = \sum_{n,m \geq 0} w_{n,m} e^{in\pi x} e^{im\pi y}$ and $p_k(x,y) = \sum_{n,m \geq 0} p_{k,n,m} e^{in\pi x} e^{im\pi y}$, where

$$w_{n,m} = \int_0^1 \int_0^1 w(x,y) e^{in\pi x} e^{im\pi y} \, dx \, dy,$$

$$p_{k,n,m} = \int_0^1 \int_0^1 p_k(x,y) e^{in\pi x} e^{im\pi y} \, dx \, dy.$$

Since $s \geq 1$ and $\{p_k\}$ is a $\delta_{H_{2,0}^s}$-name of w, it follows that

$$\|\partial_x p_k - \partial_x w\|_2^2$$
$$= \left\|\sum_{n,m} in\pi(p_{k,n,m} - w_{n,m}) e^{in\pi x} e^{im\pi y}\right\|_2^2 = \pi^2 \sum_{n,m} n^2 |p_{k,n,m} - w_{n,m}|^2$$
$$= \pi^2 \sum_{n,m} \frac{n^2}{(1+n^2+m^2)^s}(1+n^2+m^2)^s |p_{k,n,m} - w_{n,m}|^2$$
$$\leq \pi^2 \sum_{n,m} (1+n^2+m^2)^s |p_{k,n,m} - w_{n,m}|^2 = \pi^2 \|p_k - w\|_{H_2^s}^2 \leq \pi^2 \cdot 2^{-2k}$$

which further implies that

$$\|\tilde{p}_k - \partial_x w\|_2 \leq \|\tilde{p}_k - \partial_x p_k\|_2 + \|\partial_x p_k - \partial_x w\|_2 \leq 2^{-k} + \pi 2^{-k}$$

Thus, by definition, $\{\tilde{p}_k\}$ is a δ_{L_2}-name of $\partial_x w$.

It is known classically that every polygonal domain in $\mathbb{R}^2$ is Lipschitz (see, for example, [9]) and $H_2^s(U)$ is continuously embedded in $C(\overline{U})$ if $s > 1$ and U is a bounded Lipschitz domain, where $\overline{U}$ is the closure of U in $\mathbb{R}^2$ and $C(\overline{U})$ is the set of all continuous functions on $\overline{U}$. Since Ω is a bounded polygonal domain, it follows that for any $s > 1$, there is a constant $C_s > 0$ such that $\|w\|_{C(\overline{\Omega})} \leq C_s \|w\|_{H_2^s(\Omega)}$, where $\|w\|_{C(\overline{\Omega})} = \|w\|_\infty = \max\{|w(x,y)| : (x,y) \in \overline{U}\}$.

Lemma 5. *For $s > 1$, multiplication Mul : $H_2^s(\Omega) \times L_2(\Omega) \to L_2(\Omega)$, $(v, w) \mapsto vw$, is $(\delta_{H_{2,0}^s} \times \delta_{L_2}, \delta_{L_2})$-computable.*

Proof. Assume that $\{p_k\}$ is a $\delta_{H_{2,0}^s}$-name of v and $\{q_k\}$ is a δ_{L_2}-name of w. For each $n \in \mathbb{N}$, pick $k(n) \in \mathbb{N}$ such that $C_s \|v\|_{H_2^s} \|w - q_{k(n)}\|_2 \leq 2^{-(n+1)}$. Since $\|v\|_{H_2^s}$ is computable from $\{p_k\}$, the function $n \mapsto k(n)$ is computable from $\{p_k\}$ and $\{q_k\}$. Next pick $m(n) \in \mathbb{N}$ such that $\|q_{k(n)}\|_{C(\overline{\Omega})} \|v - p_{m(n)}\|_{H_2^s} \leq 2^{-(n+1)}$. It is clear that $m(n)$ is computable from $k(n)$, $\{q_k\}$, and $\{p_k\}$. The sequence $\{p_{m(n)} q_{k(n)}\}_n$ is then a δ_{L_2}-name of vw, for it is a sequence of polynomials of rational coefficients and $\|vw - p_{m(n)} q_{k(n)}\|_2 \leq \|v\|_{C(\overline{\Omega})} \|w - q_{k(n)}\|_2 + \|q_{k(n)}\|_{C(\overline{\Omega})} \|v - p_{m(n)}\|_{H_2^s} \leq 2^{-n}$.

5.2 Some Classical Properties of Fractional Powers of $\mathbb{A}$

It is known that fractional powers of the Stokes operator $\mathbb{A}$ are well defined; cf. [10, SECTION 2.6]. In the following, we summarize some classical properties of the Stokes operator and its fractional powers; these properties will be used in later proofs.

Fact 5. *Let $\mathbb{A}$ be the Stokes operator.*

(1) *For every $0 \leq \alpha \leq 1$, let $D(\mathbb{A}^\alpha)$ be the domain of $\mathbb{A}^\alpha$; this is a Banach space with the norm $\|u\|_{D(\mathbb{A}^\alpha)} := \|\mathbb{A}^\alpha u\|_{L_{2,0}^\sigma(\Omega)} = \|\mathbb{A}^\alpha u\|_2$. In particular, $D(\mathbb{A}^\alpha)$ is continuously embedded in $H_2^{2\alpha}$, that is, for every $u \in D(\mathbb{A}^\alpha)$,*

$$\|u\|_{H_2^{2\alpha}} \leq \|u\|_{D(\mathbb{A}^\alpha)} = C \|\mathbb{A}^\alpha u\|_2 \tag{25}$$

where C is a constant independent of α. Moreover, we have $D(\mathbb{A}^{1/2}) = L_{2,0}^\sigma(\Omega) \bigcap \{u \in (H^1(\Omega))^2; u = 0 \text{ on } \partial\Omega\}$.

(2) *For every nonnegative α the estimate*

$$\|\mathbb{A}^\alpha e^{-t\mathbb{A}} u\|_2 \leq C_\alpha t^{-\alpha} \|u\|_2, \quad t > 0 \tag{26}$$

is valid for all $u \in L_{2,0}^\sigma(\Omega)$, where C_α is a constant depending only on α. In particular, $C_0 = 1$. Moreover, the estimate implies implicitly that for every $u \in L_{2,0}^\sigma(\Omega)$, $e^{-t\mathbb{A}} u$ is in the domain of $\mathbb{A}$, and thus $e^{-t\mathbb{A}} u$ vanishes on the boundary of Ω for $t > 0$.

(3) *If $\alpha \geq \beta > 0$, then $D(\mathbb{A}^\alpha) \subseteq D(\mathbb{A}^\beta)$.*

(4) *For $0 < \alpha < 1$, if $u \in D(\mathbb{A})$, then*

$$\mathbb{A}^\alpha u = \frac{\sin \pi\alpha}{\pi} \int_0^\infty t^{\alpha-1} \mathbb{A}(t\mathbb{I} + \mathbb{A})^{-1} u \, dt$$

(5) *$\|\mathbb{A}^{-1/4} \mathbb{P}(u, \nabla)v\|_2 \leq M \|\mathbb{A}^{1/4} u\|_2 \|\mathbb{A}^{1/2} v\|_2$ is valid for all u, v in the domain of $\mathbb{A}^{3/5}$, where M is a constant independent of u and v.*

Proof. See Lemmas 2.1, 2.2 and 2.3 in [3] for (1) and (2) except for $C_0 = 1$; $C_0 = 1$ is proved in Lemma 3. See Theorems 6.8 and 6.9 in Sect. 2.6 of [10] for (3) and (4); Lemma 3.2 in [3] for (5).

We record, without going into the details, that the constants C, M, and C_α ($0 \le \alpha \le 1$) appeared in Fact 5 are in fact computable (some general discussions on the computability of Sobolev embedding constants and interpolation constants together with other constants in the PDE theory are forthcoming).

5.3   Proof of Proposition 4

In order to show that the iteration sequence is effectively convergent, we need to establish several estimates on various functions such as $\|\mathbb{A}^\beta u_m(t)\|_2$ and $\|\mathbb{A}^\beta(u_{m+1}(t) - u_m(t))\|_2$ for β being some positive numbers. Subsequently, as a prerequisite, $u_m(t)$ must be in the domain of $\mathbb{A}^\beta$; thus the functions $u_m(t)$ are required to have higher smoothness than the given initial function $\boldsymbol{a}$ according to Fact 5-(1). This is indeed the case: For functions $u_m(t)$ obtained by the iteration (23), it is known classically that if $u_m(0) \in L_2(\Omega)$ then $u_m(t) \in H_2^{2\alpha}(\Omega)$ for $t > 0$, where $0 \le \alpha \le 1$. In other words, $u_m(t)$ undergoes a jump in smoothness from $t = 0$ to $t > 0$ (due to the integration). In the following lemma, we present an algorithmic version of this increase in smoothness.

Lemma 6. *Let $\alpha = 3/5$. Then for the iteration (23)*

$$\boldsymbol{u}_0(t) = e^{-t\mathbb{A}}\boldsymbol{a}, \qquad \boldsymbol{u}_{m+1}(t) = \boldsymbol{u}_0(t) - \int_0^t e^{-(t-s)\mathbb{A}}\mathbb{B}u_m(s)\,ds$$

the mapping $\mathbb{S}_H : \mathbb{N} \times L_{2,0}^\sigma(\Omega) \times (0, \infty) \to H_{2,0}^{2\alpha}(\Omega) \times H_{2,0}^{2\alpha}(\Omega)$, $(m, \boldsymbol{a}, t) \mapsto \boldsymbol{u}_m(t)$, *is well-defined and* $(\nu \times \delta_{L_{2,0}^\sigma} \times \rho, \delta_{H_{2,0}^{2\alpha}})$*-computable.*

We emphasize that the lemma holds true for $t > 0$ only. Also the choice of $\alpha = 3/5$ is somewhat arbitrary; in fact, α can be selected to be any rational number strictly between $\frac{1}{2}$ and $\frac{3}{4}$. The requirement $\alpha < \frac{3}{4}$ guarantees that $D(\mathbb{A}^\alpha) \subset H_{2,0}^{2\alpha}(\Omega) \times H_{2,0}^{2\alpha}(\Omega)$ because $2\alpha < 3/2$ (see (24)). The other condition $\alpha > \frac{1}{2}$ ensures that Lemma 5 can be applied for $2\alpha > 1$.

Proof. We induct on m. Note that for any $t > 0$ and any $a \in L_{2,0}^\sigma(\Omega)$, the estimates (25) and (26) imply that

$$\|e^{-t\mathbb{A}}a\|_{H_2^{2\alpha}} \le C\|\mathbb{A}^\alpha e^{-t\mathbb{A}}a\|_2 \le CC_\alpha t^{-\alpha}\|a\|_2$$

Combining this inequality with the following strengthened version of (19): for any $a \in \mathcal{P}$,

$$\sum_{n,m \ge 1} (1 + n^2 + m^2)^2 |a_{n,m}|^2 < \infty$$

(the inequality is valid since a is C^∞), a similar argument used to prove Proposition 3 works for $m = 0$. Moreover, by type conversion (Fact 2), $a \in L^\sigma_{2,0}(\Omega) \mapsto u_0 \in C((0,\infty), H^{6/5}_{2,0}(\Omega) \times H^{6/5}_{2,0}(\Omega))$ is $(\delta_{L^\sigma_{2,0}}, [\rho \to \delta_{H^{6/5}_{2,0}}])$-computable.

Assume that $(j, a) \mapsto u_j$ is $(\nu, \delta_{L^\sigma_{2,0}}, [\rho \to \delta_{H^{6/5}_{2,0}}])$-computable for $0 \le j \le m$, where $a \in L^\sigma_{2,0}(\Omega)$, and $u_j \in C((0,\infty), (H^{6/5}_{2,0}(\Omega))^2)$. We show how to compute a $\delta_{H^{6/5}_{2,0}}$-name for $u_{m+1}(t) = e^{-t\mathbb{A}} a - \int_0^t e^{-(t-s)\mathbb{A}} \mathbb{B} u_m(s) ds$ on inputs $m + 1$, a and $t > 0$. Let us first look into the nonlinear term $\mathbb{B} u_m$. It is clear that $\mathbb{B} u_m(s)$ lies in $L^\sigma_{2,0}(\Omega)$ for $s > 0$. Moreover, it follows from Lemmas 4 and 5, and Proposition 2 that the map $(u_m, s) \mapsto \mathbb{B} u_m(s)$ is $([\rho \to \delta_{H^{2\alpha}_{2,0}}], \rho, \delta_{L^\sigma_{2,0}})$-computable for all $s \in (0, t]$. Now since $\mathbb{B} u_m(s)$ is in $L^\sigma_{2,0}(\Omega)$ for $s > 0$, it follows from the case where $m = 0$ that $(u_m, s) \mapsto e^{-(t-s)\mathbb{A}} \mathbb{B} u_m(s)$ is $([\rho \to \delta_{H^{2\alpha}_{2,0}}], \rho, \delta_{H^{6/5}_{2,0}})$-computable for $0 < s < t$.

Next let us consider the integral $\int_0^t e^{-(t-s)\mathbb{A}} \mathbb{B} u_m(s)\, ds$; we wish to compute a $\delta_{H^{6/5}_{2,0}}$-name of the integral from a and $t > 0$. We make use of the following fact: For $\theta \ge 1$, the integration operator from $C([a, b]; H^\theta_{2,0}(\Omega) \times H^\theta_{2,0}(\Omega))$ to $H^\theta_{2,0}(\Omega) \times H^\theta_{2,0}(\Omega)$, $F \mapsto \int_a^b F(t)(x) dt$, is computable from a, b, and F. This fact can be proved by a similar argument as the one used in the proof of Lemma 3.7 [24]. However, since the function $e^{-(t-s)\mathbb{A}} \mathbb{B} u_m(s)$ is not necessarily in $(H^{6/5}_2(\Omega))^2$ when $s = 0$ or $s = t$, the stated fact cannot be directly applied to the given integral. To overcome the problem of possible singularities at the two endpoints, we use a sequence of closed subintervals $[t_n, t - t_n]$ to approximate the open interval $(0, t)$, where $t_n = t/2^n$, $n \ge 1$. Then it follows from the stated fact and the induction hypotheses that a $\delta_{H^{6/5}_{2,0}}$-name, say $\{p_{n,K}\}$, of $u^n_{m+1}(t) = e^{-t\mathbb{A}} a - \int_{t_n}^{t - t_n} e^{-(t-s)\mathbb{A}} \mathbb{B} u_m(s) ds$ can be computed from inputs n, u_m, and $t > 0$, which satisfies the condition that $\|u^n_{m+1}(t) - p_{n,K}\|_{H^{6/5}_2} \le 2^{-K}$. Thus if we can show that the integrals $\int_0^{t_n} e^{-(t-s)\mathbb{A}} \mathbb{B} u_m(s) ds$ and $\int_{t-t_n}^t e^{-(t-s)\mathbb{A}} \mathbb{B} u_m(s) ds$ tend to zero effectively in $H^{6/5}_2 \times H^{6/5}_2$-norm as $n \to \infty$, then we can effectively construct a $\delta_{H^{6/5}_{2,0}}$-name of $u_{m+1}(t)$ from $\{p_{n,K}\}_{n,K}$.

It remains to show that both sequences of integrals tend to zero effectively in $H^{6/5}_2 \times H^{6/5}_2$-norm as $n \to \infty$. Since a similar argument works for both sequences, it suffices to show that the sequence $\text{Int}_n := \int_0^{t_n} e^{-(t-s)\mathbb{A}} \mathbb{B} u_m(s) ds$ tends to zero effectively as $n \to \infty$. We are to make use of Fact 5-(1), (2), (5) for showing the effective convergence. The following two claims comprise the proof.

Claim I. Let $\beta = \frac{1}{2}$ or $\frac{1}{4}$. Then the map $(a, t, m, \beta) \mapsto M_{\beta,m}$ is computable, where $M_{\beta,m}$ is a positive number satisfying the condition

$$\|\mathbb{A}^\beta u_m(s)\|_2 \le M_{\beta,m} s^{-\beta} \quad \text{for all } 0 < s < t \tag{27}$$

(note that $M_{\beta,m}$ is independent of s).

Proof. Again we induct on m. For $m = 0$, let $M_{\beta,0} = C_\beta \|a\|_2$, where C_β is the constant in estimate (26) with α replaced by β and u by a. Then $M_{\beta,0}$ is computable from a and β, and $\|A^\beta u_0(s)\|_2 \le C_\beta s^{-\beta}\|a\|_2 = M_{\beta,0}s^{-\beta}$ for any $s > 0$. Assume that $M_{\beta,k}$, $0 \le k \le m$, has been computed from k, β, a, and $t > 0$. We show how to compute $M_{\beta,m+1}$. Since $u_{m+1}(s)$ has a singularity at $s = 0$, it may not be in $H_2^{2\beta}(\Omega) \times H_2^{2\beta}(\Omega)$ at $s = 0$ (recall that $D(A^{1/2}) = L_{2,0}^\sigma(\Omega) \bigcap \{u \in H_2^1(\Omega) \times H_2^1(\Omega) : u = 0 \text{ on } \partial\Omega\}$). Let us first compute a bound (in L_2-norm) for $A^\beta \int_\epsilon^s e^{-(t-r)A}\mathbb{B}u_m(r)dr$, where $0 < \epsilon < s$. It follows from the induction hypothesis, Fact 5-(1), (2), (5), and Theorems 6.8 and 6.13 in [10] that

$$\left\|A^\beta \int_\epsilon^s e^{-(s-r)A}\mathbb{B}u_m(r)dr\right\|_2$$

$$= \left\|\int_\epsilon^s A^{\beta+1/4}e^{-(s-r)A}A^{-1/4}\mathbb{B}u_m(r)dr\right\|_2$$

$$\le C_{\beta+1/4}\int_\epsilon^s (s-r)^{-(\beta+1/4)}\|A^{-1/4}\mathbb{B}u_m(r)\|_2 dr$$

$$\le C_{\beta+1/4}M\int_\epsilon^s (s-r)^{-(\beta+1/4)}\|A^{1/4}u_m(r)\|_2\|A^{1/2}u_m(r)\|_2 dr$$

$$\le C_{\beta+1/4}MM_{1/4,m}M_{1/2,m}\int_\epsilon^t (s-r)^{-(\beta+1/4)}r^{-3/4}dr \qquad (28)$$

Subsequently, we obtain that

$$\|A^\beta u_{m+1}(s)\|_2$$

$$= \left\|A^\beta u_0(s) - \int_0^s A^\beta e^{-(s-r)A}\mathbb{B}u_m(r)dr\right\|_2$$

$$\le M_{\beta,0}s^{-\beta} + \left\|\lim_{\epsilon\to 0}\int_\epsilon^s A^\beta e^{-(s-r)A}\mathbb{B}u_m(r)dr\right\|_2$$

$$\le M_{\beta,0}s^{-\beta} + C_{\beta+\frac{1}{4}}MM_{\frac{1}{4},m}M_{\frac{1}{2},m}\int_0^s (s-r)^{-(\beta+\frac{1}{4})}r^{-3/4}dr$$

$$= M_{\beta,0}s^{-\beta} + C_{\beta+\frac{1}{4}}MM_{\frac{1}{4},m}M_{\frac{1}{2},m}B\left(\frac{3}{4} - \beta, \frac{1}{4}\right)s^{-\beta} \qquad (29)$$

where $B(\frac{3}{4} - \beta, 1/4)$ is the integral $\int_0^1 (1 - \theta)^{(\frac{3}{4}-\beta)-1}\theta^{\frac{1}{4}-1}d\theta$, which is the value of the Beta function $B(x, y) = \int_0^1 (1 - \theta)^{1-x}\theta^{1-y}d\theta$ at $x = \frac{3}{4} - \beta$ and $y = \frac{1}{4}$. It is clear that $B(\frac{3}{4} - \beta, 1/4)$ is computable. Thus if we set

$$M_{\beta,m+1} = M_{\beta,0} + C_{\beta+\frac{1}{4}}MM_{\frac{1}{4},m}M_{\frac{1}{2},m}B\left(\frac{3}{4} - \beta, \frac{1}{4}\right) \qquad (30)$$

then $M_{\beta,m+1}$ is computable and satisfies the condition that $\|A^\beta u_{m+1}(s)\|_2 \le M_{\beta,m+1}s^{-\beta}$ for all $0 < s < t$. The proof of Claim I is complete.

Claim II. $\left\|\int_0^{t_n} e^{-(t-s)A}\mathbb{B}u_m(s)ds\right\|_{H_2^{6/5}} \to 0$ effectively as $n \to \infty$.

Proof. Once again, to avoid singularity of $u_m(s)$ at $s = 0$, we begin with the following estimate: Let $0 < \epsilon < t_n$. Then it follows from Fact 5-(1), (2), (5), (27),

(30), and a similar calculation as performed in Claim I that

$$\left\| \int_\epsilon^{t_n} e^{-(t-s)\mathbb{A}} \mathbb{B} u_m(s)ds \right\|_{H_2^{6/5}}$$

$$\leq C\|\mathbb{A}^{3/5} \int_\epsilon^{t_n} e^{-(t-s)\mathbb{A}} \mathbb{B} u_m(s)ds\|_2$$

$$\leq CC_{17/20} MM_{\frac{1}{4},m} M_{\frac{1}{2},m} \int_\epsilon^{t_n} (t-s)^{-17/20} s^{-3/4} ds$$

$$\leq CC_{17/20} MM_{\frac{1}{4},m} M_{\frac{1}{2},m} (t-t_n)^{-17/20} \cdot 4(t_n^{1/4} - \epsilon^{1/4})$$

which then implies that

$$\left\| \int_0^{t_n} e^{-(t-s)\mathbb{A}} \mathbb{B} u_m(s)ds \right\|_{H_2^{6/5}}$$

$$= \left\| \lim_{\epsilon \to 0} \int_\epsilon^{t_n} e^{-(t-s)\mathbb{A}} \mathbb{B} u_m(s)ds \right\|_{H_2^{6/5}}$$

$$\leq \lim_{\epsilon \to 0} CC_{17/20} MM_{\frac{1}{4},m} M_{\frac{1}{2},m} (t-t_n)^{-17/20} \cdot 4(t_n^{1/4} - \epsilon^{1/4})$$

$$= CC_{17/20} MM_{\frac{1}{4},m} M_{\frac{1}{2},m} (t-t_n)^{-17/20} \cdot 4t_n^{1/4}$$

It is readily seen that $\int_0^{t_n} e^{-(t-s)\mathbb{A}} \mathbb{B} u_m(s)ds\|_{H_2^{6/5}} \to 0$ effectively as $n \to \infty$ (recall that $t_n = t/2^n$). The proof for the claim II, and thus for the lemma is now complete.

Remark 1. In our effort to compute an upper bound for $\|\mathbb{A}^\beta u_{m+1}(s)\|_2$, we start with the integral $\int_\epsilon^s e^{-(s-r)\mathbb{A}} \mathbb{B} u_m(r)dr$ because the integral might have a singularity at 0; then we take the limit as $\epsilon \to 0$ to get the desired estimate (see computations of (28) and (29)). The limit exists because the bound, $C_{\beta+\frac{1}{4}} MM_{\frac{1}{4},m} M_{\frac{1}{2},m} B\left(\frac{3}{4} - \beta, \frac{1}{4}\right)$, is uniform in r for $0 < r \leq s$. In the rest of the paper, we will encounter several similar computations. In those later situations, we will derive the estimates starting with $\int_0^t$ instead of $\int_\epsilon^t$. There will be no loss in rigor because the integral is uniformly bounded with respect to the integrating variable, say t, for $t > 0$.

Corollary 1. *For any $a \in L_{2,0}^\sigma(\Omega)$ and $t > 0$, let $\{u_m(t)\}$ be the sequence generated by the iteration scheme (23) based on a. Then $u_m(t) \in Dom(\mathbb{A}^{3/5}) \subset Dom(\mathbb{A}^{1/2}) \subset Dom(\mathbb{A}^{1/4})$.*

Proof. The corollary follows from Lemma 6 and Fact 5-(3).

Corollary 2. *The map from $\mathcal{P}$ to $L_2(\Omega)$, $u \mapsto \|\mathbb{A}^\alpha u\|_2$, is $(\delta_{L_{2,0}^\sigma}, \rho)$-computable, where $\alpha = 1/8, 1/4,$ or $1/2$.*

Proof. We prove the case when $\alpha = 1/4$; the other two cases can be proved in exactly the same way. Since $\mathcal{P}$ is contained in the domain of $\mathbb{A}$, it follows from

Theorem 6.9, Sect. 2.6 [10] that for every $\boldsymbol{u} \in \mathcal{P}$, $\mathbb{A}^{1/4}\boldsymbol{u} = \frac{\sin \pi/4}{\pi} \int_0^\infty t^{-3/4}\mathbb{A}(t\mathbb{I} + \mathbb{A})^{-1}\boldsymbol{u}dt$. By definition of $\mathcal{P}$, if $\boldsymbol{u} \in \mathcal{P}$, then $\boldsymbol{u}$ is C^∞ with compact support in Ω, and $\mathbb{A}\boldsymbol{u} = -\mathbb{P} \triangle \boldsymbol{u} = - \triangle \boldsymbol{u}$. Express each component of $\boldsymbol{u} = (u^1, u^2)$ in terms of the orthogonal basis $\{e^{in\pi x}e^{im\pi y}\}_{n,m}$ of $L_2(\Omega)$ in the form of $u^i = \sum_{n,m\geq 0} u^i_{n,m}e^{i\pi nx}e^{i\pi my}$, where $u^i_{n,m} = \int_{-1}^1 \int_{-1}^1 u^1(x,y)e^{i\pi nx}e^{i\pi my}dxdy$. Then a straightforward calculation shows that

$$\frac{\sin \pi/4}{\pi} \int_0^\infty t^{-3/4}\mathbb{A}(t\mathbb{I} + \mathbb{A})^{-1}u^i dt$$

$$= \frac{\sin \pi/4}{\pi} \sum_{n,m\geq 0} \left(\int_0^\infty t^{-3/4}\frac{(\pi n)^2 + (\pi m)^2}{t + (\pi n)^2 + (\pi m)^2}dt \right) u^i_{n,m}e^{i\pi nx}e^{i\pi my}.$$

Since the integral is convergent and computable, it follows that $\mathbb{A}^{1/4}\boldsymbol{u}$ is computable from $\boldsymbol{u}$ and, consequently, $\|\mathbb{A}\boldsymbol{u}\|_2$ is computable.

Proof (Proof of Proposition 4). For each $\boldsymbol{a} \in L_{2,0}^\sigma$, let $\{\boldsymbol{a}_k\}$, $\boldsymbol{a}_k = (a_k^1, a_k^2) \in \mathcal{P}$, be a $\delta_{L_{2,0}^\sigma}$-name of $\boldsymbol{a}$; i.e. $\|\boldsymbol{a} - \boldsymbol{a}_k\|_2 \leq 2^{-k}$. Let $\widetilde{C} := c_1 M B_1$, where M is the constant in Fact 5(4), $c_1 = \max\{C_{1/4}, C_{1/2}, C_{3/4}, 1\}$, and

$$B_1 = \max\{B(1/2, 1/4), B(1/4, 1/4), 1\}$$

with $B(a,b) = \int_0^1 (1-t)^{a-1}t^{b-1}dt$, $a, b > 0$, being the beta function. Then M and c_1 are computable by assumption while B_1 is computable for the beta functions with rational parameters are computable. Note that $c_1 B_1 \geq 1$. Let

$$v_m(t) = u_{m+1}(t) - u_m(t) = \int_0^t e^{-(t-s)\mathbb{A}}(\mathbb{B}u_m(s) - \mathbb{B}u_{m-1}(s))ds, \quad m \geq 1 \quad (31)$$

Our goal is to compute a constant ϵ, $0 < \epsilon < 1$, such that near $t = 0$,

$$\|v_m(t)\|_2 \leq L\epsilon^{m-1} \quad (32)$$

where L is a constant. Once this is accomplished, the proof is complete.

It follows from Corollary 1 that Fact 5-(5) holds true for all $u_m(t)$ and $v_m(t)$ with $t > 0$. It is also known classically that

$$\|\mathbb{A}^{-\frac{1}{4}}(\mathbb{B}u_{m+1}(t) - \mathbb{B}u_m(t))\|_2$$

$$= \|\mathbb{A}^{-\frac{1}{4}}\mathbb{B}u_{m+1}(t) - \mathbb{A}^{-\frac{1}{4}}\mathbb{B}u_m(t)\|_2$$

$$\leq M \left(\|\mathbb{A}^{\frac{1}{4}}v_m(t)\|_2\|\mathbb{A}^{\frac{1}{2}}u_{m+1}(t)\|_2 + \|\mathbb{A}^{\frac{1}{4}}u_m(t)\|_2\|\mathbb{A}^{\frac{1}{2}}v_m(t)\|_2 \right) \quad (33)$$

(see, for example, [5]). The equality in the above estimate holds true because $\mathbb{A}^{-1/4}$ is a (bounded) linear operator. The estimate (33) indicates that, in order to achieve (32), there is a need in establishing some bounds on $\|\mathbb{A}^\beta u_m(t)\|_2$ and $\|\mathbb{A}^\beta v_m(t)\|_2$ which become ever smaller as m gets larger uniformly for values of t near zero. The desired estimates are developed in a series of claims beginning with the following one.

Claim 1. Let $\beta = \frac{1}{4}$ or $\frac{1}{2}$; let

$$\tilde{K}^a_{\beta,0}(T) = \max_{0 \leq t \leq T} t^\beta \|\mathbb{A}^\beta e^{-t\mathbb{A}} a\|_2$$

and

$$k^a_0(T) = \max\{\tilde{K}^a_{\frac{1}{4},0}(T), \tilde{K}^a_{\frac{1}{2},0}(T)\}$$

Then there is a computable map from $L^\sigma_{2,0}(\Omega)$ to $(0,1)$, $a \mapsto T_a$, such that

$$k^a_0(T_a) < \frac{1}{8\tilde{C}}$$

Proof. First we note that $t^\beta \|\mathbb{A}^\beta e^{-t\mathbb{A}} a\|_2 = 0$ for any $a \in L^\sigma_{2,0}(\Omega)$ if $t = 0$; cf. Theorem 6.1 in [3]. Furthermore, it follows from (17) that the operator norm of $e^{-t\mathbb{A}}$, $\|e^{-t\mathbb{A}}\|_{op}$, is bound above by 1 for any $t > 0$. Since $e^{-t\mathbb{A}}$ is the identity map on $L^\sigma_{2,0}(\Omega)$ when $t = 0$, we conclude that $\max_{0 \leq t \leq T} \|e^{-t\mathbb{A}}\|_{op} \leq 1$ for any $T > 0$. Now for any $a \in L^\sigma_{2,0}(\Omega)$, it follows from Fact 5-(2) and Theorems 6.8 and 6.13 in Sect. 2.6 of [10] ($\mathbb{A}^\alpha$ and $e^{-t\mathbb{A}}$ are interchangeable) that

$$\tilde{K}^a_{\beta,0}(T) = \max_{0 \leq t \leq T} t^\beta \|\mathbb{A}^\beta e^{-t\mathbb{A}} a\|_2$$

$$= \sup_{0 \leq t \leq T} t^\beta \|\mathbb{A}^\beta e^{-t\mathbb{A}} a\|_2$$

$$\leq \sup_{0 < t \leq T} t^\beta \|\mathbb{A}^\beta e^{-t\mathbb{A}} (a - a_k)\|_2 + \sup_{0 < t \leq T} t^\beta \|\mathbb{A}^\beta e^{-t\mathbb{A}} a_k\|_2$$

$$\leq C_\beta \|a - a_k\|_2 + T^\beta \max_{0 \leq t \leq T} \|e^{-t\mathbb{A}}\|_{op} \|\mathbb{A}^\beta a_k\|_2$$

$$\leq c_1 2^{-k} + \max\{T^{1/4}, T^{1/2}\} \max\{\|\mathbb{A}^{1/4} a_k\|_2, \|\mathbb{A}^{1/2} a_k\|_2\}$$

We note that although a is not necessarily in the domain of $\mathbb{A}$ but $a_k \in \mathcal{P}$ and $\mathcal{P}$ is contained in the domain of $\mathbb{A}$; thus $\mathbb{A}^\beta a_k$ is well defined. Furthermore, it follows from Corollary 2 that $\|\mathbb{A}^\beta a_k\|_2$ is computable. Clearly one can compute a positive integer $\hat{k}$ such that

$$2^{-\hat{k}} < \frac{1}{16 c_1 \tilde{C}}$$

then compute a positive number T_a such that

$$\max\{T_a^{1/4}, T_a^{1/2}\} \max\{\|\mathbb{A}^{1/4} a_{\tilde{k}}\|_2, \|\mathbb{A}^{1/2} a_{\tilde{k}}\|_2\} < \frac{1}{16\tilde{C}}$$

The computations are performed on the inputs a and the constants c_1, M, and B_1. Consequently, $k^a_0(T_a) < 1/(8\tilde{C})$. The proof of Claim 1 is complete.

We recall that, for a given $a \in L^\sigma_{2,0}(\Omega)$, the iteration scheme (23) is based on the "seed" function $u_0(t) = e^{-t\mathbb{A}} a$. Claim 1 asserts that the seed function has the property that $\max_{0 \leq t \leq T} t^\beta \|\mathbb{A}^\beta u_0(t)\|_2$ is bounded by $\tilde{K}^a_{\beta,0}$, uniformly in t. We extend this property to the iteration sequence $\{u_m(t)\}$ in the next claim.

Claim 2. Let $\beta = \frac{1}{4}$ or $\frac{1}{2}$. Then there is a computable map $\mathbb{N} \times L_{2,0}^{\sigma} \to (0,\infty)$, $(m, \boldsymbol{a}) \mapsto K_{\beta,m}^a$, such that

$$\max_{0 \le t \le T_a} t^{\beta} \|\mathbb{A}^{\beta} u_m(t)\|_2 \le K_{\beta,m}^a \tag{34}$$

Proof. We induct on m. For $m = 0$, let $K_{\beta,0}^a = 1/(8\widetilde{C})$. Then (34) follows from Claim 1. It is clear that $K_{\beta,0}^a$ is computable.

For $m \ge 1$ and $t > 0$, $K_{\beta,m+1}^a$ is computed by the recursive formula:

$$K_{\beta,m+1}^a = K_{\beta,0}^a + C_{\beta+\frac{1}{4}} MB(1 - \beta - \frac{1}{4}, \frac{1}{4}) K_{\frac{1}{4},m}^a K_{\frac{1}{2},m}^a \tag{35}$$

The recursive formula is derived similarly as that of (29). Since the upper bound is uniformly valid for all $0 < t \le T_a$, it follows that it is also valid for $t = 0$. The proof of Claim 2 is complete.

In the next claim, we show that the sequences $\{K_{\beta,m}^a\}$, $\beta = 1/4$ or $1/2$, are bounded above with an upper bound strictly less than $1/(2\widetilde{C})$.

Claim 3. Let $k_m^a = \max\{K_{\frac{1}{4},m}^a, K_{\frac{1}{2},m}^a\}$ and let $K = \frac{4k_0^a(\sqrt{2}-1)}{\sqrt{2}}$. Then $k_m^a \le K < \frac{1}{2\widetilde{C}}$ for all $m \ge 1$.

Proof. It follows from Claim 2 that $k_0^a = \frac{1}{8\widetilde{C}}$ and $k_{m+1}^a \le k_0^a + \widetilde{C}(k_m^a)^2$ (recall that $\widetilde{C} = c_1 MB_1$). To get a bound on k_m^a, let's write $k_m^a = k_0^a w_m$. Then w_m satisfies the following inequality:

$$k_0^a w_{m+1} \le k_0^a + \widetilde{C}(k_0^a)^2 w_m^2$$

which implies that

$$w_{m+1} \le 1 + \widetilde{C} k_0^a w_m^2 = 1 + \frac{1}{8} w_m^2$$

Then a direct calculation shows that

$$w_m \le \frac{4(\sqrt{2}-1)}{\sqrt{2}}$$

Thus

$$k_m^a = k_0^a w_m \le \frac{4k_0^a(\sqrt{2}-1)}{\sqrt{2}} = \frac{\sqrt{2}-1}{2\sqrt{2}\widetilde{C}} < \frac{1}{2\widetilde{C}}$$

And so if we pick $K = \frac{4k_0^a(\sqrt{2}-1)}{\sqrt{2}}$, then $k_m^a \le K < \frac{1}{2\widetilde{C}}$ for all $m \ge 1$. The proof of Claim 3 is complete.

Next we present an upper bound for $t^{\alpha} \|\mathbb{A}^{\alpha} v_m(t)\|_2$, $m \ge 1$. Recall that $v_m(t) = u_{m+1}(t) - u_m(t)$.

Claim 4. For $t \in [0, T_a]$, $0 \le \alpha < \frac{3}{4}$, and $m \ge 1$,

$$t^\alpha \|A^\alpha v_m(t)\|_2 \le 2KC_{\alpha+\frac{1}{4}}(2\widetilde{C}K)^{m-1}B(1 - \alpha - \frac{1}{4}, \frac{1}{4}) \tag{36}$$

Proof. First we observe that (36) is true for $t = 0$. Next we assume that $0 < t \le T_a$. Once again we induct on m. At $m = 1$: We recall from the definition of c_1 and B_1 that $\frac{1}{2c_1 B_1} \le \frac{1}{2}$. Also it follows from (33), Claims 2 and 3 that $\|A^{\frac{1}{2}} u_1(t)\|_2 \le K^\alpha_{\frac{1}{2}, 1} t^{-\frac{1}{2}} \le Kt^{-\frac{1}{2}}$, $\|A^{\frac{1}{4}} u_0(t)\|_2 \le Kt^{-\frac{1}{4}}$, $\|A^{\frac{1}{4}} v_0(t)\|_2 \le 2Kt^{-\frac{1}{4}}$, and $\|A^{\frac{1}{2}} v_0(t)\|_2 \le 2Kt^{-\frac{1}{2}}$. Making use of these inequalities we obtain the following estimate:

$$
\begin{aligned}
t^\alpha &\|A^\alpha v_1(t)\|_2 \\
&= t^\alpha \|A^\alpha (u_2(t) - u_1(t))\|_2 \\
&= t^\alpha \left\| A^\alpha \int_0^t e^{-(t-s)A}(\mathbb{B}u_1(s) - \mathbb{B}u_0(s))ds \right\|_2 \\
&\le t^\alpha C_{\alpha+\frac{1}{4}} \int_0^t (t-s)^{-\alpha-\frac{1}{4}} \|A^{-\frac{1}{4}}\mathbb{B}u_1(s) - A^{-\frac{1}{4}}\mathbb{B}u_0(s)\|_2 ds \\
&\le t^\alpha C_{\alpha+\frac{1}{4}} \int_0^t (t-s)^{\alpha-\frac{1}{4}} M \bigg(\|A^{\frac{1}{4}} v_0(s)\|_2 \cdot \|A^{\frac{1}{2}} u_1(s)\|_2 \\
&\qquad\qquad\qquad\qquad + \|A^{\frac{1}{4}} u_0(s)\|_2 \cdot \|A^{\frac{1}{2}} v_0(s)\|_2 \bigg) ds \\
&\le t^\alpha C_{\alpha+\frac{1}{4}} M 2K^2 \int_0^t (t-s)^{-\alpha-\frac{1}{4}} s^{-\frac{3}{4}} ds \\
&= 2KC_{\alpha+\frac{1}{4}} MKB(1 - \alpha - \frac{1}{4}, \frac{1}{4}) \\
&< 2KC_{\alpha+\frac{1}{4}} \frac{M}{2c_1 MB_1} B(1 - \alpha - \frac{1}{4}, \frac{1}{4}) \quad \text{(recall that } K < \frac{1}{2\widetilde{C}} = \frac{1}{2c_1 MB_1}) \\
&< 2KC_{\alpha+\frac{1}{4}} (2\widetilde{C}K)^0 B(1 - \alpha - \frac{1}{4}, \frac{1}{4})
\end{aligned}
$$

Thus (36) is true for $m = 1$.

Now assuming that (36) holds for all $1 \le j \le m$, we show that (36) is also true for $m+1$. First it follows from Claims 2 and 3, and the induction hypothesis that for any $s \in (0, T_a)$,

$$
\begin{aligned}
\|A^{\frac{1}{4}} v_m(s)\|_2 \cdot \|A^{\frac{1}{2}} u_{m+1}(s)\|_2 &\le 2KC_{\frac{1}{4}+\frac{1}{4}}(2\widetilde{C}K)^{m-1}B(1 - \frac{1}{4} - \frac{1}{4}, \frac{1}{4})s^{-\frac{1}{4}} \cdot Ks^{-\frac{1}{2}} \\
&\le 2Kc_1(2\widetilde{C}K)^{m-1}B_1 Ks^{-\frac{3}{4}}
\end{aligned}
$$

Similarly,

$$\|A^{\frac{1}{2}} v_m(s)\|_2 \cdot \|A^{\frac{1}{4}} u_m(s)\|_2 \le 2Kc_1(2\widetilde{C}K)^{m-1}B_1 Ks^{-\frac{3}{4}}$$

Thus,

$$\|A^{\frac{1}{2}}u_{m+1}(s)\|_2 \cdot \|A^{\frac{1}{4}}v_m(s)\|_2 + \|A^{\frac{1}{2}}v_m(s)\|_2 \cdot \|A^{\frac{1}{4}}u_m(s)\|_2$$
$$\leq 2Kc_1(2\widetilde{C}K)^{m-1}B_1 \cdot 2Ks^{-\frac{3}{4}}$$

These inequalities imply the desired estimate:

$$t^\alpha \|A^\alpha v_{m+1}(t)\|_2$$
$$\leq t^\alpha C_{\alpha+\frac{1}{4}} \int_0^t (t-s)^{-\alpha-\frac{1}{4}} \|A^{-\frac{1}{4}}(\mathbb{B}u_{m+1}(s) - \mathbb{B}u_m(s))\|_2 ds$$
$$\leq t^\alpha C_{\alpha+\frac{1}{4}} \int_0^t (t-s)^{-\alpha-\frac{1}{4}} M\Big(\|A^{\frac{1}{2}}u_{m+1}(s)\|_2 \cdot \|A^{\frac{1}{4}}v_m(s)\|_2$$
$$+ \|A^{\frac{1}{2}}v_m(s)\|_2 \cdot \|A^{\frac{1}{4}}u_m(s)\|_2\Big) ds$$
$$\leq t^\alpha C_{\alpha+\frac{1}{4}} \int_0^t (t-s)^{-\alpha-\frac{1}{4}} M \cdot 2Kc_1(2\widetilde{C}K)^{m-1}B_1 \cdot 2Ks^{-\frac{3}{4}} ds$$
$$= t^\alpha 2KC_{\alpha+\frac{1}{4}} \cdot 2c_1 MB_1 K(2\widetilde{C}K)^{m-1} \int_0^t (t-s)^{-\alpha-\frac{1}{4}} s^{-\frac{3}{4}} ds$$
$$= 2KC_{\alpha+\frac{1}{4}}(2\widetilde{C}K)^m B(1-\alpha-\frac{1}{4}, \frac{1}{4})$$

The proof for Claim 4 is complete.

We now set $\alpha = 0$, $\epsilon = 2\widetilde{C}K$, and $L = 2KC_{\frac{1}{4}}B\left(\frac{3}{4}, \frac{1}{4}\right)$. Since $K < \frac{1}{2\widetilde{C}}$ by Claim 3, it follows that $0 < \epsilon < 1$ and

$$\|u_{m+1}(t) - u_m(t)\| \leq L\epsilon^{m-1}$$

Consequently, the iterated sequence $\{u_m(t)\}$ converges effectively to $u(t)$ and uniformly on $[0, T_a]$.

We mention in passing the following fact that can be proved by similar computations of Claims 1–3: On input (a, m, n), a positive number $T(a, m, n)$ can be computed such that $k_0^a(T(a, m, n)) < (8\widetilde{C})^{-1} \cdot 2^{-n}$, $T(a, m, n+1) < T(a, m, n)$, and $\max_{0 \leq t \leq T(a,m,n)} t^\beta \|A^\beta u_m(t)\|_2 \leq L_{\beta,m}^a \cdot 2^{-n}$, where $L_{\beta,m}^a$ is a constant independent of t and n, and computable from a and m.

5.4   Proof of Proposition 5

We now come to the proof of Proposition 5. We need to show that the map $\mathbb{S} : \mathbb{N} \times L_{2,0}^\sigma \times [0, \infty) \to L_{2,0}^\sigma$, $(m, a, t) \mapsto u_m(t)$, is $(\nu \times \delta_{L_{2,0}^\sigma} \times \rho, \delta_{L_{2,0}^\sigma})$-computable. By a similar argument as we used for proving Lemma 6, we are able to compute $u_m(t)$ on the input (m, a, t), where $m \in \mathbb{N}$, $a \in L_{2,0}^\sigma(\Omega)$, and $t > 0$. We note that $u_m(0) = u_0(0) = a$ for all $m \in \mathbb{N}$. Thus, to complete the proof, it suffices to show that there is a modulus function $\eta : \mathbb{N} \times \mathbb{N} \to \mathbb{N}$, computable from a, such that $\|u_{m+1}(t) - a\|_2 \leq 2^{-k}$ whenever $0 < t < 2^{-\eta(m+1,k)}$. Now for the details.

Given a and k. Refereeing to the last paragraph of the previous subsection and Fact 5-(2), (5), we obtain the following estimate: for $0 < t < T(a, m, n)$

$$\left\| \int_0^t e^{-(t-s)\mathbb{A}} \mathbb{B} u_m(s) ds \right\|_2$$

$$= \left\| \mathbb{A}^{1/4} \int_0^t e^{-(t-s)\mathbb{A}} \mathbb{A}^{-1/4} \mathbb{B} u(s) ds \right\|_2$$

$$\leq C_{1/4} M \int_0^t (t-s)^{-1/4} \| \mathbb{A}^{1/4} u_m(s) \|_2 \cdot \| \mathbb{A}^{1/2} u_m(s) \|_2 ds$$

$$\leq C_{1/4} M \int_0^t (t-s)^{-1/4} \cdot s^{-1/4} \cdot L^a_{1/4,m} \cdot 2^{-n} \cdot s^{-1/2} \cdot L^a_{1/2,m} \cdot 2^{-n} ds$$

$$\leq C_{1/4} M L^a_{1/4,m} L^a_{1/2,m} 2^{-2n} \int_0^t (t-s)^{-1/4} s^{-3/4} ds$$

$$= C_{1/4} M L^a_{1/4,m} L^a_{1/2,m} B(3/4, 1/4) \cdot 2^{-2n}$$

Thus if $\| e^{-t\mathbb{A}} a - a \|_2 \leq 2^{-(k+1)}$ and

$$2^{-2n} C_{1/4} M L^a_{1/4,m} L^a_{1/2,m} B(3/4, 1/4) \leq 2^{-(k+1)},$$

then

$$\| u_{m+1}(t) - a \|_2 \leq \| e^{-t\mathbb{A}} a - a \|_2 + \left\| \int_0^t e^{-(t-s)\mathbb{A}} \mathbb{B} u_m(s) ds \right\|_2 \leq 2^{-k}$$

Since $e^{-t\mathbb{A}} a$ is computable in t by Proposition 3 and $a = e^{-0\mathbb{A}} a$, there is a computable function $\theta_1 : \mathbb{N} \to \mathbb{N}$ such that $\| e^{-t\mathbb{A}} a - a \|_2 \leq 2^{-(k+1)}$ whenever $0 < t < 2^{-\theta_1(k)}$. Let $\theta_2 : \mathbb{N} \times \mathbb{N} \to \mathbb{N}$ be a computable function satisfying $C_{1/4} M L^a_{1/4,m} L^a_{1/2,m} B(3/4, 1/4) \cdot 2^{-2\theta_2(m,k)} \leq 2^{-(k+1)}$. Let $\eta(m+1, k)$ be a positive integer such that $2^{-\eta(m+1,k)} \leq \min\{2^{-\theta_1(k)}, T(a, m, \theta_2(m, k))\}$. Then η is the desired modulus function. The proof of Proposition 5 is complete.

Propositions 4 and 5 show that the solution u of the integral equation (4) is an effective limit of the computable iterated sequence $\{u_m\}$ starting with $u_0 = a$ on $[0, T_a]$; consequently, u itself is also computable. Thus we obtain the desired preliminary result:

Theorem 6. *There is a computable map $T : L^\sigma_{2,0}(\Omega) \to (0, \infty)$, $a \mapsto T(a)$, such that $u(t)$, the solution of the integral equation (4), is computable in $L^\sigma_{2,0}$ from a and t for $a \in L^\sigma_{2,0}$ and $t \in [0; T(a)]$.*

5.5   The Inhomogeneous Case and Pressure

It is known [5, THEOREM 2.3] that, also in the presence of an inhomogeneity $g \in C([0; T], L^\sigma_{2,0}(\Omega))$, the iterate sequence (5) converges to a unique solution u of Eq. (2) near $t = 0$. Similarly to (the proofs of) Propositions 5, 4, and [24, LEMMA

3.7], this solution is seen to be computable. Moreover, $g = \mathbb{P}f$ is computable from $f \in \left(L_2(\Omega)\right)^2$ according to Proposition 2. Finally the right-hand side of Eq. (6) equals

$$\left(\mathbb{I} - \mathbb{P}\right)\left[f + \triangle u - (u \cdot \nabla)u\right] =: h$$

which, by the definition of $\mathbb{P}$ projecting onto the solenoidal subspace, is conservative (=rotation-free/a pure divergence). Hence the path integral $\int_0^x h(y) \cdot d\gamma(y)$ does not depend on the chosen path from 0 to x and well-defines $P(x)$. This concludes our proof of Theorem 1.

A   Proof of Proposition 1

(a) For a divergence-free and boundary-free polynomial, its coefficients must satisfy a system of linear equations. In the following, we derive explicitly this system of linear equations in the 2-dimensional case, i.e. $\Omega = (-1,1)^2$. Let $p = (p_1, p_2) = \left(\sum_{i,j=0}^{N} a_{i,j}^1 x^i y^j, \sum_{i,j=0}^{N} a_{i,j}^2 x^i y^j\right)$ be a divergence-free and boundary-free polynomial of real coefficients. (If the degree of p_1 or p_2 is less than N, then zeros are placed for the coefficients of missing terms). Then, by definition,

$$
\begin{aligned}
\nabla \cdot p &= \frac{\partial p_1}{\partial x} + \frac{\partial p_2}{\partial y} \\
&= \sum_{1 \leq i \leq N, 0 \leq \leq N} i a_{i,j}^1 x^{i-1} y^j + \sum_{0 \leq i \leq N, 1 \leq \leq N} j a_{i,j}^2 x^i y^{j-1} \\
&= \sum_{0 \leq i,j \leq N-1} [(i+1)a_{i+1,j}^1 + (j+1)a_{i,j+1}^2]x^i y^j \\
&\quad + \sum_{0 \leq i \leq N-1} (i+1)a_{i+1,N}^1 x^i y^N + \sum_{0 \leq j \leq N-1} (j+1)a_{N,j+1}^2 x^N y^j \\
&\equiv 0 \quad \text{on } \Omega
\end{aligned}
$$

which implies that all coefficients in $\nabla \cdot p$ must be zero; or equivalently, Eq. (7) holds true. Turning to the boundary conditions, along the line $x = 1$, since

$$p(1,y) = \left(\sum_{j=0}^{N}\left(\sum_{i=0}^{N} a_{i,j}^1\right)y^j, \sum_{j=0}^{N}\left(\sum_{i=0}^{N} a_{i,j}^2\right)y^j\right)$$

is identically zero, it follows that $\sum_{i=0}^{N} a_{i,j}^1 = \sum_{i=0}^{N} a_{i,j}^2 = 0$ for $0 \leq j \leq N$. There are similar types of restrictions on the coefficients of p along the lines $x = -1$, $y = 1$, and $y = -1$. In summary, p vanishes on $\partial\Omega$ if and only if for all $0 \leq j,i \leq N$, both (8) and (9) hold true.

In the 3-dimensional case, a similar calculation shows that a polynomial triple $p(x,y,z) = \left(p_1(x,y,z), p_2(x,y,z), p_3(x,y,z)\right)$ is divergence-free and boundary-free if and only if its coefficients satisfies a system of linear equations with integer coefficients.

(b) In [8] it is shown that for any real number $s \geq 3$ and for any function $\boldsymbol{w} \in \mathcal{N}_{div}^s \cap H_{2,0}^{1,\sigma}(\Omega)^d$, the following holds:

$$\inf_{\boldsymbol{p} \in \mathcal{N}_{div}^1 \cap \mathcal{P}_N^0(\Omega)^d} \|\boldsymbol{w} - \boldsymbol{p}\|_{H_2^s(\Omega)^d} \leq CN^{-2}\|\boldsymbol{w}\|_{H_2^s(\Omega)^d}$$

where $\Omega = (-1,1)^d$,

$$\mathcal{N}_{div}^s = \{\boldsymbol{w} \in H_2^s(\Omega)^d \mid \nabla \cdot \boldsymbol{w} = 0\}, \quad \mathcal{P}_N^0(\Omega) = \mathcal{P}_N(\Omega) \bigcap H_{2,0}^{1,\sigma}(\Omega),$$

$\mathcal{P}_N$ is the set of all d-tuples of real polynomials with d variables and degree less than or equal to N with respect to each variable, $H_{2,0}^{1,\sigma}(\Omega)$ is the closure in $H_2^1(\Omega)$ of $C_0^\infty(\Omega)$, and C is a constant independent of N. This estimate implies that every function $\boldsymbol{w} \in L_{2,0}^\sigma$ can be approximated with arbitrary precision by divergence-free and boundary-free real polynomials as follows: for any $n \in \mathbb{N}$, since $\{\boldsymbol{u} \in C_0^\infty(\Omega)^d : \nabla \cdot \boldsymbol{u} = 0\}$ is dense in $L_{2,0}^\sigma$, there is a divergence-free C^∞ function $\boldsymbol{u}$ with compact support in Ω such that $\|\boldsymbol{w} - \boldsymbol{u}\|_{L_2} \leq 2^{-(n+1)}$. Then it follows from the above inequality that there exists a positive integer N and a divergence-free and boundary-free polynomial $\boldsymbol{p}$ of degree N with real coefficients such that $\|\boldsymbol{u} - \boldsymbol{p}\|_{L_2} \leq \|\boldsymbol{u} - \boldsymbol{p}\|_{H^3(\Omega)^d} \leq 2^{-(n+1)}$. Consequently, $\|\boldsymbol{w} - \boldsymbol{p}\|_{L_2} \leq \|\boldsymbol{w} - \boldsymbol{u}\|_{L_2} + \|\boldsymbol{u} - \boldsymbol{p}\|_{L_2} \leq 2^{-n}$.

It remains to show that $\mathbb{Q}_0^\sigma[\mathbb{R}^2]$, the divergence-free and boundary-free polynomial tuples with *rational* coefficients, is dense (in L_2-norm) in the set of all polynomial tuples with *real* coefficients which are divergence-free on Ω and boundary-free on $\partial\Omega$. To this end we note that, according to part (a), the divergence-free and boundary-free polynomials can be characterized, independent of their coefficient field, in terms of a homogeneous system of linear equations with integer coefficients. Then it follows from the lemma below that the set of the rational solutions of this system is dense in the set of its real solutions. And since Ω is bounded (=relatively compact), the approximations to its coefficients of a polynomial yields (actually uniform) the approximations to the polynomial itself:

$$\sup_{\boldsymbol{x} \in \Omega} |p_k(\boldsymbol{x})| \leq \sum_{i,j=0}^N |a_{i,j}^k| \cdot M^{i+j} \quad \text{for } \Omega \subseteq [-M, +M]^2 \quad \text{and } k = 1,2$$

Lemma 7. *Let $A \in \mathbb{Q}^{m \times n}$ be a rational matrix. Then the set $\text{kernel}_{IQ}(A) := \{\boldsymbol{x} \in \mathbb{Q}^n : A \cdot \boldsymbol{x} = \boldsymbol{0}\}$ of rational solutions to the homogeneous system of linear equations given by A is dense in the set $\text{kernel}_\mathbb{R}(A)$ of real solutions.*

Proof. For $d := \dim(\text{kernel}_\mathbb{R}(A))$, Gaussian Elimination yields a basis $B = (\boldsymbol{b}^1, \ldots, \boldsymbol{b}^d)$ of $\text{kernel}(A)$; in fact it holds $B \in \mathbb{Q}^{n \times d}$ and

$$\text{kernel}_\mathbb{F}(A) = \text{image}_\mathbb{F}(B) := \{\lambda_1 \boldsymbol{b}^1 + \cdots + \lambda_d \boldsymbol{b}^d : \lambda_1, \ldots, \lambda_d \in \mathbb{F}\}$$

for *every* field $\mathbb{F} \supseteq \mathbb{Q}$: Observe that the elementary row operations Gaussian Elimination employs to transform A into echelon form containing said basis B consist only of arithmetic (=field) operations! (We deliberately do not require B to be orthonormal; cf. [29, §3]). Now $\text{image}_\mathbb{Q}(B)$ is obviously dense in $\text{image}_\mathbb{R}(B)$.

B   Proof of Lemma 1

Note that $\gamma_n * \mathbb{T}_k \boldsymbol{p} = (\gamma_n * \mathbb{T}_k p_1, \gamma_n * \mathbb{T}_k p_2)$. For each $\boldsymbol{p} \in \mathbb{Q}_0^\sigma[\mathbb{R}^2]$ and $n \geq k$, since

$$
\begin{aligned}
\frac{\partial(\gamma_n * \mathbb{T}_k p_1)}{\partial x}(x,y) &= \frac{\partial}{\partial x} \int_{-1}^{1} \int_{-1}^{1} \gamma_n(x-s, y-t) \cdot \mathbb{T}_k p_1(s,t)\, ds\, dt \\
&= \int_{-1+2^{-k}}^{1-2^{-k}} \left[\int_{-1+2^{-k}}^{1-2^{-k}} \frac{\partial \gamma_n}{\partial x}(x-s, y-t) \cdot \mathbb{T}_k p_1(s,t)\, ds \right] dt \\
&= \int_{-1+2^{-k}}^{1-2^{-k}} \left[\int_{-1+2^{-k}}^{1-2^{-k}} -\frac{\partial \gamma_n}{\partial s}(x-s, y-t) \cdot \mathbb{T}_k p_1(s,t)\, ds \right] dt \quad (37)
\end{aligned}
$$

for $\mathbb{T}_k p_1 = 0$ in the exterior region of Ω_k including its boundary $\partial \Omega_k$. Note that $\frac{\partial \gamma_n}{\partial s}$ is continuous on $\mathbb{R}^2$; $\frac{\partial \gamma_n}{\partial s}(x-s, y-t) \cdot \mathbb{T}_k p_1(s,t)$ is continuous on $[-1,1]^2$ for any given $x, y \in \mathbb{R}$; $\frac{\partial \mathbb{T}_k p_1}{\partial s}(s,t)$ is continuous in $(-1+2^{-n}, 1-2^{-n})$ and $\mathbb{T}_k p_1$ is continuous on $[-1+2^{-n}, 1-2^{-n}]$ for any given $t \in [-1;1]$. Thus, we can apply the integration by parts formula to the integral

$$
\int_{-1+2^{-k}}^{1-2^{-k}} -\frac{\partial \gamma_n}{\partial s}(x-s, y-t) \cdot \mathbb{T}_k p_1(s,t)\, ds
$$

as follows:

$$
\begin{aligned}
\int_{-1+2^{-k}}^{1-2^{-k}} &-\frac{\partial \gamma_n}{\partial s}(x-s, y-t) \cdot \mathbb{T}_k p_1(s,t)\, ds \\
&= -\gamma_n(x-s, y-t) \cdot \mathbb{T}_k p_1(s,t)\Big|_{-1+2^{-k}}^{1-2^{-k}} \\
&\quad + \int_{-1+2^{-k}}^{1-2^{-k}} \gamma_n(x-s, y-t) \cdot \frac{\partial \mathbb{T}_k p_1}{\partial s}(s,t)\, ds \\
&= \int_{-1+2^{-k}}^{1-2^{-k}} \gamma_n(x-s, y-t) \cdot \frac{\partial \mathbb{T}_k p_1}{\partial s}(s,t)\, ds \quad (38)
\end{aligned}
$$

Then it follows from (37) and (38) that for any $(x,y) \in \Omega$,

$$
\frac{\partial \gamma_n * \mathbb{T}_k p_1}{\partial x}(x,y) = \int_{-1+2^{-k}}^{1-2^{-k}} \int_{-1+2^{-k}}^{1-2^{-k}} \gamma_n(x-s, y-t) \cdot \frac{\partial \mathbb{T}_k p_1}{\partial s}(s,t)\, ds\, dt.
$$

A similar calculation yields that for any $(x,y) \in \Omega$,

$$
\frac{\partial \gamma_n * \mathbb{T}_k p_2}{\partial y}(x,y) = \int_{-1+2^{-k}}^{1-2^{-k}} \int_{-1+2^{-k}}^{1-2^{-k}} \gamma_n(x-s, y-t) \cdot \frac{\partial \mathbb{T}_k p_2}{\partial t}(s,t)\, ds\, dt.
$$

Thus, for any $(x, y) \in \Omega$ and $n \geq k$,

$$
\nabla \cdot (\gamma_n * \mathbb{T}_k \boldsymbol{p})(x, y) = \frac{\partial \gamma_n * \mathbb{T}_k p_1}{\partial x}(x, y) + \frac{\partial \gamma_n * \mathbb{T}_k p_2}{\partial y}(x, y)
$$

$$
= \int_{-1+2^{-k}}^{1-2^{-k}} \int_{-1+2^{-k}}^{1-2^{-k}} \gamma_n(x - s, y - t) \cdot \left[\frac{\partial \mathbb{T}_k p_1}{\partial s} + \frac{\partial \mathbb{T}_k p_2}{\partial t} \right](s, t) \, ds \, dt
$$

$$
= 0
$$

for $\mathbb{T}_k \boldsymbol{p} = (\mathbb{T}_k p_1, \mathbb{T}_k p_2)$ is divergence-free on Ω_k. This proves that for any $\boldsymbol{p} \in \mathbb{Q}_0^{\sigma}[\mathbb{R}^2]$ and $n \geq k$, $\gamma_n * \mathbb{T}_k \boldsymbol{p}$ is divergence-free on Ω.

C   Proof of Lemma 2

Since for each $\boldsymbol{p} \in \mathbb{Q}_0^{\sigma}[\mathbb{R}^2]$ and $k \in \mathbb{N}$, $\gamma_n * \mathbb{T}_k \boldsymbol{p} \to \mathbb{T}_k \boldsymbol{p}$ effectively and uniformly on Ω_k as $n \to \infty$, it suffices to show that $\{\mathbb{T}_k \boldsymbol{p} : k \in \mathbb{N}, \boldsymbol{p} \in \mathbb{Q}_0^{\sigma}[\mathbb{R}^2]\}$ is dense in $L_{2,0}^{\sigma}(\Omega)$. On the other hand, since $\mathbb{Q}_0^{\sigma}[\mathbb{R}^2]$ is dense in $L_{2,0}^{\sigma}(\Omega)$, we only need to show that for each $\boldsymbol{p} \in \mathbb{Q}_0^{\sigma}[\mathbb{R}^2]$ and $m \in \mathbb{N}$, there is a $k \in \mathbb{N}$ such that $2^{-m} \geq \|\boldsymbol{p} - \mathbb{T}_k \boldsymbol{p}\|_{\infty} = \max\{|p_1(\boldsymbol{x}) - \mathbb{T}_k p_1(\boldsymbol{x})|, |p_2(\boldsymbol{x}) - \mathbb{T}_k p_2(\boldsymbol{x})| : \boldsymbol{x} \in \bar{\Omega}\}$.

Since p_i is uniformly continuous on $\bar{\Omega}$, there exists a $k \in \mathbb{N}$ such that $|p_i(x, y) - p_i(x', y')| \leq 2^{-m}$ whenever $|x - x'|, |y - y'| \leq 2^{-k+1}$, and, in particular, for $x' = \frac{x}{1-2^{-k}}$ and $y' = \frac{y}{1-2^{-k}}$. Also, since $p_i(x, y) = 0$ for $(x, y) \in \partial\Omega$, $|p_i(x, y)| \leq 2^{-m-1}$ for all $(x, y) \in \Omega \backslash \Omega_k$. This establishes $|p_i(x, y) - \mathbb{T}_k p_i(x, y)| \leq 2^{-m}$ on $\bar{\Omega}$.

References

1. Beggs, E., Costa, J.F., Tucker, J.V.: Axiomatising physical experiments as oracles to algorithms. Philos. Trans. R. Soc. A: Math. Phys. Eng. Sci. **370**, 3359–3384 (2012)
2. Boyer, F., Fabrie, P.: Mathematical Tools for the Study of the Incompressible Navier-Stokes Equations and Related Models. Spring Applied Mathematical Sciences. Springer, Heidelberg (2013). https://doi.org/10.1007/978-1-4614-5975-0
3. Giga, Y.: Weak and strong solutions of the Navier-Stokes initial value problem. Publ. RIMS Kyoto Univ. **19**, 887–910 (1983)
4. Giga, Y.: Time and spatial analyticity of solutions of the Navier-Stokes equations. Commun. Partial Differ. Equ. **8**, 929–948 (1983)
5. Giga, Y., Miyakawa, T.: Solutions in L^r of the Navier-Stokes initial value problem. Arch. Ration. Mech. Anal. **89**(3), 267–281 (1985)
6. Girault, V., Raviart, P.A.: Finite Element Methods for Navier-Stokes Equations. Springer Series in Computational Mathematics, vol. 5. Springer, New York (1986). https://doi.org/10.1007/978-3-642-61623-5
7. Kawamura, A., Steinberg, F., Ziegler, M.: Complexity of Laplace's and Poisson's equation. Bull. Symb. Logic **20**(2), 231 (2014). Full version to appear in Mathem. Structures in Computer Science (2016)
8. Landriani, G.S., Vandeven, H.: Polynomial approximation of divergence-free functions. Math. Comput. **52**, 103–130 (1989)

9. Mclean, W.: Strongly Elliptic Systems and Boundary Integral Equations. Cambridge University Press, London (2000)
10. Pazy, A.: Semigroups of Linear Operators and Applications to Partial Differential Equations. Springer, New York (1983). https://doi.org/10.1007/978-1-4612-5561-1
11. Pour-El, M.B., Richards, J.I.: The wave equation with computable initial data such that its unique solution is not computable. Adv. Math. **39**(4), 215–239 (1981)
12. Pour-El, M.B., Richards, J.I.: Computability in Analysis and Physics. Springer, New York (1989)
13. Pour-El, M.B., Zhong, N.: The wave equation with computable initial data whose unique solution is nowhere computable. Math. Logic Q. **43**(4), 499–509 (1997)
14. Patel, M.K., Markatos, N.C., Cross, M.: A critical evaluation of seven discretization schemes for convection-diffusion equations. Int. J. Numer. Meth. Fluids **5**(3), 225–244 (1985)
15. Brattka, V., Presser, G.: Computability on subsets of metric spaces. Theor. Comput. Sci. **305**, 43–76 (2003)
16. Smith, W.D.: On the uncomputability of hydrodynamics. NEC preprint (2003)
17. Soare, R.I.: Computability and recursion. Bull. Symb. Logic **2**, 284–321 (1996)
18. Sohr, H.: The Navier-Stokes Equations: An Elementary Functional Analytic Approach. Birkhäuser Advanced Texts. Birkhäuser, New York (2001)
19. Sun, S.M., Zhong, N., Ziegler, M.: On computability of Navier-Stokes' equation. In: Beckmann, A., Mitrana, V., Soskova, M. (eds.) CiE 2015. LNCS, vol. 9136, pp. 334–342. Springer, Cham (2015). https://doi.org/10.1007/978-3-319-20028-6_34
20. Tao, T.: Finite time blowup for an averaged three-dimensional Navier-Stokes equation. J. Am. Math. Soc. **29**, 601–674 (2016)
21. Temam, R.: Navier-Stokes Equations: Theory and Numerical Analysis. North-Holland Publishing Company, New York (1977)
22. Weihrauch, K.: Computable Analysis: An Introduction. Springer, New York (2000). https://doi.org/10.1007/978-3-642-56999-9
23. Weihrauch, K., Zhong, N.: Is wave propagation computable or can wave computers beat the Turing machine? Proc. Lond. Math. Soc. **85**(2), 312–332 (2002)
24. Weihrauch, K., Zhong, N.: Computing the solution of the Korteweg-de Vries equation with arbitrary precision on Turing machines. Theor. Comput. Sci. **332**, 337–366 (2005)
25. Weihrauch, K., Zhong, N.: Computing Schrödinger propagators on Type-2 Turing machines. J. Complex. **22**(6), 918–935 (2006)
26. Weihrauch, K., Zhong, N.: Computable analysis of the abstract Cauchy problem in Banach spaces and its applications I. Math. Logic Q. **53**, 511–531 (2007)
27. Wiegner, M.: The Navier-Stokes equations – a never-ending challenge? Jahresbericht der Deutschen Mathematiker Vereinigung (DMV) **101**(1), 1–25 (1999)
28. Zhong, N.: Computability structure of the Sobolev spaces and its applications. Theor. Comput. Sci. **219**, 487–510 (1999)
29. Ziegler, M., Brattka, V.: Computability in linear algebra. Theor. Comput. Sci. **326**, 187–211 (2004)
30. Ziegler, M.: Physically-relativized Church-Turing hypotheses: physical foundations of computing and complexity theory of computational physics. Appl. Math. Comput. **215**(4), 1431–1447 (2009)

AutoOverview: A Framework for Generating Structured Overviews over Many Documents

Jie Wang(✉)(iD)

University of Massachusetts, Lowell, MA 01854, USA
wang@cs.uml.edu
http://www.cs.uml.edu/~wang

In a conversation with Prof. Ker-I Ko about 10 years ago during a visit to Tsinghua University in Beijing, I indicated a desire to venture into a new field that would allow me to integrate algorithm designs, software development, system construction, data modeling, data management, and web technologies into a long-term project, so that a group of PhD students with various backgrounds and interests could work on different parts of the project for their dissertations. Ker-I was supportive and offered his insights. I am honored to dedicate this article on text mining and document engineering in memory of him.

Abstract. This article is an exposition of a recent study on automatic generation of a structured overview (SOV) over a very large corpus of documents, where an SOV is organized as sections and subsections according to the latent hierarchy of topics contained in the documents. We present a new framework called AutoOverview that includes and extends our previous scheme called NDORGS (best paper runner-up in ACM DocEng'2019) [47]. Different from the standard NLP task of generating a coherent summary typically over a handful of documents, AutoOverview needs to balance between two competitive objectives of accuracy and efficiency over thousands of documents. It incorporates hierarchical topic clustering, single-document summarization, multiple-document summarization, title generation, and other text mining techniques into a single platform. To assess the quality of an SOV generated over many documents, while it is possible to rely on human annotators to judge its readability, the sheer size of the inputs would make it formidable for human judges to determine if an SOV has covered all major points contained in the original texts. To overcome this obstacle, we present a text mining mechanism to evaluate topic coverage of

© Springer Nature Switzerland AG 2020
D.-Z. Du and J. Wang (Eds.): Ko Festschrift, LNCS 12000, pp. 113–150, 2020.
https://doi.org/10.1007/978-3-030-41672-0_8

the SOV against the topics contained in the original documents. We use multi-attribute decision making to help determine a suitable suite of algorithms to implement AutoOverview and the values of parameters for achieving a satisfactory SOV with respect to both accuracy and efficiency. We use NDORGS as an implementation example to address these issues and present evaluation results over a corpus of over 2,000 classified news articles and a corpus of over 5,000 unclassified news articles in a span of 10 years obtained from a search of the same keyword.

Keywords: Single document summarization · Multiple document summarization · Hierarchical topic clustering · Title generation · Multi-attribute optimization

1   Introduction

Generating an accurate, well-structured overview over a very large corpus of documents enables readers to quickly grasp the key points contained in a formidable amount of textual data, helping decision makers to make informed decisions and learners to learn new materials, among other things. This task is typically carried out by experts. However, when hundreds and thousands of fairly long articles are presented, even experienced experts would find this task hard to accomplish in a short period of time. For example, let us imagine that Alice, a politician, needs for her campaign to obtain a solid understanding of public sentiments contained in several thousand news articles on issues of global economy and regional conflicts, to help her prepare a speech for a meeting next week. Impossible to read through such a large volume of articles in a short period of time, sampling a small number of articles would be the only thing doable. However, these articles are not equally important and so sampling without knowing the underlying topic distribution will inevitably miss a few key points and capture a few minor points not needed. Fortunately, a clever use of text mining and natural language processing (NLP) techniques can come to the rescue. For example, instead of sampling articles uniformly at random from the corpus of articles, Alice may first cluster the corpus according to the underlying (latent) topics, compute a document score for each document that indicates its salience, and then sample a few articles in each cluster with high document scores. She may even generate a structured overview (SOV) of a moderate size over these articles.

AutoOverview is a text mining framework for generating an SOV over a large corpus of related documents with multiple topics in reasonable time. By "large" we mean hundreds of documents and by "very large" we mean thousands of documents, or even tens of thousands of documents, where the size of a document may range from a few pages to a few dozens of pages. By "reasonable time" we mean a few hours of CPU processing time on a normal desktop server with a moderate size of RAM. By "structured" we mean that an overview is organized in a hierarchy of two (i.e., sections and subsections) or more levels according to latent topics and subtopics contained in the original texts for easier reading, where each section and subsection must have, respectively, an appropriate title. Moreover, an SOV itself must be of a reasonable size. For a corpus of about 5,000 documents, for instance, an SOV with 20 or fewer pages would be desirable. Moreover, an SOV should also include figures to highlight frequencies and trends of interesting entities contained in the corpus.

Multiple-document summarization (MDS) is a standard NLP task, which typically takes a handful of short articles as input and outputs a short, unstructured summary. For example, the MDS systems presented in papers [5, 7, 51] are designed to handle the sizes of DUC datasets [10], where the DUC-02, DUC-03, and DUC-04 datasets provide benchmark summaries for MDS tasks, with each MDS task consisting of 10 or fewer documents as input. Directly applying these MDS systems for generating a summary for a corpus of thousands of documents could possibly generate a proportionately longer and disorganized summary. A more recent algorithm named T-CMDA [30] was devised to generate an English Wikipedia article on a specific topic over many documents. However, T-CMDA is still not suitable for generating a well-organized overview for a large corpus of documents containing many subtopics.

AutoOverview uses MDS as one of the building blocks. To generate a coherent summary that preserves the major points of the input documents, even if there are only a handful of them, a typical MDS algorithm is both CPU intensive and RAM intensive. Thus, using such an MDS imposes two constraints:

1. Each document should not be too long.
2. The number of documents should not be too big.

Violating these two constraints may cause a system crash or unacceptable delay of generating a summary.

To deal with the first constraint, we may use an appropriate summary to represent the document. Note that in an SOV of a reasonable size over a large corpus of documents, less important points should be excluded. As long as summaries contain all the major points of the original documents, generating an SOV over these summaries is expected to preserve major information, even though a major point included in an individual summary may still be considered minor for the final SOV. Thus, it suffices to first apply a single-document summarization (SDS) algorithm to obtain a summary for each document and then apply an MDS algorithm on these summaries to generate a new combined summary. Moreover, working on summaries also has an added benefit of improving efficiency.

To deal with the second constraint, we may organize the given corpus of documents into a hierarchy of two or more levels of clusters according to the underlying hierarchy of topics and subtopics, and force each cluster to have a workable size. Hierarchical topic clustering also provides a needed structure for the SOV, which helps improve the efficiency of AutoOverview.

After generating a hierarchy of topic clusters, we will then use a title generation algorithm to generate a suitable title for each section and subsection.

To assess the quality of an SOV, we rely on human annotators to judge its readability and use text mining techniques to evaluate information coverage and topic diversity. We take a holistic approach that weighs in readability, running time, information coverage, and topic diversity and use TOPSIS (Technique for Order Preference by Similarity to an Ideal Solution) [19] to determine the best ratio of the length of an SDS summary over that of the original document.

This article is organized as follows: We present in Sect. 2 the framework of AutoOverview and in Sect. 3 the evaluation method of SOVs. We then describe in Sect. 4 topic clustering algorithms, in Sect. 5 single-document-summarization (SDS) algorithms, and MDS algorithms, and in Sect. 6 title generation algorithms. We elaborate the first concrete implementation of AutoOverview and provide experimental results over two large corpora of thousands of documents, one is classified and the other unclassified. We conclude the paper in Sect. 9 with final notes.

2   AutoOverview: A General Framework

AutoOverview is a general framework consisting of the following seven components:

1. Text Wrangling (TWG),
2. Hierarchical Topic Clustering (HTC),
3. Document Summarizing (DOS),
4. Statistics and Trends of Entities (STE),
5. Cluster Summarizing (CLS),
6. Cluster Titling (CLT),
7. Assembling (ASG).

Among these components, HTC, DOS, and STE can be executed in parallel. The architecture and data flow of AutoOverview is shown in Fig. 1.

2.1   Text Wrangling

The TWG component is responsible for determining what language an input document is written in, eliminating irrelevant texts, and removing unsuitable documents. Irrelevant texts include URLs, duplicate documents, unrecognized symbols, and texts in other languages different from the underlying language the document is written in. Unsuitable documents include excessively long documents and interviews. An article of more than 100 pages, for example, would be considered excessive. Interview articles would need a separate, special treatment, for interviews may have loose structures in a number of different directions that may not be logically connected.

After text wrangling, the processed documents are inputs to HTC, DOS, and STE. For simplicity, in what follows, we will still use "documents" to denote the documents that have been processed by the TWG component, unless otherwise stated.

2.2   Hierarchical Topic Clustering

Multiple topics are anticipated over a large corpus of documents under a broader, common theme. Each topic may further contain subtopics.

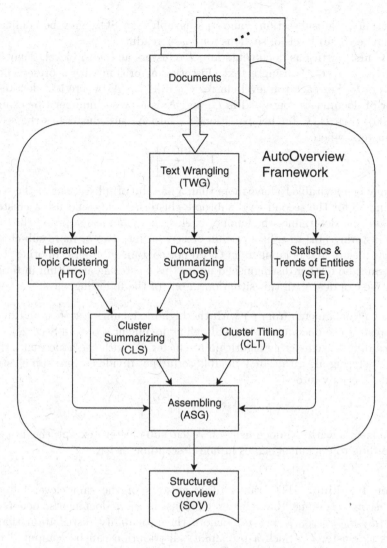

Fig. 1. AutoOverview architecture and data flow.

Two-Level Clustering. The HTC component is designed to discover the latent hierarchy of topics of two or three levels. For simplicity, we present in this article a two-level topic clustering. Topic clustering of three (or more) levels is similar.

A rule of thumb in writing an overview is not to have too many sections at each level, particularly at the top level, for top-level sections represent major topics. Too many top-level sections in an SOV may cause the reader to loose focus. A desirable number of top-level sections is not to exceed 10. Also, while an article could be divided into four levels with the following LaTeX tags: "section", "subsection", "subsubsection", and "paragraph", an SOV may be confined to two (maybe three) levels of sections for easier reading.

HTC first partitions documents into K clusters at the top level, denoted by $\mathcal{C} = \{C_1, C_2, \ldots, C_K\}$, using a text clustering algorithm with a preset number of clusters K. For each top-level cluster C_i, if $|C_i| > N$, where $|C_i|$ denotes the number of documents contained in C_i and N is a preset number (for example, $N = 100$), then HTC further partitions C_i into K_i sub-clusters as the second-level clusters, where

$$K_i = 1 + \left\lfloor \frac{|C_i|}{N} \right\rfloor.$$

It may be inevitable to have more than a handful of subsections at the second level, and so for the second-level subtopic clustering, we could use a clustering algorithm that determines the number of clusters dynamically during clustering. If the clustering fails to split a certain cluster C_i into two or more sub-clusters, then this means that documents in C_i do not contain subtopics. In other words, they cannot be further distinguished under this clustering algorithm in terms of topics. We can deal with this situation in one of the following ways:

1. Select N documents from C_i with the highest document scores, and discard the rest of the documents. We will define document scores in Sect. 2.4.
2. Use a different clustering algorithm to cluster C_i, with the hope that a different clustering algorithm may be able to further divide C_i into sub-clusters.
3. Split C_i evenly into

$$K_{ij} = 1 + \left\lfloor \frac{|C_{ij}|}{N} \right\rfloor$$

sub-clusters with N documents in each sub-cluster (except the last one) according to document scores in non-descending order.

Cluster Ranking. HTC ranks topic clusters of the same level based on cluster scores. Assume that cluster C_i consists of n documents, denoted by $C_i = \{d_{i1}, d_{i2}, \ldots, d_{in}\}$. Let p_{ij} denote the probability distribution that d_{ij} belongs to cluster C_i. Such a probability distribution can be computed using continuous topic modeling algorithms such as Latent Dirichlet Allocation (LDA) [3]. We define the score of cluster C_i using the following empirical formula:

$$s(C_i) = \frac{1}{n^2} \sum_{j=1}^{n} 2^{p_{ij}}.$$

The size n of the cluster may be viewed as a distance between any two documents contained in it, and taking a reciprocal of n^2 instead of n is somewhat similar to the Newton's law of universal gravitation that contains a reciprocal of r^2 with a distance r of two objects. Experiments show that using a reciprocal of n^2 provides a better result than using a reciprocal of n.

2.3   Document Summarizing

The DOS component generates, for each document, a summary of an appropriate length using an existing SDS algorithm. Extracting summaries is necessary for speeding up the process and is sufficient for generating a good overview, as only the most important content of an article will ultimately contribute to the overview (also noted in [30]). More information of SDS algorithms will be presented in Sect. 5.1.

The length of a summary is an important parameter, which is often measured by the percentage of the length of the original document. Let λ-summary denote a summary of a document d with a length equal to $\lambda|d|$, where $\lambda \in (0,1)$ represents a percentage and $|d|$ the length of d in terms of words or in terms of characters.

Note that we may also carry out HTC on λ-summaries for appropriate values of λ, which may reduce running time.

2.4   Statistics and Trends of Entities

The STE component uses an NLP tool of name entity recognition to identify name entities in the input documents. It then extracts, for each document, the title, the publication time, and the name of the press. For each document d, it removes the stop words contained in it. Let w be a name entity and $n_d(w)$ denote the number of times w appears in d. Let d' denote the bag of words of d after removing stop words. Compute the term frequency for each name entity w as follows:

$$\mathrm{TF}_d(w) = \frac{n_d(w)}{|d'|},$$

where $|d'|$ is the cardinal number of d'.

To compute trends of a name entity with respect to certain aspects of interests including the time when a document is published, the time an event takes place, or other types of classifications, STE groups the documents with respect to these aspects. It then computes the TFIDF value for each name entity w with respect to the aforementioned aspects of interests. That is, Let w be a term appearing in a document d, which is in a cluster G. Let

$$\mathrm{IDF}_G(w) = \log \frac{|\{d : d \in G\}|}{|\{d \in G : w \in d\}|}.$$

The TFIDF value of w with respect to (d, G) is computed as follows:

$$\mathrm{TFIDF}_{d,G}(w) = \mathrm{TF}_d(w) \cdot \mathrm{IDF}_G(w).$$

The TFIDF value of w indicates the significance of w in d with respect to G.

Measuring Trends. Let

$$\mathrm{TF}_G(w) = \sum_{d \in G} \mathrm{TF}_d(w),$$

$$\mathrm{TFIDF}_G(w) = \sum_{d \in G} \mathrm{TFIDF}_{d,G}(w).$$

These two values are used to indicate trends with respect to the underlying aspect of interests. Other text mining tools may also be used to measure trends.

Document Scoring. Let d be a document after removing stop words and C be a cluster containing d. Let $|d|$ denote the number of words in the bag of words of d. Then the document score of d with respect to C is defined by

$$s_C(d) = \frac{1}{d} \sum_{w \in d} \mathrm{TFIDF}_{d,C}(w).$$

2.5   Cluster Summarizing

The CLS component generates a coherent summary of an appropriate length for all SDS summaries in a given cluster using an MDS algorithm. This would be a second-level cluster unless a first-level cluster C does not have a second-level cluster, namely, $|\{d : \in C\}| \leq N$. In particular, for each second-level cluster, CLS takes the corresponding SDS summaries of original documents as input and uses an appropriate MDS to produce a summary of a suitable length over these SDS summaries. Section 5.3 includes more information on MDS algorithms.

2.6   Cluster Titling

The CLT component is responsible for generating section titles and subsection titles for the overview based on topic clusters and subtopic clusters. A hierarchy of concise section and subsection headings that capture the most important point of the underlying cluster is one of the most important components in a clear and well-organized overview. More information of title generation will be presented in Sect. 6.

2.7   Assembling

The ASG component is responsible for putting everything together and generating the final SOV. This includes filling in the hierarchy of topic clusters generated by the HTC component with a title generated by the CLT component for each top-level cluster (assuming it has sub-clusters), a content summary and a title for each sub-cluster generated by, respectively, the CLS component and the CLT component, and the trending graphs or tables for the entire SOV generated by the STE component. If top-level cluster does not have sub-clusters, then fill it in with a content summary in addition to a title. As a cluster with a higher cluster score represents a more significant topic, ASG lists clusters (and sub-clusters) as sections (and sub-sections) at the same level according to cluster scores in descending order.

3   Evaluation Methods

When implementing AutoOverview, the choice of algorithms for carrying out each component affects the quality of an SOV and the time complexity of generating it. Quality and efficiency are competitive objectives: Achieving a higher quality would need more computations. These criteria must be considered holistically when evaluating an SOV.

The quality of an SOV is determined by readability, information coverage, and topic diversity. While readability should be judged by human annotators, asking human annotators to judge how well information is covered is impractical because of the sheer number of documents they would need to comprehend. Thus, we would need to devise text mining algorithms to determine topic coverage.

3.1   Readability

We rely on human annotators to evaluate readability. An SOV is readable if the following eight categories all have good ratings:

1. Sentences in the SOV are coherent.
2. The SOV does not include useless or confusion text.
3. The SOV does not contain redundancy information.
4. Common nouns, proper nouns, and pronouns are well referenced in the SOV.
5. The entity re-mentions are not overly explicit.
6. Grammars are correct.
7. The SOV is well formatted.
8. Section and sub-section titles are appropriate.

The first seven categories are the DUC-04 evaluation schema [9].

3.2   Information Coverage

Let D be a corpus of text documents. Suppose that we have a gold-standard partition of D into K clusters $\mathcal{C} = \{C_1, C_2, \ldots, C_K\}$, and a clustering algorithm generates K clusters, denoted by $\mathcal{A} = \{A_1, A_2, \ldots, A_K\}$. We rearrange these clusters so that the symmetric difference of C_i and A_i, denoted by $\Delta(C_i, A_i)$, is minimum, where $\Delta(X, Y) = |X \cup Y| - |X \cap Y|$. That is, for all $1 \leq i \leq K$,

$$\Delta(C_i, A_i) = \min_{1 \leq j \leq k} \Delta(C_i, A_j).$$

This problem has an efficient algorithm. Even finding a permutation σ of $1, 2, \ldots, K$ such that $\sum_{i=1}^{K} \Delta(C_i, A_{\sigma(i)})$ is minimum is still polynomial time [16].

However, if we have one more clustering algorithm and we want to align three partitions such that the summartion of pairwise symmetric differences of three partitions is minimum, then the problem is NP-hard, but has an polynomial-time approximation with a 4/3-garantee [2].

We define CSD F1-score for $\mathcal{A}$ and $\mathcal{C}$ as follows, where CSD stands for Clusters Symmetric Difference:

$$F_1(\mathcal{A}, \mathcal{C}) = \frac{1}{K} \sum_{i=1}^{K} F_1(A_i, C_i),$$

$$F_1(A_i, C_i) = \frac{2P(A_i, C_i)R(A_i, C_i)}{P(A_i, C_i) + R(A_i, C_i)},$$

with P and R being precision and recall defined by

$$P(A_i, C_i) = \frac{|A_i \cap C_i|}{|A_i|},$$

$$R(A_i, C_i) = \frac{|A_i \cap C_i|}{|C_i|}.$$

Note that $F_1(A_i, C_i)$ can also be written as

$$F_1(A_i, C_i) = \frac{2|A_i \cap C_i|}{|A_i| + |C_i|}. \tag{1}$$

Clearly, $F_1(A, C) \leq 1$; the higher the value, the better.

Let U be a set of top k words from the original corpus and V a set of top k words from an SOV. Let

$$S_k(U, V) = \frac{|U \cap V|}{k} \tag{2}$$

denote the information coverage score of U and V. Then $S_k(U, V) \leq 1$; the higher the value, the better.

3.3   Topic Diversity

We generate clusters for the original corpus and for an SOV by treating each sentence in the SOV as a document. We then evaluate the top words among these clusters using CSD F1-scores to measure topic diversity.

3.4   Overall Quality

We evaluate the overall performance of an SOV using the following criteria (listed in the order of preference): human evaluation, time efficiency, information coverage, and topic diversity. We then use Saaty's pairwise comparison 9-point scale [41] and the Technique for Order Preference by Similarity to an Ideal Solution (TOPSIS) [19] to evaluate the overall quality of an overview.

4   Topic Clustering

A topic clustering algorithm partitions a set of documents into clusters with each cluster representing a topic. LDA [3] can be used to cluster text documents. Spectral Clustering (SC) [36] is another popular topic clustering method.

Other clustering methods, such as PW-LDA [25], Dirichlet multinomial mixture [52], and neural network models [49], are targeted at corpora of short documents such as abstracts of scientific papers, which are not suited for our task. The documents we are dealing with are much longer. Even if we use summaries to represent the original documents, a summary may still be significantly longer than a typical abstract. We note that the k-NN graph model [32] may also be used for topic clustering.

4.1   LDA Clustering

LDA assumes that the given corpus of documents $D = \{d_1, d_2, \ldots, d_L\}$ are all having the same multiple (latent) topics, labeled as $1, 2, \ldots, K$ for convenience, but each document follows a different topic distribution. In this model, topics are distributed differently in each document and words are distributed differently under each topic.

For topic clustering methods based on word counts in each document, LDA included, it is sufficient to consider only essential words contained in the document. Essential words are words that pass a part-of-speech filter, a stop-word filter, and a stemmer that reduces inflected words to the word stem. In what follows, unless otherwise stated, when words are mentioned, they are essential words. Let $W = \{w_1, w_2, \ldots, w_M\}$ denote the vocabulary of D and $\mathcal{W}$ the collection of bags-of-words of all documents.

It is convenient to view each document-to-topic distribution as a biased die of K faces and each topic-to-word distribution as a biased die of M faces. LDA assumes that there are infinite K-face dice and infinite M-face dice. LDA selects a document-to-topic die following a Dirichlet distribution with parameters α and a topic-to-word die following a Dirichlet distribution with parameters β.

LDA computes two matrices $\Theta = (\theta_{ik})_{L \times K}$ and $\Phi = (\varphi_{kj})_{K \times M}$ using Gibbs sampling to maximize the likelihood of $\mathcal{W}$, where the i-th row in Θ is a topic distribution for document d_i and the k-th row is a word distribution for topic k; namely,

$$\forall i = 1, 2, \ldots, L : \sum_{k=1}^{K} \theta_{ik} = 1, \ \theta_{ik} \geq 0;$$

$$\forall k = 1, 2, \ldots, K : \sum_{j=1}^{M} \varphi_{kj} = 1, \ \varphi_{kj} \geq 0.$$

Note that the observable probability $p(w_j|d_i)$ of word w_j in document d_i can be viewed as a dot product of the i-th row in matrix Θ and the j-th column in matrix Φ.

We can use the document-to-topic matrix (or the topic-to-word matrix) to cluster documents. For example, we may use K-means to cluster the set of rows in the document-to-topic matrix into K' clusters, where K' may or may not be equal to K. We may also simply use the largest value in the corresponding row of each document in the document-to-topic matrix to determine its cluster; namely, the document belongs to the cluster with the largest value.

LDA is a probabilistic algorithm, and so it may produce somewhat different clusters on the same corpus of documents on different runs.

4.2   Spectral Clustering

SC [46] clusters a given set of data points using eigenvalues of a similarity matrix (aka. affinity matrix) of the data points into K clusters, where K is a preset positive integer. SC can handle data points that do not satisfy convexity. In particular, we treat each document as a data point, represented as a list of words. Let $\text{WMD}(d_i, d_j)$ denote the Word Mover's Distance [23] of two documents d_i and d_j. Recall that L is the total number of documents in the corpus D of documents. The similarity matrix is an $L \times L$ matrix, where the entry s_{ij} of the matrix corresponds to a similarity between two documents d_i and d_j defined in Eq. (3). Under the WMD metric, a smaller value between two documents means that they are more similar, while a larger value means that they are less similar. This can be transformed to a similarity metric using the RBF kernel as follows:

$$\text{sim}_G(d_i, d_j) = e^{-\gamma \cdot \text{WMD}(d_i,d_j)^2}, \tag{3}$$

where γ may be set to 1. SC then uses K-means to generate clusters over eigenvectors corresponding to the K smallest eigenvalues.

The computation of WMD has a cubical time complexity in terms of the number of unique words contained in the two input documents. In practice, we may use the Linear-Complexity Relaxed WMD [1] that runs in linear time on average with only a limited loss in accuracy compared with WMD.

Similar to LDA clustering, SC also needs to predetermine the number of topics. SC is a deterministic and faster clustering algorithm, which also needs to preset the number of topics.

4.3   Affinity Propagation Clustering

Recall that both LDA clustering and spectral clustering must fix a number of clusters before clustering. This could be a concern in practice. Although Hierarchical Dirichlet Process (HDP) [44] can be used to dynamically determine the number of clusters during clustering, its complexity hinders its application in AutoOverview. Affinity propagation (AP) clustering [11] overcomes this problem. AP not only can dynamically determines the number of clusters during clustering, it is also straightforward to implement with quadratic-time complexity. AP is an exemplar-based clustering algorithm such as K-means [17] and K-medoids [21] except that AP does not need to preset the number of clusters.

Let $d_1, d_2, \ldots, d_L$ be the documents to be clustered under the similarity measure of $\text{sim}_G(d_i, d_j)$ defined in Eq. (3). Each document is a potential exemplar in K-means. AP proceeds by updating two $L \times L$ matrices $\boldsymbol{R} = (r_{ij})$ (the responsibility matrix) and $\boldsymbol{A} = (a_{ij})$ (the availability matrix) until they converge for all i and j:

1. Initially, let $r_{ij} \leftarrow 0$ and $a_{ij} \leftarrow 0$.
2. Let $r_{ij} \leftarrow \text{sim}_G(d_i, d_j) - b_{ij}$, where

$$b_{ij} = \max_{j' \neq j} \{ \text{sim}_G(d_i, d_{j'}) + a(i, j') \}.$$

3. If $i \neq j$, then let

$$a_{ij} \leftarrow \min\{0, r_{jj}\} + \sum_{i' \notin \{i,j\}} \max\{0, r_{i'j}\}.$$

Otherwise, let

$$a_{ij} \leftarrow \sum_{i' \neq j} \max\{0, r_{i'j}\}.$$

If $r_{ii} + a_{ii} > 0$, then select d_i as an exemplar. Document d_j belongs to the cluster of d_i if d_j has the largest similarity with d_i among all other exemplars.

5   Text Summarization and Title Generation

Text summarization algorithms include SDS, MDS, hierarchical summarization, and structural summarization.

5.1   Single-Document Summarization

Single-document summarization is a classic NLP task that has been studied intensively and extensively for more than six decades since 1958 [31]. The task of SDS may be formulated as a multi-objective maximization problem. Let d denote a document consisting of n sentences indexed as $S_1, S_2, \ldots, S_n$ in the order they appear, each with a length l_i and a score s_i, along with a maximum length constraint Q, where l_i is the number of characters contained in S_i. Let $F_d(D)$ denote a diversity coverage measure and x_i a 0–1 variable such that $x_i = 1$ if sentence S_i is selected, and 0 otherwise. Then the SDS task is modeled as follows, which is an NP-hard problem:

$$\text{maximize} \sum_{i=1}^{n} s_i x_i \text{ and } F_d(D),$$

$$\text{subject to} \sum_{i=1}^{n} l_i x_i \leq Q \text{ and } x_i \in \{0, 1\}.$$

There are two types of SDS algorithms in terms of content presentations, namely, extractive summaries and abstractive summaries. Extractive summaries are formed by extracting sentences from the original document, while abstractive summaries are formed by rewriting sentences. The latter is closer to what a summary is expected to look like by human readers, and is also much more difficult to produce. There are also two types of SDS algorithms in terms of methodology, namely, supervised learning and unsupervised learning.

Some of the recent publications of SDS algorithms include [6,33,34,53,54]. At the time when this article is written, the best unsupervised, extractive summarization is substantially better than the best supervised summarization and the best abstractive summarization in terms of accuracy, efficiency, scalability, and flexibility across different languages. The recent Semantic WordRank (SWR) algorithm [54] is currently the state of the art in all aspects. SWR is operated on a semantic word graph with a few other adjustments.

5.2 SWR

Semantic Word Graphs. A semantic word graph of a given document d is a weighted graph $G = (V, E)$ of essential words in d. Compute word embeddings of all words on a Wikipedia dump using an existing NLP tool. Two words are connected if at lease one of the following two conditions holds:

1. They co-occur within a small window of Δ successive words in the document (e.g, Δ is often set to 2 to capture two-word phrases).
2. The cosine similarity of their embedding representations exceeds a threshold value δ (e.g. $\delta = 0.6$).

For each edge (u, v), if only one type of connection exists, then treat the weight of the other type 0. Assign the co-occurrence count of u and v as the initial weight to the co-occurrence connection and the cosine similarity of the word embedding vectors of u and v as the initial weight to the semantic connection. Normalize the initial weights of co-occurrence connections; namely, divide the initial co-occurrence weight by the total initial co-occurrence weight. Normalize the initial weights of semantic connections; namely, divide the initial semantic weight by the total initial semantic weight. Let $w_c(u, v)$ and $w_s(u, v)$ denote, respectively, the normalized weight for the co-occurrence connection and the semantic connection of u and v. Finally, assign $w(u, v) = w_c(u, v) + w_s(u, v)$ as the weight to the edge (u, v).

Article-Structure-Biased PageRank. Article structures define how information is presented. For example, news articles are typically structured as an inverted pyramids [38], with critical information presented at the beginning, followed by additional information. Similar to the position-biased PageRank algorithm [13]. SWR uses a position-biased PageRank algorithm as follows:

$$W(v_i) = \sum_{v_j \in Adj(v_i)} \frac{d \cdot w_{ji}}{\sum_{v_k \in Adj(v_j)} w_{jk}} W(v_j) + \frac{\sum_{k:v_i \in S_k} (1 - d) \cdot \mathrm{LS}_k(v_i)}{\sum_{j,k:v_j \in S_k} \mathrm{LS}_k(v_j)}, \quad (4)$$

where LS_i is the location score of the i-th sentence s_i according the importance of sentence locations and $LS_i(v) = LS_i$ for $v \in S_i$. For example, for a document with the inverted pyramid structure, a location score for word $w \in s_i$ may be defined by $LS_i(w) = 1/i$. Compute Eq. (4) with an arbitrary initial value for each node, and iterates it until it converges.

$W(v_i)$, referred to as *salient score* of v_i, represents its importance relative to the other words in the document.

Sentence Scoring with Softplus Adjustment. Let S be a sentence. To score S, one may simply sum up the salient score $W(v)$ of each word v contained in S. This has a drawback. To see this, suppose that S_1 and S_2 are two sentences with similar scores under this scoring, and contain about the same number of words. If the distribution of word scores for words contained in S_1 follows the Pareto Principle, namely, a few words have very high scores and the rest have very low scores close to 0, while S_2 has roughly a uniform word score distribution, where the high scores of a few words in S_1 are much larger than the (almost uniform) scores of words in S_2, then the few words in S_1 with very high scores would make S_1 appear more important than S_2. Using direct summation of salient word scores, it is possible to end up with the opposite outcome.

Using the Softplus function $sp(x) = \ln(1 + e^x)$ to elevate a score helps overcome this drawback. Used as an activation function in neural networks, $sp(x)$ offers a significant elevation of x when x is a small positive number. If x is large, then $sp(x) \approx x$. Apply the Softplus function to each word, and sum up the elevated values to be the salient score of S, denoted by $\mathrm{sal}_{sp}(S)$. Namely,

$$\mathrm{sal}_{sp}(S) = \sum_{v_i \in S} \ln(1 + e^{W(v_i)}).$$

Greedy Sentence Selection and Ranking. SWR uses spectral clustering to cluster sentences into K clusters. Empirical studies suggest that setting $\lambda = 0.3$ would be the best for an λ-summary to contain almost all significant points contained in the original document. On the other hand, to avoid having too many clusters that could deteriorate performance, it is necessary to set an upper bound U. For typical news articles, for example, an upper bound $U = 8$ would be appropriate. Thus, let

$$K = \min\{\lfloor 0.3n \rfloor, U\}.$$

Each sentence S_i is now associated with the following four values:

1. Sentence index i.
2. Salient score $s_i = \mathrm{sal}_{sp}(S_i)$.
3. Sentence length l_i.
4. Cluster index j of the cluster S_i belongs to.

SWR selects sentences as a summary using a greedy strategy in a round robin fashion as follows:

1. Let $\mathcal{S}$ denote the set of selected sentences. Initially, $\mathcal{S} \leftarrow \emptyset$.
2. For each sentence S_i, compute the value per unit length to obtain a unit score $s_i' = s_i/l_i$.
3. For each cluster c_j, sort the sentences contained in it in descending order according to their unit scores.
4. While there are still sentences that have not been selected, do the followings:
   (a) Sort the remaining clusters in descending order according to the highest unit score contained in a cluster. For example, if the highest unit score in cluster c_i is smaller than the highest unit score in cluster c_j, then c_j comes before c_i in the sorted clusters.
   (b) Select the sentence from the remaining sentences with the highest unit score, one from each cluster in the order of sorted clusters, and add it to $\mathcal{S}$. That is,
   $$\mathcal{S} \leftarrow \mathcal{S} \cup \{S_{i_1}, S_{i_2}, \ldots, S_{i_k}\},$$
   where S_{ij} are the selected sentences and k is the number of remaining clusters that are nonempty.
   (c) Remove the selected sentences from their corresponding clusters.
5. Rank sentences according to the order they are selected.

Evaluation. Evaluating SWR against the SummBank benchmarks [39] indicates that SWR is a good choice for implementing AutoOverview. Each Summ-Bank benchmark consists of the following data:

1. A corpus of documents.
2. Sentences in each document are individually ranked by three human judges.
3. A combined ranking of the three judges for each sentence.

To evaluate SWR, the threshold value of similarities and the size of sliding window of co-occurrences must be determined. To demonstrate robustness and avoid overfitting, SWR is run on the DUC-01 dataset to obtain the values of these parameters that maximize the average ROUGE-1 score [29]. Testing on the DUC-02 benchmarks, SWR achieves higher ROUGE-1, ROUGE-2, and ROUGE-SU4 scores over all previous SDS algorithms. On the SummBank benchmarks, SWR outperforms each individual judge against the other two judges under the ROUGE measures, while comparing favorably with the combined ranking of all judges on ranking sentences up to the top 30% rank, which is what AutoOverview will be using. The combined ranking represents the collective judgments of the three judges.

5.3   Multi-Document Summarization

An MDS algorithm takes several documents as input and generates a summary of these documents as output. Most MDS algorithms are algorithms of selecting sentences. Sentences may be ranked using features of term frequencies, sentence positions, and keyword co-occurrences [18,33], among a few other things. Methods include graph-based lexical centrality LexRank [12], centroid-based clustering [40], Support Vector Regression [28], syntactic linkages between text [48], and Integer Linear Programming [15,26]. Selected sentences may be reordered to improve coherence using probabilistic methods [24,35].

GFLOW. Focused on sentence coherency, GLFOW [7] is an unsupervised graph-based method that selects and reorders sentences to balance coherence and salience over an approximate discourse graph (ADG) of sentences. An ADG is a weighted, directed graph on sentences, modeling sentence discourse relations among documents based on (1) deverbal noun reference, (2) event/entity continuation, (3) discourse markers, (4) sentence inference, and (5) co-reference mentions. Two sentences are connected with a direction if they have one of the aforementioned sentence relations. Edge weight is calculated based on the number of sentence relations between two sentences.

GFLOW solves uses a random-walk heuristic to find an approximation to the following NP-hard ILP problem over an ADG:

$$\text{maximize} \quad Sal(X) + \alpha Coh(X) - \beta|X|$$
$$\text{subject to} \quad \sum_{x_i \in X} l(x_i) < B,$$
$$\forall x_i, x_j \in X : rdt(x_i, x_j) = 0,$$

where variable X is a summary, $|X|$ is the number of sentences in the given summary, $Sal(X)$ is the salience score of X, $Coh(X)$ is the coherence score of X, $rdt(x_i, x_j)$ is a redundancy measure between two sentences x_i and x_j, and $l(x_i)$ is the length of sentence x_i.

Random Walks. Start from a node with the highest salience score with no incoming edges, and proceeds to the next node with the highest out-degree. Repeat this walk until it cannot go any further or the total number of sentences exceeds a given limit. Sum up the salience scores on this path. Start from another node and repeat the same procedure. Choose one path with largest salience score, or a few paths with the highest total salience score such that the total number of sentences is below the limit.

Choosing GFLOW to implement AutoOverview would be a natural choice for its strength on generating a coherent summary with minimum redundancy. Since we are using 0.3-summaries obtained by SWR, the loss of most important points of the original documents will not be a major concern.

Other Recent MDS Algorithms. Based on GFLOW, a supervised neural network model [51] was devised that combines Personalized Discourse Graph (PDG), Gated Recurrent Units (GRU), and Graph Convolutional Network (GCN) [22] to rank and select sentences. TCSum [5] is another neural network model that leverages text classification to improve the quality of multi-document summaries. However, neural network methods require large-scale training data to obtain a good result.

T-DMCA [30] is a large-scale summarization method to generate an English Wikipedia article. It combines extractive summarizations and abstractive summarizations trained on a large-scale Wikipedia dataset to summarize the text. While T-CMDA is capable of creating summaries with specified topics as Wikipedia article, it can hardly generate an SOV for a large corpus of documents containing multiple topics.

5.4 Hierarchical and Structural Summarization

There are algorithms for generating a hierarchical summary of a single document [4] and Otterbacher et al. [37] and an algorithm that summarizes the news tweets into a flexible, topic-oriented hierarchy based on Twitter-LDA [14]. SUMMA [8] is a system that creates coherent summaries hierarchically in the order of time, locations, or events. These methods focus on single documents or short texts, or require documents be written with a certain predefined structure template, making them unsuited for our task.

Structured summarization algorithms first identify topics of the input documents. For example, using a high-level structure of human-authored documents, one can generate a topic-structure multi-paragraph overview with domain-specific templates [42]. A summary template generation system was proposed to cluster sentences and words and generate sentence patterns to represent topics based on an entity-aspect LDA model [27]. Autopedia [50] is a Wikipedia article generation framework that selects Wikipedia templates as article structures.

6 Title Generation

A suitable title for a block of texts must convey the central meanings of these texts, and be succinct and catchy. Automatic title generation (ATG) for a given block of texts can be viewed as a task of generating a phrase or a short sentence to represent the central meanings of the texts. Shao and Wang [43] presented a two-phase algorithm to generate a title. In the first phase the algorithm identifies a few sentences with high rankings, with a structure suitable to being a title after some modifications. These sentences are referred to as title candidates. In the second phase, the algorithm constructs a dependency tree[1] for each title candidate using a dependency parser, and trims off possible branches using a set of empirical rules they defined. The shortest output is used as the final title. This process is called Dependency-Tree Automatic Title Generator (DTATG). Sentence trimming has been used to generate titles using context-free grammar trees [45,55]. But context-free grammar trees are tedious to work with, and working with dependency trees is better for our purpose. Experiments confirms that DTATG can generate titles comparable to titles generated by human writers on most single documents.

---

[1] For example, see http://nlpprogress.com/english/dependency_parsing.html.

6.1   DTATG

DTATG generates a title for a given block of texts as follows:

1. Rank each sentence using a suitable measure such as the sentence ranking according to the greedy selection process of SWR (see the greedy sentence selection and ranking algorithm in Sect. 5.2). Select a small number of sentences with the highest rankings, which are referred to as the central sentences. In practice, selecting three or four central sentences would be sufficient.
2. Construct a dependency tree for each central sentence using a dependency parser such as the Stanford Dependency Parser[2], starting from the sentence with the highest ranking.
3. Remove certain branches of the dependency tree based on a set of empirical rules.
4. If a trimmed sentence passes the title tests, then use it as a title.

6.2   Dependency Trees and Trimming

A dependency tree is constructed based on part-of-speech tag for each word in a sentence and grammatical relations between words in the form of triplets as follows:

$$(relation, governor, dependent).$$

Trimming Rules. For the purpose of generating a title, the following set of empirical rules specifies what words may be trimmed and what words should be kept, so that a trimmed sentence would look like a title.

1. "May-be-trimmed" rules:
   (a) The first adverbial phrase and the last adverbial phrase may be trimmed.
   (b) The phrase of "X says" and "X said", where X is a noun or pronoun, may be trimmed.
   (c) If there are two clauses connected with "and" and the first clause consists of a subject and a verb, then the second clause and the "and" may be trimmed.
   (d) If there is a clause starting with "that" and the clause has a subject and a verb, then this "that" and the words before it may be trimmed.
   (e) If there are more than one *nmod* relations next to each other, then all the *nmod* relations except the last one may be trimmed.
2. "To-be-kept" rules:
   (a) If the relation is *nsubj* or *nsubjpass*, then both the governor and the dependent are to be kept.
   (b) If the relation is *dobj* or *iobj*, then both the governor and the dependent are to be kept.

---

[2] Available at http://nlp.stanford.edu/software/stanford-dependencies.shtml.

(c) If the relation is *compound*, then both the governor and the dependent are to be kept.
(d) If the relation is *root*, then the dependent is to be kept.
(e) If the relation is *nmod*, then the dependent and the preposition in *nmod* are to be kept.
(f) If the relation is *nummod*, then both the governor and the dependent are to be kept.

For example, the sentence "Market concerns about the deficit has hit the greenback" after trimming is "Market concerns about deficit hit greenback", which would be a suitable title.

6.3 DTATG-generated Titles

A good title must be concise and catchy to capture the reader's attention. In other words, a good title should pass the following title tests:

Conciseness Test

1. A title should not exceed 15 words.
2. A title must not have clauses.
3. A title should have the following structure: Subject + Verb + Object or Subject + Verb. Subject must be specific: it can be a noun but not a pronoun.

Fluency Test. A title should contain no grammatical errors.

Topic-Relevance Test. A title must convey at least one main meaning of the document.

Evaluation. DTATG-generated titles are evaluated by five human annotators on a corpus of 2,225 classified BBC news articles published in the years of 2004 to 2005 with classification labels of business, entertainment, politics, sports, and technology, hereafter referred to as BBC News. The evaluations confirm that DTATG generates suitable titles on most cases.

7 NDORGS: The First Implementation of AutoOverview

Implementing AutoOverview involves a selection of a suite of algorithms: an HTC algorithm, a DOS algorithm, a CLS algorithm, and a CLT algorithm. It also needs to determine the best value of parameter λ for λ-summaries. NDORGS [47] is the first implementation of AutoOverview.

7.1   Data Sets

Two datasets are used to evaluate the quality of SOVs generated by NDORGS. One is the same BBC News dataset used to test DTATG (The BBC News dataset can be found at http://mlg.ucd.ie/datasets/bbc.html, which was made available by D. Greene and P. Cunningham). The other is a corpus of 5,300 unclassified articles extracted from Factiva [20] under the keyword search of "Marxism" from the year of 2008 to the year of 2017, hereafter referred to as Factiva-Marx. The latter dataset was used to test SOVs for a project of analyzing public sentiments about Marxism.

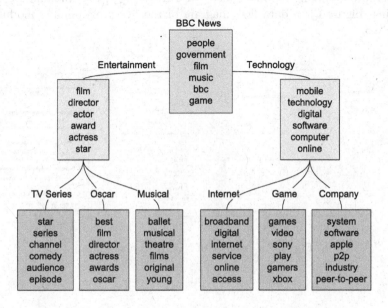

Fig. 2. A subtree of topic clusters in the BBC News dataset.

Figure 2 is an example of a subtree from the hierarchical topic clustering over BBC News, where each node represents a topic with six most frequent words under that topic. The root illustrates the most frequent words of the corpus. The first level topic cluster contains two subtopics: *Entertainment* and *Technology*. When words such as "series", "comedy", and "episode" under topic *Entertainment* are discovered for a substantial number of times, a subtopic of *TV Series* may be detected. The hierarchically structured topic clusters provide detail information about topic relationships contained in a large corpus of documents.

Table 1 shows some of the statistics of BBC News and Factiva-Marx, where NumD denotes "the total number of documents", NumW denotes "the total number of words" and AvgNumW/D denotes "the average number of words per document".

Table 1. Size comparisons between different datasets.

Dataset	NumD	NumW	AvgNumW/D	Vocab size
BBC News	2,225	8.5×10^5	380	6.56×10^4
Factiva-Marx	5,300	1.09×10^7	2,100	3.89×10^5

7.2   Programming Modules

NDORGS is focused on generating the text portion of an SOV. The STE component is carried out separately. It also combines the TWG and DOS components into one programming module called Pre-Processing (PPG) module for easier handling. Figure 3 is a data flow diagram of the five programming modules of NDORGS.

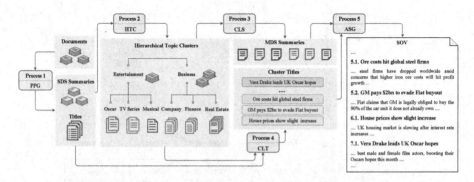

Fig. 3. An example of NDORGS processing.

7.3   Settings and Parameters

We determine empirically the values of parameters that lead to the best overall performance for both BBC News and Factiva-Marx. For each dataset, NDORGS produces three SOVs[3] corresponding to three λ-summaries with $\lambda \in \{0.1, 0.2, 0.3\}$.

1. The PPG module generates λ-summaries for each document with the length ratio $\lambda = 0.1$, 0.2, and 0.3 using the SWR algorithm described in Sect. 5.1.
2. The HTC module creates two-level topic clusters using the LDA clustering algorithm on 0.3-summaries, where the number of clusters at the first level is set to $K = 9$, as suggested in Sect. 7.4 (see Fig. 4). To generate the second-level clusters, we set $N = 200$ as an upper bound of each cluster. The number of second-level clusters is dynamically determined. Note if a cluster C_i with $|C_i| > N$ but cannot be further divided into multiple sub-clusters, the HTC

---

[3] The six SOVs generated by NDORGS are available at http://www.ndorg.net.

module follows method 3 in Sect. 2.2. Namely, split C_i evenly into $K_{ij} = 1 + \lfloor |C_{ij}|/N \rfloor$ sub-clusters with N documents in each sub-cluster (except the last one) according to document scores in non-descending order.

3. The CLS module uses GFLOW to produce a cluster summary for each cluster. The length l_i of an MDS summary of (nonempty) cluster C_i is determined by

$$l_i = \begin{cases} 150 \cdot \lfloor |C_i|/10 \rfloor + 300, & \text{if } |C_i| < 70, \\ 200 \cdot \lfloor |C_i|/10 \rfloor, & \text{if } |C_i| \geq 70. \end{cases}$$

This is an empirical setting, and may be changed in practice.

4. The CLT module applies DTATG to generate a title for each cluster and sub-cluster.

5. The ASG module reorders clusters at the same level according to cluster scores S_C defined in Sect. 2.2. For each level of clusters, if there are clusters containing less than 70 documents, then ASG merges these clusters' summaries into a separate section under the title of "Other Topics", where each cluster's summary is listed as a bullet item, in descending order of cluster scores.

7.4   Text Clustering Evaluations for Deciding K

Let HLDA-D, HSC-D, HLDA-S, and HSC-S denote, respectively, the algorithms of applying HTC using LDA clustering and HTC using spectral clustering on original documents and 0.3-summaries generated by SWR.

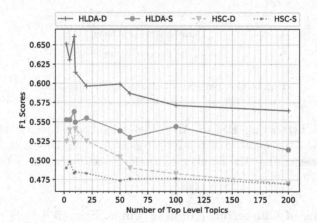

Fig. 4. CSD F1-scores on labeled BBC News articles.

Comparisons of Clustering Quality. Figure 4 compares the CSD F1-scores of HLDA-D, HLDA-S, HSC-D, and HSC-S over the labeled BBC News articles. We can see that HLDA-D is better than HLDA-S, which is better than HSC-D,

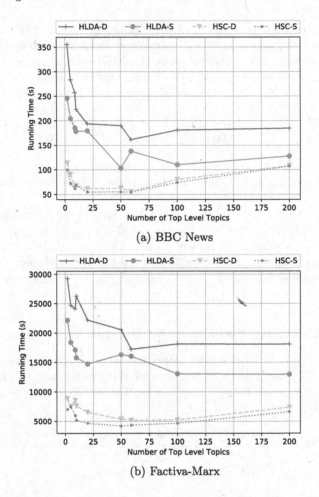

(a) BBC News

(b) Factiva-Marx

Fig. 5. Comparisons of clustering running time

and HSC-D is better than HSC-S. All of these algorithms have the highest CSD F1-scores when the number of top-level topics $K = 9$. This is in line with a general experience that the number of top-level sections in an SOV should be bounded by 10.

Comparisons of Clustering Running Time. Figure 5 depicts the running time of clustering the BBC News and Factiva-Marx datasets by different algorithms into two-level clusters on a Dell desktop with a quad-core Intel Xeon 2.67 GHz processor and 12 GB RAM. The top-level clusters numbers K are ranged from 2 to 200.

We can see that for both corpora, HSC-S is the fastest, HSC-D is slightly slower, HLDA-D is the slowest, and HLDA-S is in between HLDA-D and HSC-D. Thus, HLDA-S would be a good choice for balancing accuracy and efficiency.

On the other hand, it can be seen that when the number of top-level clusters is small, the two-level clustering running time is high. This is because, for a given corpus, having a smaller number of top-level clusters would imply a larger number of second-level clusters, and so would require significantly more time for clustering at the second level. The turning points are around $K = 20$.

7.5   Evaluations of Overall Quality of SOVs

Readability. Human annotators with high ratings were recruited from Amazon Mechanical Turk (AMT) to evaluate six SOVs generated by NDORGS on, respectively, 0.1, 0.2, and 0.3-summaries of documents in the two corpora [47], with each SOV being evaluated by four human annotators. The evaluation scores are provided in Table 2.

Figure 6 shows the average scores of human annotators using a 4-point system, with 4 being the best. It can be seen from Fig. 6a that, for BBC News, the SOVs generated on 0.3-summaries outperforms reports generated on 0.2-summaries and 0.1-summaries in all categories except "OverlyExplicit". From Fig. 6b it can be seen that, for Factiva-Marx, the SOVs generated on 0.2-summaries is better than reports generated on 0.3-summaries and 0.1-summaries in most of the categories; they are "UselessText", "Referents", "OverlyExplicit", "Grammatical", and "Formatting". Note that a larger value of λ yields a better SOV on BBC News, while a smaller value of λ yields a better SOV on Factiva-Marx. Note that the Factiva-Marx corpus contains almost three times more documents than the BBC News corpus, and each document in Factiva-Marx contains on average over ten times larger number of tokens than that in a document from BBC News. This may indicate that for a larger corpus of longer documents, using a smaller value of λ might be better.

Information Coverage. Information coverage is evaluated by comparing the top words in an SOVs and the top words in the corresponding corpus. Top words in BBC News and Factiva-Marx, are listed below, respectively, for comparisons, where the first item depicts the top 50 words in the original corpus, and the second, third, and fourth items depict, respectively, the top 50 words in the report generated on 0.1-summaries, 0.2-summaries, and 0.3-summaries, listed in descending order of keyword scores. The words in **bold** are the common top words that occur across all four rows. The words with <u>underlines</u> are the top words that occur in the first row and two of the other three rows. The words in *italics* are the top words that occur in the first row and just one of the other three rows.

Top word comparisons for BBC News

1. **people**, told, **best**, **government**, *time*, **year**, **number**, **three**, **film**, **music**, <u>bbc</u>, *set*, **game**, going, <u>years</u>, **labour**, good, well, *top*, **british**, *european*, <u>win</u>, *market*, **won**, <u>company</u>, public, <u>second</u>, play, *mobile*, *work*, **firm**, <u>blair</u>, **games**, minister, <u>expected</u>, <u>england</u>, *chief*, technology, party, sales, *news*, *plans*, *including*, *help*, *election*, digital, players, director, *economic*, big

Table 2. Human evaluation Scores, where $C_1, C_2, \ldots, C_7$ represent, respectively, Coherence, UselessText, Redundancy, Referents, OverlyExplicit, Grammatical, and Formatting

Corpus	SOV	Human evaluation score						
		C_1	C_2	C_3	C_4	C_5	C_6	C_7
BBC News	$\lambda = 0.1$	3	1	2	2	2	2	1
		4	4	4	4	4	4	4
		3	3	2	1	3	1	2
		1	1	2	2	2	4	4
	$\lambda = 0.2$	4	3	3	3	0	3	2
		2	3	3	3	4	4	4
		1	1	2	3	2	2	4
		3	3	3	3	3	3	2
	$\lambda = 0.3$	4	3	2	4	4	4	4
		4	1	3	2	1	1	3
		3	2	2	3	1	3	3
		4	4	4	4	4	4	4
Factiva-Marx	$\lambda = 0.1$	3	3	3	1	3	2	3
		4	1	4	2	4	4	4
		2	2	3	2	3	1	2
		3	2	2	3	1	2	3
	$\lambda = 0.2$	3	2	2	3	3	4	2
		2	2	2	3	3	3	4
		2	4	3	3	3	3	3
		4	4	4	4	4	2	4
	$\lambda = 0.3$	4	3	4	4	3	3	3
		3	3	3	2	3	3	2
		2	2	2	2	2	2	2
		3	3	3	3	3	3	2

2. **people**, **best**, **number**, **government**, **film**, **year**, **three**, **game**, howard, **music**, london, **british**, face, biggest, net, action, **firm**, deal, rise, national, foreign, singer, michael, leader, oil, blair, dollar, stock, star, cup, online, future, **games**, 2004, *work*, **won**, list, international, coach, win, mark, tory, **labour**, brown, general, prices, *market*, car, *help*, users

3. **year**, **people**, **number**, **three**, **best**, **british**, **film**, company, **won**, **labour**, **music**, net, bbc, **government**, leader, shares, *european*, earlier, chart, third, **games**, state, win, coach, expected, second, months, political, house, *economic*, **game**, years, team, start, manchester, england, *election*, *chief*, international, michael, profit, champion, award, star, announced, service, future, **firm**, *top*, *news*

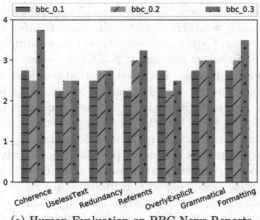

(a) Human Evaluation on BBC News Reports

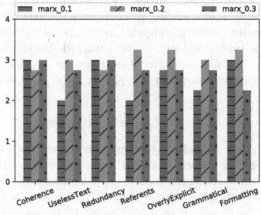

(b) Human Evaluations on Marx Reports

Fig. 6. Human evaluations. *(a) BBC-0.1, BBC-0.2, and BBC-0.3: reports with summary length ratio = 0.1, 0.2, and 0.3 over BBC News. (b) Marx-0.1, Marx-0.2, and Marx-0.3: reports with summary length ratio = 0.1, 0.2, and 0.3 over Factiva-Marx.*

4. **people**, england, **year**, **film**, **labour**, boss, **firm**, despite, **number**, **three**, wales, **british**, nations, **best**, company, **music**, blair, *set*, record, oil, *time*, years, **won**, prices, *plans*, net, online, *including*, films, bbc, court, **games**, **game**, brown, david, **government**, expected, club, action, beat, total, group, unit, firms, rules, *mobile*, second, analysts, future, computer

Top word comparisons for Factiva-Marx

1. **party**, **chinese**, **china**, **political**, **people**, **communist**, **economic**, **national**, **state**, **government**, **years**, **social**, *great*, time, **rights**, development, international, **president**, **central**, **war**, **north**, university,

power, **united**, work, country, **foreign**, *global*, military, history, south, **marxism**, **human**, **western**, *soviet*, well, system, **mao**, *american*, **news**, public, *cultural*, long, states, *countries*, three, left, **media**, british, including

2. **party**, **china**, **chinese**, **communist**, **political**, **years**, **economic**, **rights**, **human**, **president**, **people**, **national**, year, **state**, leaders, **government**, **central**, *countries*, **news**, **social**, country, leader, time, **foreign**, **power**, **north**, nuclear, top, **marxism**, ideological, led, **media**, **war**, beijing, **western**, **united**, development, *soviet*, **mao**, states, history, university, capitalism, official, market, officials, march, korea, democracy, south

3. **china**, **party**, **communist**, **chinese**, **political**, **economic**, **years**, **rights**, **president**, **people**, **central**, **state**, **social**, **united**, **north**, beijing, **western**, **news**, **media**, **mao**, cpc, **war**, **human**, anniversary, public, members, country, jinping, leader, states, **government**, south, **marxism**, democratic, **national**, **power**, year, **foreign**, *american*, education, international, july, nuclear, day, book, leadership, committee, leaders, copyright, study

4. **china**, **party**, **communist**, **chinese**, **economic**, **years**, **people**, **political**, **human**, **news**, **state**, **social**, **government**, **central**, **national**, leader, **president**, **media**, *cultural*, **rights**, **mao**, **power**, development, year, international, university, leaders, history, **united**, beijing, copyright, socialist, *global*, *great*, top, nation, universities, **western**, revolution, nuclear, **foreign**, public, agency, **marxism**, time, members, congress, **war**, change, **north**

Listed below are the summary of the comparison results:

1. For BBC News, over 70% of the top words in the corpus are also top words in the three overview reports combined, over one-third of the top words in the corpus are top words in the report on 0.1-summaries, over one half of the top words in the corpus are top words in the report on 0.2-summaries as well as in the report on 0.3-summaries, and over 80% of the top 10 words in the corpus are top words in each report.

2. For Factiva-Marx, 82% of the top words in the corpus are also top words in the three overview reports combined, 70% of the top words in the corpus are top words in the report on 0.1-summaries as well as in the report on 0.3-summaries, 64% of the top words in the corpus are top words in the report on 0.2-summaries, and the top 12 words in the corpus are top words in each summary. These results indicate that NDORGS is capable of capturing important information of a large corpus.

Information-Coverage Score. The information coverage scores (see Eq. 2 in Sect. 3.2 for definition) for SOVs over BBC News and Factiva-Marx are given in Table 3. It can be seen that the report generated on 0.2-summaries achieves the highest information coverage score over BBC News, and the SOV generated on 0.1-summaries or 0.3-summaries achieves the highest information coverage score over Factiva-Marx.

Table 3. Information-coverage scores.

Dataset	$\lambda = 0.1$	$\lambda = 0.2$	$\lambda = 0.3$
BBC News	0.38	0.54	0.52
Factiva-Marx	0.70	0.64	0.70

Topic Diversity. Topic diversity scores are shown in Table 4. It can be seen that SOVs generated on 0.2-summaries outperform those generated on 0.1-summaries and 0.3-summaries for both BBC News and Factiva-Marx.

Table 4. Topic-diversity scores.

Dataset	$\lambda = 0.1$	$\lambda = 0.2$	$\lambda = 0.3$
BBC News	0.1278	0.1444	0.1278
Factiva-Marx	0.1056	0.1167	0.1111

Time Efficiency. Figure 7 illustrates the running time of NDORGS on BBC News and Factiva-Marx, with the following results:

1. NDORGS incurs, respectively, about 80% and 56% more time to generate an SOV on 0.3-summaries and 0.2-summaries than 0.1-summaries (see Fig. 7(a)).
2. NDORGS incurs, respectively, over 2 times and 1.8 times longer to generate an SOV on 0.3-summaries and 0.2-summaries than 0.1-summaries (see Fig. 7(b)).
3. NDORGS achieves the best time efficiency on 0.1-summaries, which is expected.
4. Working on a larger summary length ratio incurs a longer running time, which is expected.

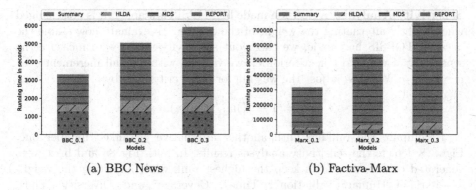

(a) BBC News   (b) Factiva-Marx

Fig. 7. Comparisons of running time.

Overall Performance. The overall performance of NDORGS is evaluated using the following criteria (listed in the order of preference): human evaluation, time efficiency, information coverage, and topic diversity. In particular, we use Saaty's pairwise comparison 9-point scale [41] and the Technique for Order Preference by Similarity to an Ideal Solution (TOPSIS) [19] to determine which value of λ has the best performance.

Let the three SOVs for the same corpus be the three alternatives, denoted by a_1, a_2, a_3. Let the human evaluation mean score, running time, information coverage score, and topics diversity score be four criteria, denoted by c_1, c_2, c_3, c_4. Next, we use Saaty's pairwise comparison 9-point scale to determine weights for each criterion. A weight vector $\boldsymbol{w} = \{w_1, w_2, w_3, w_4\}$ is then computed using the Analytic Hierarchy Process (AHP) procedure [41], where w_i is the weight for criterion c_i. A weighted normalization decision matrix $\boldsymbol{T}$ is then generated from the normalized matrix $\boldsymbol{R}$ and the weight vector $\boldsymbol{w}$. The alternatives $a_1, a_2,$ and a_3 are ranked using Euclidean distance and a similarity method (see Table 5). It can be seen that the SOVs generated on 0.2-summaries achieve the best overall performance on both BBC News and Factiva-Marx.

Table 5. Overall performance.

Rank	Model	Readability	Time	Coverage	Diversity
3	BBC-0.1	3.57	3310	0.38	0.1278
1	**BBC-0.2**	3.71	5060	0.54	0.1444
2	BBC-0.3	4.03	5930	0.52	0.1278
2	Marx-0.1	3.57	317023	0.70	0.1056
1	**Marx-0.2**	4.03	539474	0.64	0.1167
3	Marx-0.3	3.75	758404	0.70	0.1111

Sensitivity Analysis. A decision made by TOPSIS is stable if it is not changed when slightly alternating the weight of the criteria. To evaluate how stable the decision TOPSIS has made, we carry out sensitivity analyses to measure the sensitivity of weights. For criterion c_i, we vary w_i with a small increment c by $w_i' = w_i + c$. We then adjust the weights for other criteria c_j by

$$w_j' = (1 - w_i')w_j/(1 - w_i).$$

Recompute the ranking until another alternative is ranked number one. Figure 8 depicts the sensitivity analyses results. In both Fig. 8a and b, reports generated on 0.2-summaries keep the highest rank while adjusting the weight of criteria of "Human Evaluation", "Time", "Coverage", and "Diversity". Thus, the decision made by TOPSIS is stable over both BBC News and Factiva-Marx.

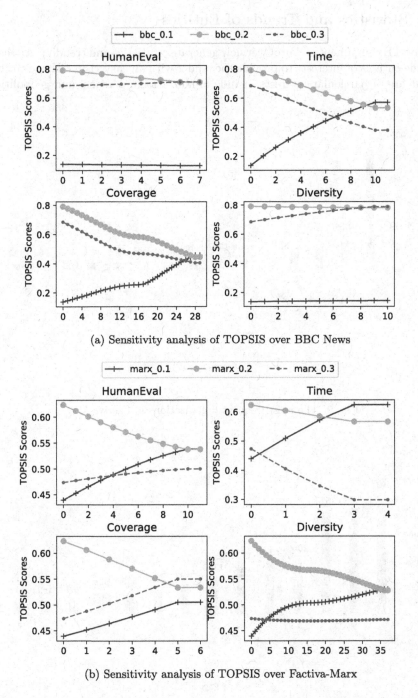

(a) Sensitivity analysis of TOPSIS over BBC News

(b) Sensitivity analysis of TOPSIS over Factiva-Marx

Fig. 8. The x-axis indicates the weight of corresponding criterion in increments/decrements of 0.02 each time, and the y-axis shows the new TOPSIS values.

144   J. Wang

8 Statistics and Trends of Entities

The STE component of AutoOverview generates statistics and trending graphs of name entities of interests to provide the reader with an easy visual. Name entities that are of particular interests include organizations, persons, and geopolitical

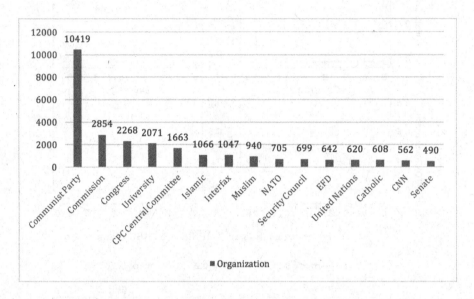

Fig. 9. The most frequent organizations in Factiva-Marx

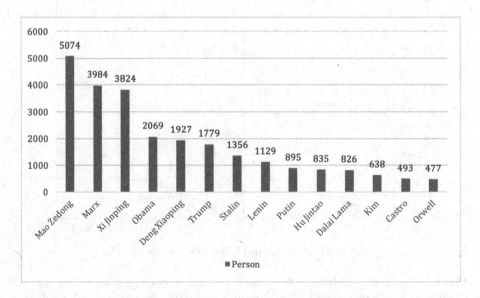

Fig. 10. The most frequent persons in Factiva-Marx

entities. An implementation of STE uses a name-entity-recognition tool (such as nltk.org) to tag name entities and compute their frequencies. Figures 9, 10, and 11 are the trending graphs generated over the Factiva-Marx dataset.

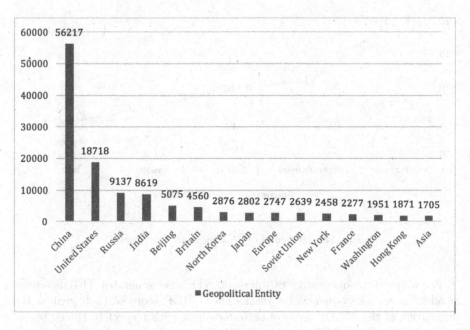

Fig. 11. The most frequent geopolitical entities in Factiva-Marx

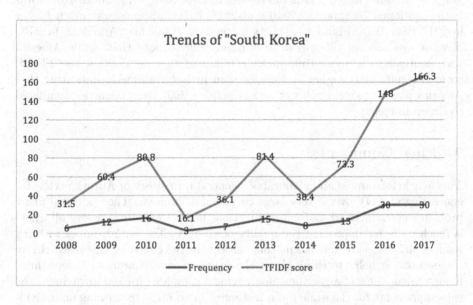

Fig. 12. The trend of "South Korea" in Factiva-Marx

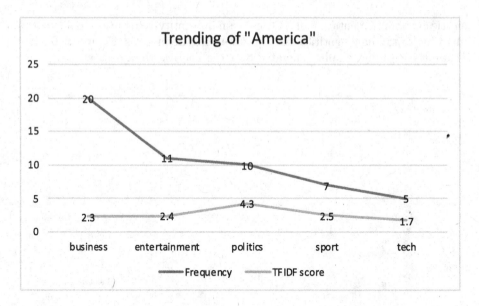

Fig. 13. The trend of "America" in BBC News

For a specific name entity of interests, STE also generates TFIDF scores, in addition to its frequency by years. The TFIDF score of each year is the summation of the TFIDF score of each document with respect to the corpus of articles in that year, which measures its significance. Figure 12 depicts a trending graph for "South Korea". It is interesting to note that, while South Korea was both mentioned 30 times in 2016 and 2017, its significance was much higher in 2017 than 2016. Figure 13 depicts a trending graph for "America" in BBC News across the six categories. It is interesting to note that, while America was mentioned one more time in Entertainment than in Politics, the TFIDF score is significantly higher in Politics than in Entertainment, indicating that America plays a more significant role in politics than other countries compared to Entertainment.

9   Final Comments

This article introduces and elaborates a general framework of AutoOverview for generating an SOV over a very large corpus of documents. There are still many problems not addressed and many directions not yet explored. Above all, there is much room for improving the quality of an SOV. The author hopes to bring readers' attentions to this fascinating subject. In the course of this research, we realized that it helps to think about algorithms as combinatorics of algorithms, namely, algorithms of algorithms, for solving a complex problem involving many sub-components for the same input instances, in addition to working on individual algorithms dealing with combinatorics of different components in an instance.

Acklowledgement. During the past five years, a number of students have worked on various parts of AutoOverview for their PhD degrees at the University of Massachusetts Lowell, although the term of AutoOverview and its current framework were not introduced until now in this article. They are (in alphabetical order) Yiqi Bai, Ming Jia, Liqun (Catherine) Shao, Jingwen (Jessica) Wang, Wenjing Yang, Cheng Zhang, and Hao Zhang, and five of them have graduated. Most of their contributions have been published elsewhere. The term of AutoOverview was inspired from a conversation with Prof. Jiawei Han of the University of Illinois at Urbana-Champaign.

References

1. Atasu, K., et al.: Linear-complexity relaxed word mover's distance with GPU acceleration. In: Proceedings of the 2017 IEEE International Conference on Big Data (BigData 2017), Boston, Massachusetts, USA, 11–14 December 2017, pp. 889–896 (2017)
2. Berman, P., DasGupta, B., Kao, M.Y., Wang, J.: On constructing an optimal consensus clustering from multiple clusterings. Inform. Process. Lett. **104**(4), 137–145 (2007)
3. Blei, D.M., Ng, A.Y., Jordan, M.I.: Latent dirichlet allocation. J. Mach. Learn. Res. **3**, 993–1022 (2003)
4. Buyukkokten, O., Garcia-Molina, H., Paepcke, A.: Seeing the whole in parts: text summarization for web browsing on handheld devices. In: Proceedings of the 10th International Conference on World Wide Web (WWW 2001), Hong Kong, China, 1–5 May 2001, pp. 652–662. ACM (2001)
5. Cao, Z., Li, W., Li, S., Wei, F.: Improving multi-document summarization via text classification. In: Proceedings of the 31st AAAI Conference on Artificial Intelligence (AAAI 2017), San Francisco, USA, 4–9 February 2017, pp. 3053–3059 (2017)
6. Cao, Z., Li, W., Li, S., Wei, F., Li, Y.: AttSum: joint learning of focusing and summarization with neural attention. In: Proceedings of the 26th International Conference on Computational Linguistics (COLING 2016), Osaka, Japan, 11–16 December 2016, pp. 547–556 (2016)
7. Christensen, J., Mausam, Soderland, S., Etzioni, O.: Towards coherent multi-document summarization. In: Proceedings of the 2013 Conference of the North American Chapter of the Association for Computational Linguistics: Human Language Technologies (NAACL-HLT 2013), Atlanta, Georgia, USA, 9–15 June 2013, pp. 1163–1173 (2013)
8. Christensen, J., Soderland, S., Bansal, G., et al.: Hierarchical summarization: scaling up multi-document summarization. In: Proceedings of the 52nd Annual Meeting of the Association for Computational Linguistics (ACL 2014), Baltimore, Maryland, USA, 22–27 June 2014, vol. 1, pp. 902–912 (2014)
9. DUC: DUC 2004 quality questions (2004). http://duc.nist.gov/duc2004/quality. questions.txt
10. DUC: Document understanding conference (2014). https://www-nlpir.nist.gov/ projects/duc/intro.html
11. Dueck, D.: Affinity propagation: clustering data by passing messages. Citeseer (2009)

12. Erkan, G., Radev, D.R.: LexRank: graph-based lexical centrality as salience in text summarization. J. Artif. Intell. Res. **22**, 457–479 (2004)
13. Florescu, C., Caragea, C.: A position-biased pagerank algorithm for keyphrase extraction. In: Proceedings of the 31st AAAI Conference on Artificial Intelligence (AAAI 2017), San Francisco, California, USA, 4–9 February 2017, pp. 4923–4924 (2017)
14. Gao, W., Li, P., Darwish, K.: Joint topic modeling for event summarization across news and social media streams. In: Proceedings of the 21st ACM International Conference on Information and Knowledge Management (CIKM 2012), Maui, Hawaii, USA, 29 October–2 November 2012, pp. 1173–1182 (2012)
15. Gillick, D., Favre, B.: A scalable global model for summarization. In: Proceedings of the Workshop on Integer Linear Programming for Natural Langauge Processing, pp. 10–18 (2009)
16. Gusfield, D.: Partition-distance: a problem and class of perfect graphs arising in clustering. Inform. Process. Lett. **82**(3), 159–164 (2002)
17. Hartigan, J.A., Wong, M.A.: Algorithm AS 136: a k-means clustering algorithm. J. Roy. Stat. Soc. Ser. C (Appl. Stat.) **28**(1), 100–108 (1979)
18. Hong, K., Conroy, J.M., Favre, B., Kulesza, A., Lin, H., Nenkova, A.: A repository of state of the art and competitive baseline summaries for generic news summarization. In: Proceedings of the 9th edition of the Language Resources and Evaluation Conference (LREC 2014), Reykjavik, Iceland, 26–31 May 2014, pp. 1608–1616 (2014)
19. Hwang, C.L., Yoon, K.: Methods for multiple attribute decision making. In: Hwang, C.L., Yoon, K. (eds.) Multiple Attribute Decision Making. Lecture Notes in Economics and Mathematical Systems, vol. 186, pp. 58–191. Springer, Heidelberg (1981). https://doi.org/10.1007/978-3-642-48318-9_3
20. Jones, D.: Factiva global news database (2018). https://www.dowjones.com/products/factiva/
21. Kaufman, L., Rousseeuw, P.J.: Finding Groups in Data: An Introduction to Cluster Analysis. Wiley, Hoboken (2009)
22. Kipf, T.N., Welling, M.: Semi-supervised classification with graph convolutional networks. In: Proceedings of the 4th International Conference on Learning Representations (ICLR 2016), San Juan, Puerto Rico, 2–4 May 2016
23. Kusner, M.J., Sun, Y., Kolkin, N.I., Weinberger, K.Q.: From word embeddings to document distances. In: Proceedings of the 32nd International Conference on Machine Learning (ICML 2015), Lille, France, 06–11 July 2015, vol. 37 (2015)
24. Lapata, M.: Probabilistic text structuring: experiments with sentence ordering. In: Proceedings of the 41st Annual Meeting on Association for Computational Linguistics (ACL 2003), Sapporo, Japan, 7–12 July 2003, vol. 1, pp. 545–552 (2003)
25. Li, C., et al.: LDA meets Word2Vec: a novel model for academic abstract clustering. In: Companion Proceedings of the Web Conference (WWW 2018), pp. 1699–1706 (2018)
26. Li, C., Qian, X., Liu, Y.: Using supervised bigram-based ILP for extractive summarization. In: Proceedings of the 51st Annual Meeting of the Association for Computational Linguistics (ACL 2013), Sofia, Bulgaria, 4–9 August 2013, vol. 1, pp. 1004–1013 (2013)
27. Li, P., Jiang, J., Wang, Y.: Generating templates of entity summaries with an entity-aspect model and pattern mining. In: Proceedings of the 48th Annual Meeting of the Association for Computational Linguistics (ACL 2010), Uppsala, Sweden, 11–16 July 2010, pp. 640–649 (2010)

28. Li, S., Ouyang, Y., Wang, W., Sun, B.: Multi-document summarization using support vector regression. In: Proceedings of DUC. Citeseer (2007)
29. Lin, C.Y.: ROUGE: a package for automatic evaluation of summaries. In: Proceedings of Workshop on Text Summarization Branches Out, Barcelona, Spain, 21–26 July 2004, pp. 74–81 (2004)
30. Liu, P.J., et al.: Generating Wikipedia by summarizing long sequences. In: Proceedings 6th International Conference on Learning Representation (ICLR 2018), Vancouva, Canada, 30 April-3 May 2018, vol. abs/1801.10198 (2018)
31. Luhn, H.P.: The automatic creation of literature abstracts. IBM J. Res. Dev. 2(2), 159–165 (1958)
32. Lulli, A., Debatty, T., Dell'Amico, M., Michiardi, P., Ricci, L.: Scalable k-NN based text clustering. In: Proceedings of 2015 IEEE International Conference on Big Data (IEEE BigData 2015), Santa Clara, California, USA, 29 October–1 November 2015, pp. 958–963 (2015)
33. Mihalcea, R., Tarau, P.: TextRank: Bringing order into texts. In: Proceedings of the 2004 Conference on Empirical Methods in Natural Language Processing (EMNLP 2004), Barcelona, Spain, 25–26 July 2004, pp. 404–411 (2004)
34. Nallapati, R., Zhai, F., Zhou, B.: SummaRuNNer: a recurrent neural network based sequence model for extractive summarization of documents. In: Proceedings of the 31st AAAI Conference on Artificial Intelligence (AAAI 2017), San Francisco, USA, 4–9 February 2017, pp. 3075–3081 (2017)
35. Nayeem, M.T., Chali, Y.: Extract with order for coherent multi-document summarization. In: Proceedings of the Workshop on Graph-based Methods for Natural Language Processing (TextGraphs 2011), Vancouver, Canada, 3 August 2017, pp. 51–56 (2017)
36. Ng, A.Y., Jordan, M.I., Weiss, Y.: On spectral clustering: analysis and an algorithm. In: Advances in Neural Information Processing Systems, pp. 849–856 (2002)
37. Otterbacher, J., Radev, D., Kareem, O.: News to go: hierarchical text summarization for mobile devices. In: Proceedings of the 29th Annual International ACM SIGIR Conference on Research and Development on Information Retrieval (SIGIR 2006), Seattle, Washington, USA, 6–11 August 2006, pp. 589–596 (2006)
38. Pottker, H.: News and its communicative quality: the inverted pyramidwhen and why did it appear? J. Stud. 4(4), 501–511 (2003)
39. Radev, D., et al.: SummBank 1.0 LDC2003T16. web download. Linguistic Data Consortium, Philadelphia (2003)
40. Radev, D.R., Jing, H., Sty, M., Tam, D.: Centroid-based summarization of multiple documents. Inf. Process. Manage. 40, 919–938 (2004)
41. Saaty, T.: The Analytical Hierarchy Process. McGraw Hill, New York (1980)
42. Sauper, C., Barzilay, R.: Automatically generating Wikipedia articles: a structure-aware approach. In: Proc of the Joint Conference of the 47th Annual Meeting of the ACL and the 4th International Joint Conference on Natural Language Processing of the AFNLP (ACL 2009), Suntec, Singapore, 2–7 August 2009, pp. 208–216 (2009)
43. Shao, L., Wang, J.: DTATG: an automatic title generator based on dependency trees. In: Proceedings of the 8th International Joint Conference on Knowledge Discovery and Information Retrieval (KDIR 2016), Porto, Portugal, 9–11 November 2016, pp. 166–173. SCITEPRESS - Science and Technology Publications, Lda, Portugal (2016). https://doi.org/10.5220/0006035101660173
44. Teh, Y.W., Jordan, M.I., Beal, M.J., Blei, D.M.: Hierarchical dirichlet processing. J. Am. Stat. Assoc. 101(476), 1566–1581 (2006)

45. Vandegehinste, V., Pan, Y.: Sentence compression for automated subtitling: a hybrid approach. In: Proceedings of the 42nd Annual Meeting of the Association for Computational Linguistic (ACL 2004), Barcelona, Spain, 21–26 July 2004, pp. 89–95 (2004)
46. Von Luxburg, U.: A tutorial on spectral clustering. Stat. Comput. **17**(4), 395–416 (2007)
47. Wang, J., Zhang, H., Zhang, C., Yang, W., Wang, J.: An effective scheme for generating an overview report over a very large corpus of documents. In: Proceedings ACM Symposium on Document Engineering (DocEng 2019), Berlin, Germany, 23–26 September 2019. (Best paper runnerup)
48. Wang, X., Nishino, M., Hirao, T., Sudoh, K., Nagata, M.: Exploring text links for coherent multi-document summarization. In: Proceedings of the 26th International Conference on Computational Linguistics (COLING 2016), Osaka, Japan, 11–16 December 2016, pp. 213–223 (2016)
49. Xu, J., et al.: Short text clustering via convolutional neural networks. In: Proceedings of the 2015 Conference of the North American Chapter of the Association for Computational Linguistics Human Language Technologies (NAACL HLT 2015), Denver, Colorado, USA, 31 May-5 June 2015, pp. 62–69 (2015)
50. Yao, C., Jia, X., Shou, S., Feng, S., Zhou, F., Liu, H.: Autopedia: automatic domain-independent Wikipedia article generation. In: Proceedings of the 20th International Conference Companion on World Wide Web (WWW 2011), Hyderabad, India, 28 March–1 April 2011, pp. 161–162 (2011)
51. Yasunaga, M., Zhang, R., Meelu, K., Pareek, A., Srinivasan, K., Radev, D.R.: Graph-based neural multi-document summarization. In: Proceedings of the SIGNLL Conference on Computational Natural Language Learning (CoNLL 2017), Vancouver, Canada, 3–4 August 2017
52. Yin, J., Wang, J.: A Dirichlet multinomial mixture model-based approach for short text clustering. In: Proceedings of the 20th ACM SIGKDD International Conference on Knowledge Discovery and Data Mining (KDD 2014), New York, NY, USA, 24–27 August 2014, pp. 233–242 (2014)
53. Yogatama, D., Liu, F., Smith, N.A.: Extractive summarization by maximizing semantic volume. In: Proceedings of the 2015 Conference on Empirical Methods in Natural Language Processing (EMNLP 2015), Lisbon, Portugal, 17–21 September 2015, pp. 1961–1966 (2015)
54. Zhang, H., Wang, J.: Semantic WordRank: generating finer single-document summarizations. In: Yin, H., Camacho, D., Novais, P., Tallón-Ballesteros, A.J. (eds.) IDEAL 2018. LNCS, vol. 11314, pp. 398–409. Springer, Cham (2018). https://doi.org/10.1007/978-3-030-03493-1_42
55. Zhang, Y., Peng, C., Wang, H.: Research on Chinese sentence compression for the title generation. In: Ji, D., Xiao, G. (eds.) CLSW 2012. LNCS (LNAI), vol. 7717, pp. 22–31. Springer, Heidelberg (2013). https://doi.org/10.1007/978-3-642-36337-5_3

Better Upper Bounds for Searching on a Line with Byzantine Robots

Xiaoming Sun[1,2]([✉]), Yuan Sun[1,2], and Jialin Zhang[1,2]

[1] CAS Key Lab of Network Data Science and Technology, Institute of Computing
Technology, Chinese Academy of Sciences, Beijing, China
{sunxiaoming,sunyuan2016,zhangjialin}@ict.ac.cn
[2] University of Chinese Academy of Sciences, Beijing, China

Abstract. Searching on a line with Byzantine robots was first posed
by Czyzowicz et al. in [13]: Suppose there are n robots searching on an
infinite line to find a target which is unknown to the robots. At the
beginning all robots stay at the origin and then they can start to search
with maximum speed 1. Unfortunately, f of them are *Byzantine fault*,
which means that they may ignore the target when passing it or lie that
they find the target. Therefore, the target is found if at least $f+1$ robots
claim that they find the target at the same location. The aim is to design
a parallel algorithm to minimize the competitive ratio $S(n, f)$, the ratio
between the time of finding the target and the distance from origin to
the target in the worst case by n robots among which f are Byzantine
fault.

In this paper, our main contribution is a new algorithm framework
for solving the Byzantine robot searching problem with (n, f) sufficiently
large. Under this framework, we design two specific algorithms to improve
the previous upper bounds in [13] when $f/n \in (0.358, 0.382) \cup (0.413, 0.5)$.
Besides, we also improve the upper bound of $S(n, f)$ for some small (n, f).
Specifically, we improve the upper bound of $S(6, 2)$ from 4 to 3.682, and
the upper bound of $S(3, 1)$ from 9 to 8.53.

Keywords: Searching on a line · Mobile robots · Parallel search ·
Competitive ratio · Byzantine fault

1 Introduction

1.1 Problem Description

In this paper, we consider the following robot searching problem from [13]. Suppose there are n robots searching on a one-dimensional axis in parallel, and their
aim is to find a target placed somewhere on the line unknown to these robots.
At the beginning all robots are at the origin of the axis. The maximum speed of
each robot is 1 per unit time, both on moving either along the positive direction
(which can be seen as moving right) or negative direction (which can be seen
as moving left). During the searching, any robot can change its direction at any

© Springer Nature Switzerland AG 2020
D.-Z. Du and J. Wang (Eds.): Ko Festschrift, LNCS 12000, pp. 151–171, 2020.
https://doi.org/10.1007/978-3-030-41672-0_9

position without loss of time. The state of a position is detected by a robot only when this robot has passed there. Robots are assumed to have full knowledge of the detected searching states at any time. The mission is to find the target as fast as possible.

If all robots are functioning normally, this problem is quite trivial. However, when we introduce *Byzantine fault* into the robot system, the problem becomes much complicated. In fact, when we build a communication network to control the actions of all robots, it is natural to consider the fault tolerance of the network. We say a robot is *Byzantine fault* or a *Byzantine robot*, if during the searching process it may ignore the target or lie that it finds the target at its position. In order to deal with Byzantine faults, we allow that robots can *communicate* and *vote*. All robots can communicate with each other in wireless mode with any distance. Communication does not cost time. When a robot claims it finds the target, it also set up a vote for this claim in the communication channel. Before the vote ends any robot passing the controversial position should vote on supporting or opposing this claim. If we know that there are f Byzantine robots in the robot system, then the vote ends immediately when at least $f + 1$ robots holding the same opinion on one side. More precisely, when there are at least $f + 1$ robots claiming that the target is found at the same position, we can finish the searching process and believe that we find the target indeed. Similarly, when there are at least $f + 1$ robots showing that a finding claim is false, we can mark all the robots which support the claim as Byzantine robots. We allow that there are more than one votes in the communication channel simultaneously.

Under the Byzantine robot model, we need some evaluation functions to judge the efficiency of a protocol among all robots. One obvious way to evaluate a searching algorithm is to consider the ratio between time and the distance from the origin in the worst case. In the next subsection, we will define such functions formally.

1.2   Evaluation Functions

For the problem that n robots among which are f byzantine fault ones search on a line, we define two kinds of functions, $S(n, f)$ and $T(\frac{f}{n})$, to represent the optimal search time among all algorithms for the searching problem. This notations is similar to the ones in [13].

First, the concept *searching time* need be clarified:

Definition 1 (Searching Time). *Given two integers n, f, a real number x and a protocol (or an algorithm) $\mathcal{A}$, we define $S_x^{\mathcal{A}}(n, f)$ as the searching time of the situation that under the algorithm $\mathcal{A}$, $(n - f)$ normal robots with f Byzantine fault robots together find the target whose coordinate is x on the axis.*

Then we can evaluate the efficiency of an algorithm by the following *competitive ratio* function:

Definition 2 (Competitive Ratio). *Define*

$$S^{\mathcal{A}}(n, f) = \sup_{|x|>0} \left\{ \frac{S_x^{\mathcal{A}}(n, f)}{|x|} \right\}$$

as the competitive ratio of the algorithm $\mathcal{A}$ with the parameter pair (n, f). In other words, this function represents the ratio of the searching time and distance in the worst case.

Remark 1. W.o.l.g, we may need to assume that the target can not be placed to close to the origin. For example, we can suppose that the distance between the target and the origin is at least 0.01.

For the whole problem, first let $\mathscr{A}$ be the set of all algorithms for this problem. Then we use

$$S(n, f) = \inf_{\mathcal{A} \in \mathscr{A}} \{S^{\mathcal{A}}(n, f)\}$$

to be the competitive ratio of the Byzantine robot searching problem with a robot system with fault ratio β.

When both n and f are both sufficiently large, we turn to consider the relationship between $\frac{f}{n}$ (we say $\frac{f}{n}$ is the *fault ratio* of a robot system with f Byzantine robots and $(n - f)$ normal robots) and the competitive ratio, which induces the *asymptotic competitive ratio* function:

Definition 3 (Asymptotic Competitive Ratio). *For $\beta \in (0, \frac{1}{2})$, define*

$$T(\beta) = \varlimsup_{n \to \infty} S(n, \beta n)$$

as the asymptotic competitive ratio of the Byzantine robot searching problem with a robot system with fault ratio β.

Remark 2. For competitive ratio, the parameter (n, f) we are interested in satisfies $2f + 1 \leq n \leq 4f + 1$, because if $n \leq 2f$ then we can not judge any position where f robots claim while others give different opinion, and if $n \geq 4f + 2$ then we just let two groups with $\frac{n}{2}$ robots search each side [13].

For asymptotic competitive ratio, we only consider the fault ratio β satisfying $\frac{1}{4} < \beta < \frac{1}{2}$ for the same reason. Furthermore, for convenience the status of a position is totally checked if at least $2\beta n$ robots have passed there, instead of $2\beta n + 1$.

1.3 Our Results

The main contribution of this work is that, for sufficiently large n and f we provide an algorithm framework, BYZANTINESEARCH(BASE, μ), to give upper bounds of asymptotic competitive ratio. Here, BASE is a searching algorithm which can only deal with the robot systems with fault ratio no more than μ. The details of this algorithm framework are in Algorithm 1, Sect. 3. As a brief description, the main process of BYZANTINESEARCH is that if the fault ratio of the current robot system is no more than μ we just call BASE, otherwise we call a recursion subroutine repetitively to reduce the fault ratio. In the recursion subroutine, when some robots claim that they find the target, we do not let others come to check immediately, but let them move along their original routes for a while then turn to check. The deferred time on checking depends on an extra parameter. This operation is the key to the trade-off among all bad cases. We merge all extra parameters in each times of recursive program into a parameter sequence. For a given (BASE, μ) we can give an optimal assignment to the parameter sequence to get the best upper bound of asymptotic competitive ratio under the algorithm BYZANTINESEARCH(BASE, μ).

As a result, we achieve the upper bounds of asymptotic competitive ratio in Theorems 1 and 2 by BYZANTINESEARCH(BASE, μ) with two different (BASE, μ) pairs:

Theorem 1. *Define a real number sequence* $\{\beta_n\}_{n \geq 1}$: $\beta_0 = \frac{5}{14}, \beta_n = \frac{1+\beta_{n-1}}{4-2\beta_{n-1}}(n \geq 1)$ *and a function sequence* $\{f_n(x)\}_{n \geq 1}$: $f_1(x) = x + 1, f_n(x) = (x+1)(1 - \frac{1}{f_{n-1}(x)})(n \geq 2)$. *Then if the fault ratio* β *satisfies* $\beta_{n-1} < \beta \leq \beta_n$ *for some* $n \geq 1$, *we have* $T(\beta) \leq 3 + 2x^*$, *where* $x^* \in (0, 3)$ *is the largest real root of the equation* $f_n(x) = \frac{x+1}{x}$.

Theorem 2. *Define a real number sequence* $\{\beta_n\}_{n \geq 1}$: $\beta_0 = \frac{13}{34}, \beta_n = \frac{1+\beta_{n-1}}{4-2\beta_{n-1}}(n \geq 1)$ *and a function sequence* $\{f_n(x)\}_{n \geq 1}$: $f_1(x) = x + 1, f_n(x) = (x+1)(1 - \frac{1}{f_{n-1}(x)})(n \geq 2)$. *Then if the fault ratio* β *satisfies* $\beta \leq \beta_0$, *we have* $T(\beta) \leq 3.682$; *otherwise if the fault ratio* β *satisfies* $\beta_{n-1} < \beta \leq \beta_n$ *for some* $n \geq 1$, *we have* $T(\beta) \leq 3 + 2x^*$, *where* $x^* \in (0, 3)$ *is the largest real root of the equation* $f_n(x) = \frac{x+1}{x-0.341}$.

Table 1. Upper bounds of $T(\beta)$ when $\frac{5}{14} < \beta < \frac{1}{2}$. (The previous upper bounds are from [13])

β	$(\frac{5}{14}, \frac{13}{34}]$	$(\frac{13}{34}, \frac{19}{46}]$	$(\frac{19}{46}, \frac{47}{110}]$	$(\frac{47}{110}, \frac{65}{146}]$
Previous upper bounds	4	5	6	7
Our upper bounds	3.682	5	5.682	6.236
Achieved by	Theorem 2	Theorem 1	Theorem 2	Theorem 1
β	$(\frac{65}{146}, \frac{157}{346}]$	$(\frac{157}{346}, \frac{211}{454}]$	$(\frac{211}{454}, \frac{503}{1070}]$	$(\frac{503}{1070}, \frac{665}{1394}]$
Previous upper bounds	8	9	9	9
Our upper bounds	6.747	7	7.369	7.494
Achieved by	Theorem 2	Theorem 1	Theorem 2	Theorem 1
β	$(\frac{665}{1394}, \frac{1573}{3274}]$	$(\frac{1573}{3274}, \frac{2059}{4246}]$	$(\frac{2059}{4246}, \frac{4847}{9950}]$	...
Previous upper bounds	9	9	9	9
Our upper bounds	7.762	7.828	8.026	...
Achieved by	Theorem 2	Theorem 1	Theorem 2	...

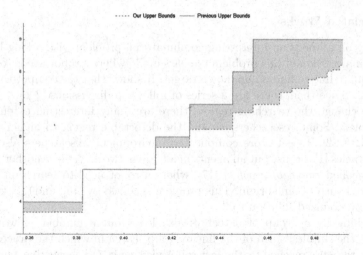

Fig. 1. The comparison between the efficiency of previous and current results. X axis represents the ratio parameter β, and Y axis represents the upper bounds of $S(\beta)$. The straight line is the previous results while the dashed line is the results achieved in this paper.

The final upper bounds are achieved by combining Theorems 1 and 2. Table 1 gives an intuitive version of the above theorems, and compares our results with the previous ones. It can be seen that when $\beta \in (\frac{5}{14}, \frac{13}{34}] \cup (\frac{19}{46}, \frac{1}{2})$, Theorems 1 and 2 make significant improvement. Moreover, Fig. 1 presents a straightforward visual feeling of the improvement.

For some small (n, f), we also make achievement on the upper bounds of $S(n, f)$. First, a corollary of Theorem 2 directly gives a better upper bound of $S(6, 2)$:

Corollary 1. $S(6, 2) \leq 3.682$.

Next, for the case where $(n, f) = (3, 1)$, we get the following theorem by a totally different algorithm, which is based on an algorithm from [11]:

Theorem 3. $S(3, 1) \leq 8.653$.

Table 2 compares these results with the previous ones.

Table 2. Upper bounds of $S(n, f)$ for some small n and f. (The previous upper bounds are from [13], and lower bounds are from [13, 19])

(n, f)	Previous upper bounds	Our upper bounds	Lower bounds
$(3, 1)$	9	8.653	5.23
$(6, 2)$	4	3.682	3

1.4   Related Works

Searching on a line is an interesting combinatorial problem with a long history. The original version of this problem appears in [3], where a robot needs to search on an infinite line to find an unknown target. It shows that the competitive ratio is exact 9. Since then, there are a series of follow-up discussions [2–6].

If we change the searching region, there are many interesting extension of this problem. Some researchers expand the detectable region to higher dimensions [16,18,22]. To get more complicated environment, researchers also introduce obstacles [1], or just put all agents in a large network [7,14]. Another variant is the so-called *cow-path problem* [17], where a robot need to search on several rays shared with a same origin. This problem is solved by [15], and [17] gives an optimal randomized solution to it.

Searching by a group of detectors also leads out a number of works. [8] considers the simplest form of the linear group search in which the process ends when the target is reached by the last robot visiting it and shows that increasing the number of robots does not help and the optimal competitive ratio is still 9. [10,14] think of the cost and restrictions of communications and information exchanges between the group members.

In group searching, it is natural to consider the safety of the whole system. The concept *Byzantine Fault Tolerance* has been widely studied in distributed computing [21]. [20] notices that it is important to keep the whole system safe even if some of the agents are broken down and send wrong messages to others, which induces the *Byzantine general problem*. On group searching with faulty members, there are also many impressive works [9,10]. Recently, researchers discuss the problem that the robots search on a line by group together while some of them are crash or Byzantine fault [11,13,19]. These is the main model discussed in this paper. See also [12] as a survey.

1.5   Organization

In Sect. 2, we will denote some necessary notations to describe our algorithm better. In Sect. 3, we will give an algorithm framework for the asymptotic competitive ratio. In Sect. 4, we provide two specific base algorithms which can be used as inputs for the framework and prove main theorems. In Sect. 5, we turn to discuss the competitive ratio for small n and f. In Sect. 6, we conclude our work and give some open problems.

2   Notations

In order to describe our algorithms clearly, we provide some new definitions.

First, we give several notations on robot systems.

Definition 4 (Robot Status Representation (RSR)). *For a robot r, its robot status representation (RSR) is a tuple (pos_r, d_r), where $pos_r \in R$ is its coordinate and $d_r \in \{-1, 0, 1\}$ is its current moving direction. Here $d_r = 1$ (−1)* *means r will move along the right (left) direction after receiving move order, and $d_r = 0$ means r will stop at its position after receiving move order.*

Definition 5 (Group Searching System (GSS)). *At any time of the searching progress, the status of all working robots can be described as a group searching system (GSS) $\mathcal{P} = (n, f, R, S)$, where n is the number of robots, f is the number of Byzantine fault robots, R is the set of all robots and S is the set of all RSRs. Note that we do not know which f robots are Byzantine fault. In a GSS, if all robots in a set A is at a same position, we say that the position of A is the position of robots in A.*

Definition 6 (Symmetric Two-group Searching System (STSS)). *A GSS $\mathcal{P}$ is a symmetric two-group searching system (STSS) if all robots in $\mathcal{P}$ can be divided into two groups A and B such that:*

- *Robots in the same group have the same RSR tuple;*
- *for any two robots $r_1 \in A$ and $r_2 \in B$, $pos_{r_1} = -pos_{r_2}$ and $d_{r_1} = -d_{r_2}$.*

We write an STSS $\mathcal{P}$ as $\mathcal{P} = (n, f, A, B, (pos, d))$, where (pos, d) is the RSR of robots in the set A. In particular, a STSS $\mathcal{P}$ is an initial STSS, if all robots are at the origin, and half of robots have directions -1 while others have directions 1. An initial STSS can also be seen as a status of a robot system at the beginning of searching.

Next, we define some basic orders on a robot system, which can be used as basic components of our algorithms.

- MOVE($\mathcal{P}$, R, t): for the GSS P, keep robots in the robot set R moving along their directions for time t with speed 1. In other words, this function will cost t units of time and for any robot $r \in R$ after this function its position will be $pos_r := pos_r + td_r$. Robots in R will check their position to find the target while moving.
- MOVEALL($\mathcal{P}$, t): for the GSS P, keep all robots moving along their directions for time t with maximum speed. All robots will check their position to find the target while moving.
- TURNAROUND($\mathcal{P}$, R): for the GSS P, change directions of robots in the robot set R. i.e., set $d_r := -d_r$ for all $r \in R$.
- TURNAROUNDALL(P): for the GSS P, change directions of all robots.
- REMOVE($\mathcal{P}$, R): for the GSS P, remove all robots in R from $\mathcal{P}$.
- JOIN($\mathcal{P}$, R_1, R_2, k): for the GSS $\mathcal{P}$, arbitrarily choose k robots from the robot set R_1, and add them to the robot set R_2. This operation only works when k is no larger than $|R_1|$ and the positions of robots in both R_1 and R_2 are all the same.
- SELECT($\mathcal{P}$, R_0, k): for the GSS $\mathcal{P}$ and a robot set R_0, this operation will randomly choose k robots in R_0, and return a robot set containing these k robots. This operation only work when $k \leq |R_0|$ and $R_0 \subseteq \mathcal{P}.R$.

Also, here are some specific variables closely related to the searching process:

- *TargetClaim*: a Boolean variable which is False initially, but turns to be True immediately after some robots claim the finding of the target, while the number of robots with opposite opinion is no more than f. This variable will turn to be False again when all robots reach a consensus that the claim on *ClaimPos* is false, and turn to be True for the next effective claim, and so on.
- *ClaimPos*: a real number representing the position where some robots claim they find the target which change the value of *TargetClaim* from False to True. (This variable is meaningless when *TargetClaim* is False.)
- *ClaimSet*: The set of robots which claim they find the target (This variable is meaningless when *TargetClaim* is False.)
- *TargetVeri*: a Boolean variable which is False initially, but turns to be True immediately after the target is actually found. (i.e., at least $f+1$ robots claim they find the target at the same place.) After it turns to be True, its value will not be changed.
- *TargetPos*: a real number representing the position of the target (This variable is meaningless when *TargetVeri* is False.)

Remark 3. Here are some extra supplementary explanation for the operations and variables in Sect. 2.2.

- It is easy to see that the time cost of a searching process is the sum of time parameters in MOVE and MOVEALL functions.
- When we use MOVE order some robots, the others stay at their positions.
- Positions on the line can be verified even when *TargetClaim* is true, but the value of *TargetPos* won't be changed when a new claim occurs but *TargetClaim* is already True.
- If more than one positions are claimed to be the location of target which makes *TargetClaim* from False to True, randomly choose one position as the value of *ClaimPos*.
- If during a MOVE process a position is verified to be the real position of the target, the MOVE function will be ended immediately and values of the two variables *TargetPos* and *TargetVeri* will be updated.

3   An Algorithm Framework for Byzantine Robot Searching Problem

In this section we provide an algorithm framework, BYZANTINESEARCH, for the Byzantine robot searching problem. The detail is stated in Algorithm 1.

Algorithm 1. BYZANTINESEARCH(BASE, μ): A searching algorithm framework

---

1: $\mathcal{P}$:=Initial STSS
2: $i := 0$, $m :=$ TIMES$(\mathcal{P}.f/\mathcal{P}.n, \mu)$
3: **while** $(\mathcal{P}.f/\mathcal{P}.n > \mu) \wedge (TargetVeri = \text{False})$ **do**
4:     $i := i + 1$
5:     RECUR$(\mathcal{P}, \lambda(i, m))$
6: **if** $TargetVeri = $ True **then**
7:     **return** $TargetPos$
8: **else**
9:     BASE$(\mathcal{P})$

This framework has two input parameters, BASE and μ. BASE is a *partial* searching algorithm, and μ is the corresponding fault ratio limit. More precisely, BASE can deal with the Byzantine searching problem with an STSS only when the fault ratio is no more than μ. In Sect. 4, we will provide two different (BASE, μ) to Algorithm 4 and prove Theorems 1 and 2.

At the beginning of BYZANTINESEARCH, the robot system turns to be an initial STSS $\mathcal{P}$. If the fault ratio is no more than μ, we can just call BASE algorithm. Otherwise, we provide a searching subroutine called RECUR. Each time after we call RECUR, either we find the target or the fault ratio of STSS $\mathcal{P}$ is reduced. Thus we can call BASE algorithm when the fault ratio of STSS $\mathcal{P}$ is reduced to be no more than μ.

In BYZANTINESEARCH, the procedure RECUR is the core component, which can reduce the fault ratio and keep the robot system still an STSS. The process details of RECUR are stated in Algorithm 2. The RECUR procedure has two inputs, an STSS $\mathcal{P}$ and a real number $\lambda \geq 1$. The basic idea is that first let $\mathcal{P}$ search until some robots claim, then all robots turn around and move again. When all robots meet at the origin, two groups swap some members. After this subroutine either the target is found or $\mathcal{P}$ is still an STSS but with lower fault ratio. However, if each time when some robots claim others move to check immediately, the time cost will be too large if the target is near $ClaimPos$. To balance this bad case, the key point of RECUR is deferred verification. When some robots claim that they find the target, robots in the other group will not turn around to check it immediately but move along their directions for a while. The moving length depends on λ, the second input. In BYZANTINESEARCH, the i-th time we call RECUR the value of the second input λ is $\lambda(i, m)$. $\{\lambda(i, m)\}(1 \leq i \leq m)$ is a real parameter sequence, whose value is not related to the correctness of BYZANTINESEARCH, but can affect the (asymptotic) competitive ratio. For a given (BASE, μ), we can give an optimal assignment to the parameter sequence to get the best upper bound of asymptotic competitive ratio under the algorithm BYZANTINESEARCH(BASE, μ), by solving a convex optimization. In the proof of Theorem 4 we will give an example on this assertion.

To give an intuitive impression, the process of Algorithm 2 is also showed in Fig. 2.

Algorithm 2. RECUR($\mathcal{P}$, λ): A group searching subroutine with an STSS $\mathcal{P}$ and a real number $\lambda \geq 1$ to reduce the value of $\mathcal{P}.f / \mathcal{P}.n$

1: **do**
2:    MOVEALL($\mathcal{P}$, 0.01)   ▷ Here 0.01 can be a arbitrarily small number.
3: **until** $TargetClaim$ = True
4: MOVEALL($\mathcal{P}$, $(\lambda - 1) \cdot |ClaimPos|$)
5: TURNAROUNDALL($\mathcal{P}$)
6: MOVEALL($\mathcal{P}$, $\lambda \cdot |ClaimPos|$) ▷ After this order, all robots are at the origin.
7: $m := |ClaimSet|$, $k := \min\{n - 2f, m/2\}$
8: **if** $ClaimSet \subseteq \mathcal{P}.A$ **then**
9:    JOIN($\mathcal{P}$, $\mathcal{P}.B$, $\mathcal{P}.A$, k)
10: **else**
11:    JOIN($\mathcal{P}$, $\mathcal{P}.A$, $\mathcal{P}.B$, k)
              ▷ After this order $\mathcal{P}$ turns to be a general GSS.
12: $R' :=$ SELECT($\mathcal{P}$, $ClaimSet$, $2k$)
13: MOVEALL($\mathcal{P}$, $\lambda \cdot |ClaimPos|$)
14: REMOVE($\mathcal{P}$, R')   ▷ After this order $\mathcal{P}$ turns to be an $STSS$ again.

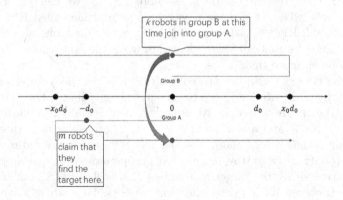

Fig. 2. Process of Algorithm 2

Now we show some basic properties of RECUR. It is easy to verify the next two facts:

Fact 1. *For any STSS $\mathcal{P}$, it will still be an STSS after* RECUR($\mathcal{P}$, λ).

Fact 2. *For any STSS $\mathcal{P}$, all points in the interval $[-\lambda \cdot |ClaimPos|$, $\lambda \cdot |ClaimPos|]$ will totally be checked after* RECUR($\mathcal{P}$, λ).

Next, Lemma 1 shows the ability of RECUR on reducing the fault ratio of $\mathcal{P}$:

Lemma 1. *For a given STSS $\mathcal{P}$, let $\beta = \frac{\mathcal{P}.f}{\mathcal{P}.n}$. Then either* RECUR($\mathcal{P}$, λ) *actually finds the target, or it can reduce β to a number no greater than $\frac{4\beta - 1}{2\beta + 1}$. (Notice that $\beta < \frac{1}{2}$, so we always have $\frac{4\beta - 1}{2\beta + 1} < \beta$.)*

Proof. For an STSS $\mathcal{P} = (n, f, A, B, (pos, d))$ with fault ration $\beta = \frac{f}{n}$, suppose there are m robots claiming the finding after line 3 in Algorithm 2. Note that we have $m \geq \frac{n}{2} - f$, otherwise the claiming will be verified as a false statement immediately by all robots at $ClaimPos$. Thus after $\text{RECUR}(\mathcal{P}, \lambda)$, if the finding claim is false, the fault ratio will turn to be

$$\frac{f-m}{n-m} \leq \frac{1 - (\frac{1}{2\beta} - 1)}{\frac{1}{\beta} - (\frac{1}{2\beta} - 1)} = \frac{4\beta - 1}{2\beta + 1},$$

or

$$\frac{f - (2n - 4f)}{n - (2n - 4f)} = \frac{5f - 2n}{4f - n} \leq \frac{5\beta - 2}{4\beta - 1} \leq \frac{4\beta - 1}{2\beta + 1}.$$

By Lemma 1, if we have a BASE algorithm which can deal with the Byzantine robots searching problem with fault ratio no more than μ, then for a given initial STSS $\mathcal{P}$ with fault ratio β, we can easily calculate the maximum possible times that we call RECUR to reduce the fault ratio of $\mathcal{P}$ no more than μ, which is the function $\text{TIMES}(\beta, \mu)$. The details of the function TIMES are stated in Algorithm 3.

Algorithm 3. TIMES(β, μ): The maximum possible recursion times in BYZANTINESEARCH

1: **if** $\beta \leq \mu$ **then**
2:     **return** 0
3: **else**
4:     **return** TIMES$((4\beta - 1)/(2\beta + 1), \mu) + 1$

---

4   Two Different Base Algorithms

In this section, we give two different (BASE, μ) pairs to prove Theorems 1 and 2 respectively.

4.1   A Base Algorithm with $\mu \leq \frac{5}{14}$

For an initial STSS $\mathcal{P}$ with fault ratio no more than $5/14$, [11] gives an algorithm in Theorem 13. Here we modify this algorithm as a base algorithm called LEFT-RIGHT. We give details of LEFTRIGHT in Algorithm 4. The main idea is that, after checking $ClaimPos$ if $TargetClaim$ is false, we still keep two groups of sufficient number robots move along different directions respectively after remove Byzantine robots. (That is why we call this algorithm "LEFTRIGHT".) Here we omit the proof of the correctness of Algorithm 4, since it is similar to the proof of Theorem 13 in [13].

Algorithm 4. LEFTRIGHT($\mathcal{P}$): A base algorithm for an STSS $\mathcal{P}$ satisfying $\frac{\mathcal{P}.f}{\mathcal{P}.n} \leq \frac{5}{14}$

1: **do**
2:     MOVEALL($\mathcal{P}$, 0.01)   ▷ Here 0.01 can be an arbitrary small positive real
    number.
3: **until** $TargetClaim$ = True   ▷ W.l.o.g, suppose $ClaimSet \subseteq \mathcal{P}.A$.
4: R_1 :=SELECT($\mathcal{P}$, $\mathcal{P}.B$, $(3/5)|\mathcal{P}.B|$)
5: R_2 :=SELECT($\mathcal{P}$, $ClaimSet$, $(2/5)|ClaimSet|$)
6: R_3 :=SELECT($\mathcal{P}$, $\mathcal{P}.A \setminus ClaimSet$, $(2/5)|\mathcal{P}.A \setminus ClaimSet|$)
7: TURNAROUND($\mathcal{P}$, $R_1 \cup R_2 \cup R_3$) ▷ After this order $\mathcal{P}$ turns to be a general
    GSS.
8: MOVE($\mathcal{P}$, $R_1 \cup R_2 \cup R_3$, $2|ClaimPos|$)
9: **if** $TargetVeri$ = False **then**
10:     REMOVE($\mathcal{P}$, $ClaimSet$)
11:     **do**
12:         MOVEALL($\mathcal{P}$, 0.01)
13:     **until** $TargetVeri$ = True
14: **return** $TargetPos$

Combining BYZANTINESEARCH and LEFTRIGHT, we have the following result leading to Theorem 1:

Theorem 4. BYZANTINESEARCH(LEFTRIGHT, 5/14) *with a specific parameter sequence* $\{\lambda(i, m)\}$ *can achieve the upper bounds of asymptotic competitive ratio in Theorem 1.*

Proof. First, it is easy to verify that the sequence $\{\beta_n\}$ in Theorem 1 is monotonically increasing and $\lim_{m\to\infty} \beta_m = \frac{1}{2}$. So for a given β with $\frac{5}{14} < \beta < \frac{1}{2}$, there exists an integer $m \geq 1$ such that $\beta_m < \beta \leq \beta_{m+1}$.

Second, suppose $\mathcal{P}$ is an initial STSS with $\frac{\mathcal{P}.f}{\mathcal{P}.n} = \beta$, by Lemma 1 and the definition of $\{\beta_n\}$ we know that the **while** loop in BYZANTINESEARCH(LEFTRIGHT, 5/14) will repeat at most m times. Denote that in the i-th ($1 \leq i \leq m$) repetition of the **while** loop the value of the parameter $\lambda(i, m)$ is λ_i, and let $\lambda_0 = 1$. Our aim is to choose specific $\{\lambda_i\}$ to reduce the asymptotic competitive ratio.

There are two cases on finding the target. First, suppose in the i-th ($1 \leq i \leq m$) repetition of the **while** loop we actually find the target. Then the worst case will be that the target is placed at $\pm(\lambda_{i-1}ClaimPos + \epsilon)$, where $\epsilon > 0$ is an arbitrary small number. Second, the target is found in the base algorithm LEFTRIGHT. Then the worst case will be that the target is placed at $\pm(\lambda_m ClaimPos + \epsilon)$, where $\epsilon > 0$ is an arbitrary small number. After calculating the searching time of all bad cases, we can express the asymptotic competitive ratio of BYZANTINESEARCH(LEFTRIGHT, 5/14) as

$$\max\{3 + 2\lambda_1, 3 + 2\lambda_2 + \frac{2}{\lambda_1}, 3 + 2\lambda_3 + \frac{2}{\lambda_2} + \frac{2}{\lambda_2\lambda_1}, ...,$$

$$3 + 2\lambda_m + \frac{2}{\lambda_{m-1}} + \frac{2}{\lambda_{m-1}\lambda_{m-2}} + ... + \frac{2}{\lambda_{m-1}\lambda_{m-2}...\lambda_1},$$

$$5 + \frac{2}{\lambda_m} + \frac{2}{\lambda_m\lambda_{m-1}} + ... + \frac{2}{\lambda_m\lambda_{m-1}...\lambda_1}\}$$

By analysis in the convex optimization, we know that if $(\lambda_1, \lambda_2, ..., \lambda_m)$ satisfies:

$$3 + 2\lambda_1 = 3 + 2\lambda_2 + \frac{2}{\lambda_1} = ... = 5 + \frac{2}{\lambda_m} + \frac{2}{\lambda_m\lambda_{m-1}} + ... + \frac{2}{\lambda_m\lambda_{m-1}...\lambda_1}, \quad (1)$$

$$\lambda_1 \geq 1, \lambda_2 \geq 1, ..., \lambda_m \geq 1$$

then $3 + 2\lambda_1$ will be the upper bound of the asymptotic competitive ratio, and this assignment to the parameter sequence $\{\lambda(i, m)\}$ is optimal for BYZANTINE-SEARCH(LEFTRIGHT, 5/14).

To solve (1) we let $x = \lambda_1$ and use the first two equations $3 + 2\lambda_1 = 3 + 2\lambda_2 + \frac{2}{\lambda_1}$. Then we get that $\lambda_2 = \frac{x^2-1}{x}$. Similarly we can express $\lambda_i(i > 2)$ as $\lambda_i = (1 + x)(1 - \frac{1}{\lambda_{i-1}})$ by the equations $3 + 2\lambda_1 = 3 + 2\lambda_i + \frac{2}{\lambda_i} + \frac{2}{\lambda_i\lambda_{i-1}} + ... + \frac{2}{\lambda_i\lambda_{i-1}...\lambda_1}$, $3 \leq i \leq m$. Combining these results to the equation $3 + 2\lambda_1 = 5 + \frac{2}{\lambda_m} + \frac{2}{\lambda_m\lambda_{m-1}} + ... + \frac{2}{\lambda_m\lambda_{m-1}...\lambda_1}$, and then we have:

$$(1 - \frac{1}{\lambda_m})(3 + 2x) = (5 + \frac{2}{\lambda_m} + \frac{2}{\lambda_m\lambda_{m-1}} + ... + \frac{2}{\lambda_m\lambda_{m-1}...\lambda_1})$$
$$- \frac{1}{\lambda_m}(3 + 2\lambda_m + \frac{2}{\lambda_{m-1}} + \frac{2}{\lambda_{m-1}\lambda_{m-2}} + ... + \frac{2}{\lambda_{m-1}\lambda_{m-2}...\lambda_1}) \quad (2)$$

Finally the equation turns to be:

$$\lambda_m = \frac{x + 1}{x} \quad (3)$$

Since all λ_i $(1 \leq i \leq m)$ can be expressed as a rational function of x, (3) is a polynomial equation about x. The last thing is to verify that there exists a real root x satisfying that $\lambda_i \geq 1, \forall 1 \leq i \leq m$. Lemmas 2 and 3 show that, the largest real root of (3) can meet our requirements.

Lemma 2. *Define a rational function sequence about x: $f_1(x) = x + 1$, $f_n(x) = (1 + x)(1 - \frac{1}{f_{n-1}(x)})$ $(n \geq 2)$. Let $\{\alpha_n\}$ be the sequence of the largest real root of the equation $f_n(x) = \frac{x+1}{x}$. Then sequence $\{\alpha_n\}$ is monotonically increasing with finite limit and its limit is no greater than 3.*

Proof. Define a rational function sequence $\{g_n(x)\}$ related to $\{f_n(x)\}$: $\forall n \geq 1$, $g_n(x) = f_n(x) - \frac{x+1}{x}$. Then $\{\alpha_n\}$ be the sequence of the biggest real root of the equation $f_n(x) = \frac{x+1}{x} \iff \{\alpha_n\}$ be the sequence of the biggest real root of the equation $g_n(x) = 0$. Note that $f_{n+1}(x) = (1+x)(1 - \frac{1}{f_n(x)})$, $n \geq 1$, which induces the recursive formula of $g_n(x)$:

$$g_{n+1}(x) = (1+x)\frac{x^2 g_n(x) - xg_n(x) - 1}{x^2 g_n(x) + x^2 + x}, \quad n \geq 1.$$

Now we can prove this lemma by induction. In fact, we will show a much stronger conclusion as the following list:

- $g_n(3) = \frac{2n+6}{3n}$.
- $\forall n \geq 2$, $g_n(x)$ is monotonically increasing and continuous when $x \geq \alpha_n$;
- $\alpha_{n-1} \leq \alpha_n < 3$;

The case of $n = 2$ is trivial. Suppose this proposition holds when $n \leq k$. Now consider $n = k + 1$. First, we have

$$g_{k+1}(3) = 4\frac{6g_k(3) - 1}{9g_k(3) + 12} = 4\frac{6\frac{2k+6}{3k} - 1}{9\frac{2k+6}{3k} + 12} = \frac{2(k+1) + 6}{3(k+1)}.$$

Second, calculate the derivation of $g_{k+1}(x)$, we have

$$g'_{k+1}(x) = \frac{1 + 2x^3 g_k(x) + 2x(1 + g_k(x)) + x^2(1 + g_k(x))^2 + x^5 g'_k(x) + x^4(g_k(x) + g_k^2(x) + g'_k(x))}{x^2(1 + x + xg_k(x))^2}.$$

Since when $x \geq \alpha_k$, we have $g_k(x) \geq g_k(\alpha_k) = 0$ and $g'_k(x) > 0$ by induction hypothesis, we have

$$x^2(1 + x + xg_k(x))^2 > 0$$

and

$$1 + 2x^3 g_k(x) + 2x(1 + g_k(x)) + x^2(1 + g_k(x))^2 + x^5 g'_k(x) + x^4(g_k(x) + g_k^2(x) + g'_k(x)) > 0,$$

so $g_{k+1}(x)$ is monotonically increasing and continuous when $x \geq \alpha_k$.
Finally, since

$$g_{k+1}(\alpha_k) = (1+x)\frac{x^2 g_k(\alpha_k) - xg_k(\alpha_k) - 1}{x^2 g_k(\alpha_k) + x^2 + x} = -\frac{1}{x} < 0,$$

there exists exactly one root x_0 of $g_{k+1}(x)$ which lies in the interval $(\alpha_k, 3)$. Because $g_{k+1}(x)$ is monotonically increasing and continuous when $x \geq \alpha_{k+1}$, we have $\alpha_{k+1} = x_0$. Thus $\alpha_k \leq \alpha_{k+1} < 3$.
Combining the above discussions, we finish the induction step of $n = k + 1$.

Lemma 3. *Define a function sequence about x:* $f_1(x) = x + 1, f_n(x) = (1 + x)(1 - \frac{1}{f_{n-1}(x)})(n \geq 2)$. *Let $\{\alpha_n\}$ be the sequence of the largest real root of the equation $f_n(x) = \frac{x+1}{x}$. Then $\forall n$, if $x \geq \alpha_n$, we get that $f_i(x) > 1$ and $f_i'(x) > 0$, $\forall 1 \leq i \leq n$.*

Proof. We still prove this proposition by induction. When $n = 1$, we have $\alpha_1 = 1$ and $\forall x \geq 1$, $f_1(x) = 1 + x > 1$ and $f_1'(x) = 1 > 0$. Now suppose that this lemma holds for all $1 \leq n \leq k$. Consider about $n = k + 1$. Since $\alpha_{k+1} \geq \alpha_k$ from Lemma 2, by induction hypothesis we just need to prove that $f_{k+1}(x) > 1$ and $f_{k+1}'(x) > 0$ when $x \geq \alpha_{k+1}$. First, we have

$$f_{k+1}'(x) = 1 - \frac{1}{f_k(x)} + (1+x)(1 + \frac{f_k'(x)}{f_k^2(x)}) > 1 - \frac{1}{1} + \frac{(1+x)f_k'(x)}{f_k^2(x)} > 0.$$

So if $x \geq \alpha_{k+1}$, $f_{k+1}(x)$ is monotonically increasing. Since α_{k+1} is a root of $f_{k+1}(x) = \frac{x+1}{x}$, we get that

$$f_{k+1}(x) \geq f_{k+1}(\alpha_{k+1}) = \frac{\alpha_{k+1} + 1}{\alpha_{k+1}} > 1,$$

which finish the induction step of $n = k + 1$.

4.2 A Base Algorithm with $\mu \leq \frac{13}{34}$

In this subsection we design a new base algorithm, LEFTMIDDLERIGHT, which accepts an STSS $\mathcal{P}$ with fault ratio no more than $13/34$ as input. The details of this algorithm are stated in Algorithm 5, which is totally different from LEFT-RIGHT.

Here we give a brief description for the process. Algorithm 5 can be divided into two parts. The first part of it is from line 1 to line 15, and the **if** component that starts at line 16 is the second part. For a given STSS $\mathcal{P}$, if some robots claim they find the target at the first part, we repartition all robots into 5 groups ($ClaimSet, A_2, A_1 \setminus A_2, B_1, B \setminus B_1$) and let 4 of them (except $ClaimSet$) move along different routes. At the end of the first part, $ClaimPos$ is checked. If the claim is true, we can finish the searching, otherwise we remove all Byzantine robots and merge remained robots into 3 groups (A_2, B_2, C_2). The position of robots in A_2 and B_2 are symmetric about the position of robots in C. (That is why we call this algorithm "LEFTMIDDLERIGHT".) Then we let robots in A_2 and B_2 move along their directions until $TargetClaim$ turns to be true again. At this time, robots in C move to $ClaimPos$ to check it.

Algorithm 5. LEFTMIDDLERIGHT($\mathcal{P}$): A base algorithm for an STSS $\mathcal{P}$ with $\frac{\mathcal{P}.f}{\mathcal{P}.n} \leq \frac{13}{34}$

1: **do**
2:    MOVEALL($\mathcal{P}$, 0.01)
3: **until** $TargetClaim$ = True   ▷ W.l.o.g, suppose $ClaimSet \subseteq \mathcal{P}.A$.
4: $m := |ClaimSet|$, $k := (1/2)n + (2/3)m - (2/3)f$
5: $x_0 := 1.341$, $x_1 := 1.797$, $x_2 := 2.41$
6: $A := \mathcal{P}.A$, $B := \mathcal{P}.B$
7: $A_1 := A \setminus ClaimSet$, $B_1 :=$ SELECT($\mathcal{P}$, B, k)
8: MOVE($\mathcal{P}$, $A_1 \cup B$, $x_0|ClaimPos|$)   ▷ After this order $\mathcal{P}$ turns to be a general GSS.
9: TURNAROUND($\mathcal{P}$, B_1)
10: MOVE($\mathcal{P}$, $A_1 \cup B$, $(x_1 - x_0)|ClaimPos|$)
11: $A_2 :=$ SELECT($\mathcal{P}$, A_1, $\frac{4}{3}(f - m)$)
12: TURNAROUND($\mathcal{P}$, A_2)
13: MOVE($\mathcal{P}$, $A_2 \cup B$, $(x_2 - x_1)|ClaimPos|$)
14: TURNAROUND($\mathcal{P}$, $B \setminus B_1$)
15: MOVE($\mathcal{P}$, $A_2 \cup B$, $(x_2 + x_1 - x_0)|ClaimPos|$)
16: **if** $TargetVeri$ = False **then**
17:    REMOVE($\mathcal{P}$, $ClaimSet$)
18:    $B_2 := (A_1 \setminus A_2) \cup B_1$, $C := B \setminus B_1$
19:    $d := (x_1 - x_0)ClaimPos$ ▷ At this moment, Robots in $B \setminus B_1$ are at d.
20:    **do**
21:      MOVE($\mathcal{P}$, $A_2 \cup B_2$, 0.01)
22:    **until** $TargetClaim$ = True   ▷ W.l.o.g, suppose $ClaimSet \subseteq A_2$.
23:    **if** $TargetVeri$ = False **then**
24:      MOVE($\mathcal{P}$, C, $d - ClaimPos$)
25:      **if** $TargetVeri$ = False **then**
26:        REMOVE($\mathcal{P}$, $ClaimSet$)
27:        **do**
28:          MOVEALL($\mathcal{P}$, 0.01)
29:        **until** $TargetClaim$ = True
30: **return** $ClaimPos$

To give an intuitive impression, the first part of Algorithm 5 is also showed in Fig. 3.

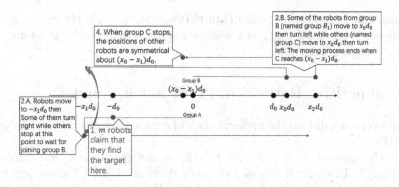

Fig. 3. Process of Algorithm 5

Lemma 4. *Given an STSS $\mathcal{P}$ with fault ratio no more than $13/34$, LEFTMIDDLERIGHT($\mathcal{P}$) can actually find the target.*

Proof. For any given STSS $\mathcal{P}$ with fault ratio no more than $\frac{13}{34}$, in order to prove that LEFTMIDDLERIGHT($\mathcal{P}$) actually finds the target, let us analyze the algorithm step by step.

In the first part, if the target is truly lying in the interval $[-x_1 d_0, -d_0] \cup [d_0, x_0 d_0]$, the searching process will stop immediately and the asymptotic competitive ratio will be no more than $1 + 2x_0$ because all the points in the interval $[-x_1 d_0, x_0 d_0]$ are checked by at least $2\beta n$ robots; If the target is lying in the interval $[x_0 d_0, x_2 d_0]$, then it will be verified immediately after the robots from group A_2 check its position; Otherwise after the first part the asymptotic competitive ratio of $\mathcal{P}$ turns to be

$$\beta' = \frac{f - m}{n - m} \le \frac{1 - (\frac{1}{2\beta} - 1)}{\frac{1}{\beta} - (\frac{1}{2\beta} - 1)} \le \frac{3}{10}.$$

Now let us consider the second part of the algorithm. If $\beta \le \frac{13}{34}$, then $\beta' \le \frac{3}{10}$ and at the beginning the group A_2 and B_2 are is symmetrical about $(x_0 - x_1)d_0$ (the position group C stays), and the ratio between the number of robots in A_2, B_2, C is $2 : 2 : 1$. We let the groups A_2 and B_2 keep on moving until $TargetClaim$ turns to be true again. We then let robots in group C move $ClaimPos$. Notice that there will be at least $\frac{3}{5}n' = 2f'$ robots checking $ClaimPos$, so the state of the position can be clearly determined. If the claim is true, then we can finish the search process. If not, then after the REMOVE function $\mathcal{P}.f/\mathcal{P}.n$ will be no greater than $(\frac{3}{10} - \frac{1}{10})/(1 - \frac{1}{10}) = \frac{2}{9} < \frac{1}{4}$, and the distribution of the robots at this time allows us to use an extra MOVE process to finish the searching.

Combining BYZANTINESEARCH and LEFTMIDDLERIGHT, we have the following theorem leading to Theorem 2, whose proof is similar to the one of Theorem 4:

Theorem 5. BYZANTINESEARCH(LEFTMIDDLERIGHT, 13/34) *with a specific parameter sequence* $\{\lambda(i,m)\}$ *can achieve the upper bounds of asymptotic competitive ratio in Theorem 2.*

5   Competitive Ratio $S(n, f)$ for Small n and f

In this section we discuss the upper bounds of competitive ratio $S(n, f)$ for small n and f.

For the case where $n = 6$ and $f = 2$, we design an algorithm similar to Algorithm 5. First, let 3 robots (we mark them as A, B, C) move left while others move right (we mark them as D, E, F). W.l.o.g, suppose A claims that it finds the target at position $-d_0 (d_0 > 0)$. If B or C support this claim, then let D and E turn round immediately to $-d_0$ to check the position while F still moves on. In this situation the competitive ratio will be no greater than 3. Otherwise B and C oppose the claim. Then we let B and C move to $-x_1 d_0$ then turn right, while D and E move to $x_0 d_0$ then turn left and F moves to $x_2 d_0$ then turns left until it stops at the location. $(x_0 - x_1)d_0$. (Here $x_0 = 1.341$, $x_1 = 1.797$, $x_2 = 2.41$.) It is not hard to verify that by this searching process, $S(6, 2) \leq 3.682$, which leads to Corollary 1.

For the case where $n = 3$ and $f = 1$, we will use some results from the problem searching on a line with *crash* robots, where crash means that some robots cannot detect the target. [13] gives an efficient algorithm to deal with this problem. For completeness, we also give the details of the proof of this algorithm.

Lemma 5 (Czyzowicz et al. [13]). *There is an algorithm for the case* $(n, f) = (3, 1)$ *that can solve the problem in the version of crash robots with competitive ratio* $(\beta + 1)^{\frac{4}{3}}(\beta - 1)^{-\frac{1}{3}} + 1$ *($\beta > 1$ is a parameter that can be determined by ourselves).*

Proof (Proof of Lemma 5). We provide an algorithm to achieve this competitive ratio. At first we let all robots move right. For robot i $(1 \leq i \leq 3)$, we define its turning-around location sequence as $x_{i,0}, x_{i,1}, x_{i,2}, \ldots$. The first turning-around location of each robot are calculated by the formulas below: $x_{1,0} = 1$, $x_{2,0} = (\frac{\beta+1}{\beta-1})^{\frac{2}{n}}\beta - \beta + 1$, $x_{3,0} = (\frac{\beta+1}{\beta-1})^{\frac{4}{n}}\beta - \beta + 1$. The later turning-around locations of each robot can be calculate by the formulas below: $\forall 1 \leq i \leq 3$, $j \geq 1$, $x_{i,j} = x_{i,j-1} \cdot (\frac{-\beta+1}{\beta-1})$. To give a direct impression, the relationships between time and locations of each robot can be viewed in the left side of Fig. 4. This strategy can give the competitive ratio $(\beta + 1)^{\frac{4}{3}}(\beta - 1)^{-\frac{1}{3}} + 1$ for $S(3, 1)$.

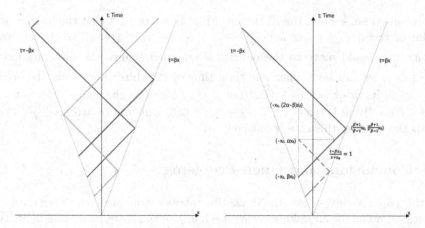

Fig. 4. Left side: The coordinate map between time and distance in the proof of Lemma 5. The three lines represent the traces of each robot. Right side: The coordinate map between time and distance in the proof of Theorem 3. The thick dashed line represents the behavior of the robots which claims that it finds the target near $-x_0$. The straight line represents the behavior of the robot whose turning point is $-x_0$, while the thin dashed line represents its changing after receiving the target detecting message.

We design Algorithm 6 to prove Theorem 3. In the Byzantine fault version, we split the search process into two parts, and express it as follows:

Algorithm 6. BYZANTINESEARCH31($\mathcal{P}$): A searching algorithm with an initial STSS $\mathcal{P}$ where $n = 3$ and $f = 1$

1: In the first part, we run the algorithm described in the proof of Lemma 5, until there is a robot claiming that it finds the target. At the end of this part, w.l.o.g suppose a robot claims that it finds the target at position $-x_0(x > 0)$.
2: In the second part, at the beginning for the other two robots if some of them have not detected the position $-x_0$ and are moving right, let them turn around immediately to move forward to $-x_0$. After checking the claim, if it is false then there remain two normal robots. The final step is letting them moving on opposite directions.

Proof (Proof of Theorem 3). There are many situations need to be discussed in the analysis of Algorithm 6. Here we provide the analysis of the worst situation, where the target is truly at $-x_0$ which is ignored by the first robot passing there, and the claiming is by the second robot reaching the position $-x_0$. In the first part of Algorithm 6, the search time is αx_0, where $\alpha = (\beta + 1)^{\frac{4}{3}}(\beta - 1)^{-\frac{1}{3}} + 1$ ($\beta > 1$ is a parameter undetermined yet). Notice that there are two robots having detected the position $-x_0$, in which one claims that there is the target while the other holds the different opinion. Now consider the only robot which has not check $-x_0$ yet. We build a coordinate map between the time and coordinate from original axis (horizontal axis represents original coordinate, and vertical axis represents time, see also the right side of Fig. 4), we can get that the intersection point of two lines $t = \beta x$ and $\frac{t - \beta x_0}{x + x_0} = 1$ is $(\frac{\beta+1}{\beta-1} x_0, \beta \frac{\beta+1}{\beta-1} x_0)$, so if there was no alert it would move to the point $\beta \frac{\beta+1}{\beta-1} x_0$ then change the direction. So if $\beta \frac{\beta+1}{\beta-1} > \alpha$ we can save time. The total time of the third robot from the origin to $-x_0$ in its orbit is $(2\alpha - \beta)x_0$. Let $\beta = 2.023$, then the competitive ratio is $(2\alpha - \beta) = 2(\beta + 1)^{\frac{4}{3}}(\beta - 1)^{-\frac{1}{3}} + 2 - \beta = 8.653$, which also satisfies $\beta \frac{\beta+1}{\beta-1} > \alpha$. Thus the proof of this case is finished.

6   Conclusions and Open Problems

In this paper we investigate the Byzantine robots search problem where n robots search on a line for an unknown target while f of them are Byzantine fault. For sufficiently large (n, f), we provide a new algorithm framework, and design two specific algorithm under this framework. With these algorithms, We improve the upper bounds of asymptotic competitive ratio $T(\beta)$ significantly with the fault ratio $\beta \in (\frac{5}{14}, \frac{13}{34}) \cup (\frac{19}{46}, \frac{1}{2})$ compared with the results in [13]. Besides, we also improve upper bounds of competitive ratio $S(n, f)$ for the cases where $(n, f) = (6, 2)$ and $(3, 1)$.

For more discussion, We conjecture that the upper bounds of asymptotic competitive ratio achieved in this paper would be optimal. Another conjecture is that $\lim_{\beta \to \frac{1}{2}} T(\beta) = 9$. In other words, when the number of byzantine robots is almost half of the total number, we can not do much better than the algorithm in [2].

Another interesting direction of this problem is how to improve the lower bounds. The hardness comes from the lack of methods to analysis the complicate behaviors of all robots at all time. For example, we even do not know whether all the robots should not stop at all time in every optimal algorithm to the searching problem with $(n, f) = (3, 1)$. However, a feasible way to the cases $(n, f) = (2f + 1, f)$ is to consider a weaker version where all the faulty robots are crash fault. Some researchers have improve the lower bounds by this idea [19], and their results match the upper bound for crash robots. But for Byzantine version, the lower bounds are still not tight.

References

1. Albers, S., Kursawe, K., Schuierer, S.: Exploring unknown environments with obstacles. Algorithmica **32**(1), 123–143 (2002)
2. Beck, A.: On the linear search problem. Israel J. Math. **2**(4), 221–228 (1964)
3. Beck, A.: More on the linear search problem. Israel J. Math. **3**(2), 61–70 (1965)
4. Beck, A., Newman, D.J.: Yet more on the linear search problem. Israel J. Math. **8**(4), 419–429 (1970)
5. Beck, A., Warren, P.: The return of the linear search problem. Israel J. Math. **14**(2), 169–183 (1973)
6. Bellman, R.: An optimal search. SIAM Rev. **5**(3), 274 (1963)

7. Casteigts, A., Flocchini, P., Quattrociocchi, W., Santoro, N.: Time-varying graphs and dynamic networks. Int. J. Parallel Emergent Distrib. Syst. **27**(5), 387–408 (2012)
8. Chrobak, M., Gąsieniec, L., Gorry, T., Martin, R.: Group search on the line. In: Italiano, G.F., Margaria-Steffen, T., Pokorný, J., Quisquater, J.-J., Wattenhofer, R. (eds.) SOFSEM 2015. LNCS, vol. 8939, pp. 164–176. Springer, Heidelberg (2015). https://doi.org/10.1007/978-3-662-46078-8_14
9. Chuangpishit, H., Czyzowicz, J., Kranakis, E., Krizanc, D.: Rendezvous on a line by location-aware robots despite the presence of byzantine faults. In: Fernández Anta, A., Jurdzinski, T., Mosteiro, M.A., Zhang, Y. (eds.) ALGOSENSORS 2017. LNCS, vol. 10718, pp. 70–83. Springer, Cham (2017). https://doi.org/10.1007/978-3-319-72751-6_6
10. Czyzowicz, J., Gasieniec, L., Kosowski, A., Kranakis, E., Krizanc, D., Taleb, N.: When patrolmen become corrupted: monitoring a graph using faulty mobile robots. Algorithmica **79**(3), 925–940 (2017)
11. Czyzowicz, J., et al.: Search on a line by byzantine robots. In: 27th International Symposium on Algorithms and Computation, vol. 27 (2016)
12. Czyzowicz, J., Georgiou, K., Kranakis, E.: Group search and evacuation. In: Flocchini, P., Prencipe, G., Santoro, N. (eds.) Distributed Computing by Mobile Entities, pp. 335–370. Springer, Cham (2019). https://doi.org/10.1007/978-3-030-11072-7_14
13. Czyzowicz, J., Kranakis, E., Krizanc, D., Narayanan, L., Opatrny, J.: Search on a line with faulty robots. In: Proceedings of the 2016 ACM Symposium on Principles of Distributed Computing, pp. 405–414. ACM (2016)
14. Dieudonné, Y., Pelc, A., Peleg, D.: Gathering despite mischief. ACM Trans. Algorithms (TALG) **11**(1), 1 (2014)
15. Gal, S.: Minimax solutions for linear search problems. SIAM J. Appl. Math. **27**(1), 17–30 (1974)
16. Hoffmann, F., Icking, C., Klein, R., Kriegel, K.: The polygon exploration problem. SIAM J. Comput. **31**(2), 577–600 (2001)
17. Kao, M.-Y., Reif, J.H., Tate, S.R.: Searching in an unknown environment: an optimal randomized algorithm for the cow-path problem. Inf. Comput. **131**(1), 63–79 (1996)
18. Kleinberg, J.M.: On-line search in a simple polygon. In: SODA, vol. 94, pp. 8–15 (1994)
19. Kupavskii, A., Welzl, E.: Lower bounds for searching robots, some faulty. In: Proceedings of the 2018 ACM Symposium on Principles of Distributed Computing, pp. 447–453. ACM (2018)
20. Lamport, L., Shostak, R., Pease, M.: The byzantine generals problem. ACM Trans. Program. Lang. Syst. (TOPLAS) **4**(3), 382–401 (1982)
21. Lynch, N.A.: Distributed Algorithms. Elsevier, Amsterdam (1996)
22. Schuierer, S.: Lower bounds in on-line geometric searching. Comput. Geom. **18**(1), 37–53 (2001)

A Survey on Double Greedy Algorithms for Maximizing Non-monotone Submodular Functions

Qingqin Nong[1], Suning Gong[1]([✉]), Qizhi Fang[1], and Dingzhu Du[2]

[1] School of Mathematical Science, Ocean University of China,
Qingdao 266100, Shandong, People's Republic of China
gsn_ouc@163.com
[2] Department of Computer Science, University of Texas, Dallas 75083, USA

Abstract. Maximizing non-monotone submodular functions is one of the most important problems in submodular optimization. A breakthrough work on the problem is the double-greedy technique introduced by Buchbinder et al. [7]. Prior work has shown that this technique is very effective. This paper surveys on double-greedy algorithms for maximizing non-monotone submodular functions from discrete domains of sets and integer lattices to continuous domains.

Keywords: Submodular · Non-monotone · Algorithm · Double greedy

1 Introduction

Submodularity captures the principle of diminishing returns in economics. Motivated by the principle of economy of scale, prevalent applications such as machine learning [3], computer vision [16], and operations research [17] have stimulated research on problems with submodular objective functions. The most early results on submodular maximization concentrate on submodular set functions. The greedy approach is a basic method for the problems: start with the empty set being the initial solution, go through all the elements one at a time, and iteratively add to the current solution set an element that results in the largest positive marginal gain of the objective function while satisfying the constraints. This approach is effective for monotone submodular maximization with different types of constraints. However, this greedy algorithm performs poorly with non-monotone submodular functions such as the Unconstrained Submodular Maximization (USM). It has applications in various practical settings such as marketing in social networks [15], revenue maximization with discrete choice [1]. Given a non-negative submodular set fucntion $f : N \to R^+$, the goal of USM is to find

This research was supported in part by the National Natural Science Foundation of China under grant numbers 11201439 and 11871442, and was also supported in part by the Natural Science Foundation of Shandong Province under grant number ZR2019MA052 and the Fundamental Research Funds for the Central Universities.

D.-Z. Du and J. Wang (Eds.): Ko Festschrift, LNCS 12000, pp. 172–186, 2020.
https://doi.org/10.1007/978-3-030-41672-0_10

a subset $S \subseteq N$ such that $f(S)$ is maximized. For the problem, consider a similar algorithm to the greedy algorithm that starts with a solution consisting of all the elements, and then goes through all the elements one by one, and removes the element from the current solution if its marginal contribution to the current solution is negative. This algorithm is called reverse greedy. It is shown that this greedy algorithm also performs poorly for USM. In a breakthrough work, Buchbinder et al. [7] combined greedy and reverse-greedy by using them concurrently. When going through the elements, information from both greedy and reverse-greedy is used to decide whether to add the current element to the solution. Inspired by the *double-greedy* framework of Buchbinder et al. [7], researchers have designed effective algorithms for a large number of non-monotone submodular maximization problems and their generalizations. In this paper we survey on the results related to double-greedy algorithms for maximizing non-monotone submoduar functions from discrete domains of sets and integer lattices to continuous domains.

2 Maximizing Non-monotone Submodular Set Functions

Definition 1 (Submodular Set Function). *Given a finite ground set N, a set function $f : 2^N \to \mathbb{R}$ is said to be submodular, if it satisfies one of the following two conditions:*

(a) $f(S) + f(T) \geq f(S \cup T) + f(S \cap T)$, for any pair of subsets $S, T \subseteq N$;
(b) $f_S(j) \geq f_T(j)$, for any $S \subseteq T \subset N$ and $j \in N \setminus T$.

Unconstrained Submodular Maximization (USM) has been studied since the 1960s. The first rigorous study of it was conducted by Feige et al. [11]. They showed that a subset S chosen uniformly at random is a $\frac{1}{4}$-approximation. They presented two local search algorithms, one with an approximation ratio of $\frac{1}{3}$ and the other with an approximation guarantee of $\frac{2}{5}$. Gharan and Vondrák [13] provided an algorithm with an improved approximation of roughly 0.41. On the negative side, Feige et al. [11] showed that for any constant $\epsilon > 0$, any algorithm achieving an approximation of $(\frac{1}{2} + \epsilon)$ requires an exponential number of oracle queries. This hardness result holds even if f is symmetric, that is, even if $f(S) = f(N \setminus S)$ for every $S \subseteq N$.

To improve the approximation guarantees of algorithms for the USM, Buchbinder et al. [7] firstly introduced double-greedy framework. Basing on this idea, they presented a simple deterministic algorithm with an approximation guarantee of $\frac{1}{3}$ and a randomized algorithm achieving a tight approximation guarantee of $\frac{1}{2}$.

Let $N = \{u_1, \ldots, u_n\}$. The $\frac{1}{3}$-approximate deterministic double-greedy algorithm runs as follows:

Lemma 1. *For every $i \in \{1, \ldots, n\}$, we have $a_i + b_i \geq 0$.*

Algorithm 1. Deterministic USM(f, N)

Input: A submodular set function $f : 2^N \to \mathbb{R}_+$ and a value oracle for f.
Output: An approximate solution to $\max_{S \subseteq N} f(S)$

1: Initialization: $X_0 \leftarrow \emptyset$, $Y_0 \leftarrow N$
2: **for** $i = 1$ to n **do**
3:     $a_i \leftarrow f(X_{i-1} \cup \{u_i\}) - f(X_{i-1})$
4:     $b_i \leftarrow f(Y_{i-1} \setminus \{u_i\}) - f(Y_{i-1})$
5:     **if** $a_i \geq b_i$ **then**
6:         $X_i \leftarrow X_{i-1} \cup \{u_i\}, Y_i \leftarrow Y_{i-1}$
7:     **else**
8:         $X_i \leftarrow X_{i-1}, Y_i \leftarrow Y_{i-1} \setminus \{u_i\}$
9:     **end if**
10: **end for**
11: **return** X_n

Proof. Observe that $X_{i-1} \subseteq Y_{i-1} \setminus \{u_i\}$ and $u_i \in Y_{i-1}$.

$$
\begin{aligned}
a_i + b_i &= [f(X_{i-1} \cup \{u_i\}) - f(X_{i-1})] + [f(Y_{i-1} \setminus \{u_i\}) - f(Y_{i-1})] \\
&= [f(X_{i-1} \cup \{u_i\}) - f(X_{i-1})] - [f(Y_{i-1}) - f(Y_{i-1} \setminus \{u_i\})] \\
&\geq 0,
\end{aligned}
$$

where the last inequality holds from the submodularity of f.

The above result indicates that the solutions X_i and Y_i returned by the algorithm improve at every step.

Corollary 1. *For every $i \in \{1, \ldots, n\}$: $f(X_i) \geq f(X_{i-1})$ and $f(Y_i) \geq f(Y_{i-1})$.*

Proof. Consider the ith loop. If $a_i \geq b_i$, then $X_i = X_{i-1} \cup \{u_i\}$ and $Y_i = Y_{i-1}$. $a_i + b_i \geq 0$ implies that $a_i \geq 0$ and thus $f(X_i) - X_{i-1} \geq 0$ and $f(Y_i) - f(Y_{i-1}) = 0$. If $a_i < b_i$, $X_i = X_{i-1}$ and $Y_i = Y_{i-1} \setminus \{u_i\}$. $a_i + b_i \geq 0$ means that $b_i \geq 0$ and thus $f(Y_i) - f(Y_{i-1}) \geq 0$ and $f(X_i) - f(X_{i-1}) \geq 0$.

Let OPT be an optimal solution and $OPT_i = (OPT \vee X_i) \wedge Y_i$ ($i = 0, 1, \ldots, n$). OPT_i is the set that coincides with X_i and Y_i on $u_1, \ldots, u_i$ and coincides with OPT on $u_{i+1}, \ldots, u_n$. One can see that $X_i \subseteq OPT_i \subseteq Y_i$ ($i = 0, 1, \ldots, n$), $OPT_n = X_n = Y_n$ and $OPT_0 = OPT$. Note that OPT_i is the crucial sequence of sets used through the analysis.

Lemma 2. *For every $i \in \{1, \ldots, n\}$, we have*

$$
[f(X_i) - f(X_{i-1})] + [f(Y_i) - f(Y_{i-1})] \geq f(OPT_{i-1}) - f(OPT_i).
$$

Proof. Assume that $a_i \geq b_i$ (the case that $a_i < b_i$ is similar). Then $X_i = X_{i-1} \cup \{u_i\}$, $Y_i = Y_{i-1}$ and the sum of the left terms of the inequality is a_i. If $u_i \in OPT_{i-1}$, then $OPT_i = OPT_{i-1}$ and thus $f(OPT_i) = f(OPT_{i-1})$. Note that $a_i \geq 0$ in this case. The inequality follows. If $u_i \notin OPT_{i-1}$, then $OPT_i =$

$OPT_{i-1} \cup \{u_i\}$. By the submodularity and the fact that $OPT_{i-1} \subseteq Y_{i-1} \setminus \{u_i\}$, we have

$$
\begin{aligned}
f(OPT_{i-1}) - f(OPT_i) &= f(OPT_{i-1}) - f(OPT_{i-1} \cup \{u_i\}) \\
&\leq f(Y_{i-1} \setminus \{u_i\}) - f(Y_{i-1}) \\
&= b_i \\
&\leq a_i.
\end{aligned}
$$

This concludes the proof.

Theorem 1. *Algorithm 1 has an approximation guarantee of $\frac{1}{3}$ for USM.*

Proof.

$$
\begin{aligned}
f(X_n) + f(Y_n) &\geq f(X_n) - f(X_0) + f(Y_n) - f(Y_0) \\
&= \sum_{i=1}^{n} [f(X_i) - f(X_{i-1}) + f(Y_i) - f(Y_{i-1})] \\
&\geq \sum_{i=1}^{n} f(OPT_{i-1}) - f(OPT_i) \\
&= f(OPT_0) - f(OPT_n),
\end{aligned}
$$

where the first inequality holds from the fact that $f(X_0) \geq 0$ and $f(Y_0) \geq 0$, and the second holds from Lemma 2. Observe that $OPT_0 = OPT$ and $X_n = Y_n = OPT_n$. The result follows.

In the case where a_i and b_i are both positive, either the decision of picking u_i or of rejecting u_i could be a good decision. By either picking it or rejecting it with some probability, the approximation guarantee can be improved to $\frac{1}{2}$. The randomized algorithm is described at the below.

Lemma 3. *For every $i \in \{2, \ldots, n\}$, we have*

$$
\mathbb{E}[f(X_i) - f(X_{i-1}) + f(Y_i) - f(Y_{i-1})] \geq 2\mathbb{E}[f(OPT_{i-1}) - f(OPT_i)].
$$

Proof. If $a_i \geq 0$ and $b_i \leq 0$, then with probability 1, $X_i = X_{i-1} \cup \{u_i\}$ and $Y_i \leftarrow Y_{i-1}$. The sum of the left terms of the inequality is a_i. If $u_i \in OPT_{i-1}$, then $OPT_i = OPT_{i-1}$ and thus $f(OPT_i) = f(OPT_{i-1})$. The inequality follows. If $u_i \notin OPT_{i-1}$, then $OPT_i = OPT_{i-1} \cup \{u_i\}$. By the submodularity and the fact that $OPT_{i-1} \subseteq Y_{i-1} \setminus \{u_i\}$, we have

$$
\begin{aligned}
f(OPT_{i-1}) - f(OPT_i) &= f(OPT_{i-1}) - f(OPT_{i-1} \cup \{u_i\}) \\
&\leq f(Y_{i-1} \setminus \{u_i\}) - f(Y_{i-1}) \\
&= b_i \\
&\leq \frac{a_i}{2}.
\end{aligned}
$$

Algorithm 2. Randomized USM(f, N)

Input: A submodular set function $f : 2^N \to \mathbb{R}_+$ and a value oracle for f.
Output: An approximate solution to $\max_{S \subseteq N} f(S)$

1: Initialize: $X_0 \leftarrow \emptyset$, $Y_0 \leftarrow N$
2: **for** $i = 1$ to n **do**
3:     $a_i \leftarrow f(X_{i-1} \cup \{u_i\}) - f(X_{i-1})$
4:     $b_i \leftarrow f(Y_{i-1} \setminus \{u_i\}) - f(Y_{i-1})$
5:     $a_i' \leftarrow \max\{a_i, 0\}$, $b_i' \leftarrow \max\{b_i, 0\}$,
6:     with probability $\frac{a_i'}{a_i' + b_i'}$: $X_i \leftarrow X_{i-1} \cup \{u_i\}, Y_i \leftarrow Y_{i-1}$
7:     with probability $\frac{b_i'}{a_i' + b_i'}$: $X_i \leftarrow X_{i-1}, Y_i \leftarrow Y_{i-1} \setminus \{u_i\}$
8: **end for**
9: **return** X_n
    (Note: assume that $\frac{a_i'}{a_i' + b_i'} = 1$ if $a_i' = b_i' = 0$)

If $b_i \geq 0$ and $a_i < 0$, then with probability 1: $X_i = X_{i-1}, Y_i = Y_{i-1} \setminus \{u_i\}$. The proof follows similarly as the previous case. By Lemma 1, $a_i + b_i \geq 0$. Thus the only case remains is $a_i > 0$ and $b_i > 0$. In this case $a_i' = a_i$ and $b_i' = b_i$. With probability $\frac{a_i}{a_i + b_i}$, $X_i = X_{i-1} \cup \{u_i\}$ and $Y_i = Y_{i-1}$, and with probability $\frac{b_i}{a_i + b_i}$, $X_i = X_{i-1}$ and $Y_i = Y_{i-1} \setminus \{u_i\}$. Thus,

$$\mathbb{E}[f(OPT_{i-1}) - f(OPT_i)$$
$$= \frac{a_i}{a_i + b_i}(f(OPT_{i-1}) - f(OPT_{i-1} \cup \{u_i\})) + \frac{b_i}{a_i + b_i}(f(OPT_{i-1}) - f(OPT_{i-1} \setminus \{u_i\})).$$

(1)

If $u_i \in OPT_{i-1}$, then $OPT_{i-1} \cup \{u_i\} = OPT_{i-1}$ and thus

$$f(OPT_{i-1}) - f(OPT_{i-1} \cup \{u_i\}) = 0.$$

By submodularity of f and the fact that $X_{i-1} \subseteq OPT_{i-1} \setminus \{u_i\}$, we have

$$f(OPT_{i-1}) - f(OPT_{i-1} \setminus \{u_i\}) \leq f(X_{i-1} \cup \{u_i\}) - f(X_{i-1}) = a_i.$$

If $u_i \notin OPT_{i-1}$, then $OPT_{i-1} \setminus \{u_i\} = OPT_{i-1}$ and thus

$$f(OPT_{i-1}) - f(OPT_{i-1} \setminus \{u_i\}) = 0.$$

By the submodularity of f and the fact that $OPT_{i-1} \subseteq Y_{i-1} \setminus \{u_i\}$, we have

$$f(OPT_{i-1}) - f(OPT_{i-1} \cup \{u_i\}) \leq f(Y_{i-1} \setminus \{u_i\}) - f(Y_{i-1}) = b_i.$$

Therefore, in both cases the sum of the right terms of Equality (1) is $\frac{a_i b_i}{a_i + b_i}$. On the other hand,

$$
\begin{aligned}
& \mathbb{E}[f(X_i) - f(X_{i-1}) + f(Y_i) - f(Y_{i-1})] \\
& = \frac{a_i}{a_i + b_i}(f(X_{i-1} \cup \{u_i\}) - f(X_{i-1})) + \frac{b_i}{a_i + b_i}(f(Y_{i-1} \setminus \{u_i\}) - f(Y_{i-1})) \\
& = \frac{a_i^2 + b_i^2}{a_i + b_i} \\
& \geq \frac{2 a_i b_i}{a_i + b_i} \\
& \geq \mathbb{E}[f(OPT_{i-1}) - f(OPT_i)].
\end{aligned}
$$

This completes the proof of the lemma.

Lemma 3 implies the following theorem.

Theorem 2. *Algorithm 2 has an approximation guarantee of $\frac{1}{2}$ for USM.*

In most cases the approximation guarantees obtained by randomized algorithms are superior to the best results obtained by the known deterministic algorithms. Buchbinder and Feldman [5] gave evidence that randomization is not necessary for obtaining good algorithms by presenting a new technique for derandomization of Algorithm 2. Its approximation guarantee remains $\frac{1}{2}$. Their idea is to maintain explicitly a (small) distribution over the states of the algorithm, and carefully update it using marginal values obtained from an extreme point solution of a suitable linear formulation.

Inspired by the double-greedy framework, Buchbinder et al. [6] presented a continuous double-greedy algorithm for maximizing non-monotone submodular set functions with cardinality constraints $\max_{|S| \leq k} f(S)$, where k is a positive integer. They showed that the approximate guarantee of the algorithm is at least $(1 + \frac{n}{2\sqrt{(n-k)k}})^{-1}$. Ene and Nguyễn [9] considered the problem $\max_{S \in \mathcal{I}} f(S)$, where f is a non-monotone submodular set function and $\mathcal{I}$ is an independent system ($\mathcal{I} \subseteq 2^N$, and $B \in \mathcal{I}$ and $A \subseteq B$ implies $A \in \mathcal{I}$) with ground set N. Given a set function $f : 2^N \to \mathbb{R}_+$, a function $F : [0,1]^N \to \mathbb{R}_+$ is the multilinear extension of f if

$$
F(\boldsymbol{x}) = \mathbb{E}[f(R(\boldsymbol{x}))] = \sum_{R \subseteq N} f(R) \cdot \prod_{u \in R} x_u \cdot \prod_{v \notin R} (1 - x_v).
$$

Ene and Nguyễn [9] described an algorithm for maximizing the multilinear extension of f. The algorithm picks the best out of the following two solutions. The first solution is constructed by running a Continuous Greedy algorithm with an additional dampening constraint. The second solution is constructed by running a double-greedy exactly when the marginal gain becomes low. They showed that the approximation guarantee of the algorithm is 0.372, improving over the $\frac{1}{e}$ approximation achieved by the unified Continuous Greedy algorithm Feldman et al. [12].

3   Maximizing Non-monotone Submodular Integer Lattice Functions

Set functions are powerful for describing problems with variable selection. But they cannot cast the case that allows multiple choices of an element in the ground set. To deal with such situations, several generalizations of submodularity have been proposed. Different from submodular set functions, the diminishing-return-style characterization is not equivalent to submodularity for integer lattice functions [18,20]. It leads to two kinds of submodularities, weak-submodular and DR-submodular, where the latter is stronger than the former.

Let $B = (B_1, B_2, \ldots, B_n) \in \mathbb{Z}_+^n$ be an integer vector and $[B] = \{x \in \mathbb{Z}_+^n : 0 \le x_i \le B_i, \forall 1 \le i \le n\}$ be the set of all smaller non-negative integer vectors. Specially, $[B_i] = \{0, 1, \ldots, B_i\}$ for any integer $B_i \in \mathbb{Z}_+$. Let $x \vee y$ be the vector whose i-th coordinate is $\max\{x_i, y_i\}$ and $x \wedge y$ be the vector whose i-th coordinate is $\min\{x_i, y_i\}$.

Definition 2 (Weak-submodular Integer Lattice Function). *An integer lattice function $f : [B] \to \mathbb{R}$ is said to be weak-submodular, if it satisfies one of the following two conditions:*

(a) $f(x) + f(y) \ge f(x \vee y) + f(x \wedge y)$, for any pair of vectors $x, y \in [B]$;
(b) $f(x + k1_i) - f(x) \ge f(y + k1_i) - f(y)$, for any $i \in \{1, \ldots, n\}$, $k \in \mathbb{Z}_+$ and any pair of $x, y \in [B]$ such that $x \le y$ and $x_i = y_i$, where 1_i is the vector with the ith component equal to 1 and each of the others equals to 0.

Definition 3 (DR-submodular Integer Lattice Function). *An integer lattice function $f : [B] \to \mathbb{R}$ is said to be DR-submodular, if*

$$f(x + k1_i) - f(x) \ge f(y + k1_i) - f(y),$$

for any $i \in \{1, \ldots, n\}$, $k \in \mathbb{Z}_+$ and any pair of $x, y \in [B]$ such that $x \le y$.

Given a non-monotone weak-submodular integer lattice function $f : [B] \to \mathbb{R}_+$, Gottschalk and Peis [14] considered the following maximization problem on a bounded integer lattice (W-MBIL):

$$\max_{x \in [B]} f(x).$$

They presented the following double-greedy algorithm for the problem and showed that its tight approximate guarantee is $\frac{1}{3}$. For a vector $x \in \mathbb{Z}^n$, let (c, x_{-i}) denote a vector that the ith component is set to c while the others remain the same as x.

Lemma 4 below shows that with the progress of Algorithm 3, both of $f(x^{(i)})$ and $f(y^{(i)})$ is non-decreasing.

Lemma 4. *For every $i \in \{1, \ldots, n\}$, we have $f(x^{(i)}) \ge f(x^{(i-1)})$ and $f(y^{(i)}) \ge f(y^{(i-1)})$.*

Algorithm 3. Double Greedy for W-MBIL
___
Input: A non-monotone weak-submodular integer lattice function $f : [B] \to \mathbb{R}_+$.
Output: An approximate solution to $\max_{x \in [B]} f(x)$.
1: Initialize: $x^{(0)} = 0$, $y^{(0)} = B$
2: **for** $i = 1$ to n **do**
3:     $c' \leftarrow \arg\max_{c \in [B_i]} f(c, x_{-i}^{(i-1)}) - f(x^{(i-1)})$; $\delta_{xi} \leftarrow f(c', x_{-i}^{(i-1)}) - f(x^{(i-1)})$
4:     $c'' \leftarrow \arg\max_{c \in [B_i]} f(c, y_{-i}^{(i-1)}) - f(y^{(i-1)})$; $\delta_{yi} \leftarrow f(c'', y_{-i}^{(i-1)}) - f(y^{(i-1)})$
5:     **if** $\delta_{xi} \geq \delta_{yi}$ **then**
6:         $x^{(i)} \leftarrow (c', x_{-i}^{(i-1)})$; $y^{(i)} \leftarrow (c', y_{-i}^{(i-1)})$
7:     **else**
8:         $x^{(i)} \leftarrow (c'', x_{-i}^{(i-1)})$; $y^{(i)} \leftarrow (c'', y_{-i}^{(i-1)})$
9:     **end if**
10: **end for**
11: **return** $x^{(n)}$
___

Proof. Consider the ith loop. Clearly, $x^{(i)}$ coincides with $y^{(i)}$ on the first i components, denoted by $c_1, c_2, \ldots, c_{i-1}$ respectively. That is,

$$x^{(i)} = (c_1, c_2, \ldots, c_{i-1}, c_i, 0, \ldots, 0),$$
$$y^{(i)} = (c_1, c_2, \ldots, c_{i-1}, c_i, B_{i+1}, \ldots, B_n).$$

Case 1. $\delta_{xi} \geq \delta_{yi}$.
It is straightforward see that $c_i = c'$ and $f(x^{(i)}) \geq f(x^{(i-1)})$ holds naturally. Next we consider $f(y^{(i)})$ and $f(y^{(i-1)})$. From the submodularity of f and the fact that $c_i = \arg\max_{c \in [B_i]} f(c, x_{-i}^{(i-1)}) - f(x^{(i-1)})$, we have

$$
\begin{aligned}
f(y^{(i-1)}) - f(y^{(i)}) &= f(B_i, y_{-i}^{(i-1)}) - f(c_i, y_{-i}^{(i-1)}) \\
&\leq f(B_i, x_{-i}^{(i-1)}) - f(c_i, x_{-i}^{(i-1)}) \\
&\leq 0.
\end{aligned}
$$

Case 2. $\delta_{xi} < \delta_{yi}$.
The proof is similar to that of case 1. In this case, $c_i = c''$ and $f(y^{(i)}) \geq f(y^{(i-1)})$ holds naturally. From the submodularity of f and the fact that $c_i = \arg\max_{c \in [B_i]} f(c, y_{-i}^{(i-1)}) - f(y^{(i-1)})$, we have

$$
\begin{aligned}
f(x^{(i)}) - f(x^{(i-1)}) &= f(c_i, x_{-i}^{(i-1)}) - f(0, x_{-i}^{(i-1)}) \\
&\geq f(c_i, y_{-i}^{(i-1)}) - f(0, y_{-i}^{(i-1)}) \\
&\geq 0.
\end{aligned}
$$

This completes the proof of the lemma.

Let OPT be the optimal solution of W-MBIL and $OPT^{(i)} = (OPT \vee x^{(i)}) \wedge y^{(i)}$. Then $x^{(i)} \leq OPT^{(i)} \leq y^{(i)}$ for each $i \in \{0, 1, \ldots, n\}$. Similar to Lemma 2, we have the following result.

Lemma 5. *For every $i \in \{1, \ldots, n\}$, we have*

$$[f(\boldsymbol{x}^{(i)}) - f(\boldsymbol{x}^{(i-1)})] + [f(\boldsymbol{y}^{(i)}) - f(\boldsymbol{y}^{(i-1)})] \geq f(OPT^{(i-1)}) - f(OPT^{(i)}).$$

Proof. For convenience, let O_i be the ith component of OPT. Consider the ith loop. Clearly, the first i components of $OPT^{(i)}$ are the same with $\boldsymbol{x}^{(i)}$ and $\boldsymbol{y}^{(i)}$ while the others are the same with OPT, namely,

$$OPT^{(i)} = (c_1, \ldots, c_{i-1}, c_i, O_{i+1}, \ldots, O_n).$$

It is easy to see that $OPT^{(i)} = (c_i, OPT_{-i}^{(i-1)})$.

Case 1. $\delta_{xi} \geq \delta_{yi}$.
Then $c_i = c'$ in this case.

Case 1.1. $c_i \leq O_i$.
From the submodularity of f, we have

$$\begin{aligned} f(OPT^{(i-1)}) - f(OPT^{(i)}) &= f(O_i, OPT_{-i}^{(i-1)}) - f(c_i, OPT_{-i}^{(i-1)}) \\ &\leq f(O_i, \boldsymbol{x}_{-i}^{(i-1)}) - f(c_i, \boldsymbol{x}_{-i}^{(i-1)}) \\ &\leq 0, \end{aligned}$$

where the second inequality follows from the fact that $c_i = \arg\max_{c \in [B_i]} f(c, \boldsymbol{x}_{-i}^{(i-1)}) - f(\boldsymbol{x}^{(i-1)})$. Combined with Lemma 4, the result follows.

Case 1.2. $c_i > O_i$.
From the submodularity of f, we have

$$\begin{aligned} f(OPT^{(i)}) - f(OPT^{(i-1)}) &= f(c_i, OPT_{-i}^{(i-1)}) - f(O_i, OPT_{-i}^{(i-1)}) \\ &\geq f(c_i, \boldsymbol{y}_{-i}^{(i-1)}) - f(O_i, \boldsymbol{y}_{-i}^{(i-1)}) \\ &= [f(\boldsymbol{y}^{(i)}) - f(\boldsymbol{y}^{(i-1)})] + [f(\boldsymbol{y}^{(i-1)}) - f(O_i, \boldsymbol{y}_{-i}^{(i-1)})] \\ &\geq [f(\boldsymbol{y}^{(i-1)}) - f(O_i, \boldsymbol{y}_{-i}^{(i-1)})] \\ &\geq [f(\boldsymbol{y}^{(i-1)}) - f(c'', \boldsymbol{y}_{-i}^{(i-1)})] \\ &\geq f(\boldsymbol{x}^{(i-1)}) - f(\boldsymbol{x}^{(i)}). \end{aligned}$$

where the second inequality holds from Lemma 4, the third inequality holds from the fact that $c'' = \arg\max_{c \in [B_i]} f(c, \boldsymbol{y}_{-i}^{(i-1)}) - f(\boldsymbol{y}^{(i-1)})$ and the last inequality from the assumption that $\delta_{xi} \geq \delta_{yi}$. Thus $f(\boldsymbol{x}^{(i)}) - f(\boldsymbol{x}^{(i-1)}) \geq f(OPT^{(i-1)}) - f(OPT^{(i)})$. Combined with Lemma 4, the result follows.

Case 2. $\delta_{xi} < \delta_{yi}$.
The proof is similar to that of case 1 and we omit it here.

By Lemma 5 and having a discussion as done in the proof of Theorem 1, one can obtain the following theorem. Gottschalk and Peis [14] also provide an example that shows the analysis is tight.

Theorem 3. *Algorithm 3 has an approximation guarantee of $\frac{1}{3}$ for W-MBIL.*

Soma and Yoshida [21] considered non-monotone DR-submodular integer lattice function maximization (DR-MBIL). They presented a $\frac{1}{2}$-approximation algorithm (Algorithm 4) with a running time of $O(||B||_1\theta+||B||_1)$, where θ is the running time of evaluating f. They also speeded up the algorithm by rounding the values of $f(x_i+1,\boldsymbol{x}_{-i})-f(\boldsymbol{x})$ and $f(y_i-1,\boldsymbol{y}_{-i})-f(\boldsymbol{y})$ to the form $\delta(1+\epsilon)^k$, where δ is the minimum positive marginal gain and k is a positive integer. They showed that the resulting algorithm is a $\frac{1}{2+\epsilon}$ -approximation algorithm with time complexity $O(\frac{n}{\epsilon}\log(\frac{\Delta}{\delta})\log||B||_\infty\theta+||B||_1\log||B||_\infty)$, where Δ is the maximum marginal gain.

Let $a = f(x_i+1,\boldsymbol{x}_{-i})-f(\boldsymbol{x})$ and $b = f(y_i-1,\boldsymbol{y}_{-i})-f(\boldsymbol{y})$. By the submodularity of f, $a+b \geq 0$. Algorithm 4 is described below.

Algorithm 4. Double Greedy for DR-MBIL

Input: A non-monotone DR-submodular integer lattice function $f : [B] \to \mathbb{R}_+$.
Output: An approximate solution to $\max_{\boldsymbol{x}\in[B]} f(\boldsymbol{x})$.
1: Initialize: $\boldsymbol{x} = \boldsymbol{0}$, $\boldsymbol{y} = \boldsymbol{B}$
2: **for** $i = 1$ to n **do**
3:    **while** $x_i < y_i$ **do**
4:       $a \leftarrow f(x_i+1,\boldsymbol{x}_{-i})-f(\boldsymbol{x})$ and $b \leftarrow f(y_i-1,\boldsymbol{y}_{-i})-f(\boldsymbol{y})$
5:       **if** $b < 0$ **then** $x_i \leftarrow x_i + 1$
6:       **else if** $a < 0$ **then** $y_i \leftarrow y_i - 1$
7:       **else** $x_i \leftarrow x_i + 1$ with probability $\frac{a}{a+b}$ and $y_i \leftarrow y_i - 1$ with probability $\frac{b}{a+b}$
8:       (Note: assume that $\frac{a}{a+b} = 1$ if $a = b = 0$)
9:       **end if**
10:   **end while**
11: **end for**
12: **return** x

---

4   Maximizing Submodular Continuous Functions

Submodularity has naturally been considered for functions defined on continuous domains [2]. Similar to the integer lattice settings, there are two kinds of submodularities in continuous domains.

Definition 4 (Weak-Submodular Continuous Function). *A function $f :$ $\mathbb{R}^n \to \mathbb{R}$ is said to be weak-submodular, if it satisfies one of the following two equivalent conditions:*

(a) $f(\boldsymbol{x}) + f(\boldsymbol{y}) \geq f(\boldsymbol{x} \vee \boldsymbol{y}) + f(\boldsymbol{x} \wedge \boldsymbol{y})$, for any pair of vectors $\boldsymbol{x}, \boldsymbol{y} \in \mathbb{R}^n$;
(b) $f(\boldsymbol{x} + k\boldsymbol{1}_i) - f(\boldsymbol{x}) \geq f(\boldsymbol{y} + k\boldsymbol{1}_i) - f(\boldsymbol{y})$, for any $i \in \{1,\dots,n\}$, $k \in \mathbb{R}_+$ and
    $\boldsymbol{x} \leq \boldsymbol{y} \in \mathbb{R}^n$ such that $x_i = y_i$.

Definition 5 (DR-submodular Continuous Function). *A function* f : $\mathbb{R}^n \to \mathbb{R}$ *is said to be DR-submodular, if for any of vectors* $\boldsymbol{x} \le \boldsymbol{y} \in \mathbb{R}^n$, $i \in \{1, \dots, n\}$ *and* $k \in \mathbb{R}_+$ *we have*

$$f(\boldsymbol{x} + k\boldsymbol{1}_i) - f(\boldsymbol{x}) \ge f(\boldsymbol{y} + k\boldsymbol{1}_i) - f(\boldsymbol{y}).$$

Bian et al. [4] generalized the simple non-optimal $\frac{1}{3}$-approximation deterministic double-greedy algorithm of [7] for set functions to DR-submodular continuous functions and obtain a $\frac{1}{3}$-approximation. Niazadeh and Roughgardenthe [19] considered the problem of maximizing a nonnegative and coordinate-wise Lipschitz continuous submodular function over a hypercube $[0, 1]^n$. They provided a randomized double-greedy $\frac{1}{2}$-approximation algorithm for weak-submodular continuous functions. For the special case of DR-submodular maximization, they presented a faster $\frac{1}{2}$-approximation algorithm that runs in almost linear time. Ene et al. [10] studied the problem of maximizing a non-negative differentiable DR-submodular submodular function over a hypercube $[0, 1]^n$. They designed a parallel double continuous greedy $\frac{1}{2}$-approximate algorithm and then offered a discrete parallel version of the algorithm. They showed that the discrete algorithm achieves a $(\frac{1}{2} - \epsilon)$-approximation and uses $O(\frac{1}{\epsilon} \log(\frac{1}{\epsilon}))$ parallel rounds of function evaluations. If f is differentiable, f is DR-submodular if and only if $\nabla f(\boldsymbol{x}) \ge \nabla f(\boldsymbol{y})$ for any pair of vectors $\boldsymbol{x} \le \boldsymbol{y} \in [0, 1]^n$. Let t be a continuous time variable. The parallel double continuous greedy algorithm runs as follows.

Algorithm 5. Parallel Double Continuous Greedy Algorithm

Input: A diferentiable DR-submodular function $f : [0, 1]^n \to \mathbb{R}_+$ and a value oracle for f and its gradient ∇f.

Output: An approximate solution to $\max\limits_{\boldsymbol{x} \in [0,1]^n} f(\boldsymbol{x})$.

1: Initialize: $t \leftarrow 0$, $\boldsymbol{x} \leftarrow \boldsymbol{0}$, $\boldsymbol{y} \leftarrow \boldsymbol{1}$ and $N = [n]$
2: **while** $t < 1$ and $N \ne \emptyset$ **do**
3:     **if** $\nabla_i f(\boldsymbol{x}) \le 0$ **then** $\boldsymbol{y} \leftarrow (x_i, \boldsymbol{y}_{-i})$ and $N \leftarrow N \setminus \{i\}$
4:     **else if** $\nabla_i f(\boldsymbol{y}) \ge 0$ **then** $\boldsymbol{x} \leftarrow (y_i, \boldsymbol{x}_{-i})$ and $N \leftarrow N \setminus \{i\}$
5:     **else if** $\nabla_i f(\boldsymbol{x}) > 0$ and $\nabla_i f(\boldsymbol{y}) < 0$ **then**
6:         $\dot{x}_i(t) = \frac{\nabla_i f(\boldsymbol{x})}{\nabla_i f(\boldsymbol{x}) - \nabla_i f(\boldsymbol{y})}$ and $\dot{y}_i(t) = \frac{\nabla_i f(\boldsymbol{y})}{\nabla_i f(\boldsymbol{x}) - \nabla_i f(\boldsymbol{y})}$
7:     **end if**
8: **end while**
9: **return** $\boldsymbol{x}$

One can see that in the process of the algorithm $\boldsymbol{x}$ and $\boldsymbol{y}$ change with time.

Lemma 6. $\boldsymbol{x} = \boldsymbol{y}$ *at the moment* $t = 1$ *of Algorithm 5.*

Proof. We only need to show that $x_i = y_i$ for each $i \in N$ at the moment $t = 1$. To see this, consider an arbitrary $i \in N$. If there is some moment such that $\nabla_i f(\boldsymbol{x}) \le 0$ or $\nabla_i f(\boldsymbol{y}) \ge 0$, $x_i = y_i$ by the algorithm. Otherwise, the increasing speed of x_i and the decreasing speed of y_i are $\frac{\nabla_i f(\boldsymbol{x})}{\nabla_i f(\boldsymbol{x}) - \nabla_i f(\boldsymbol{y})}$ and $\frac{\nabla_i f(\boldsymbol{y})}{\nabla_i f(\boldsymbol{x}) - \nabla_i f(\boldsymbol{y})}$, respectively. Thus $|\dot{x}_i(t)| + |\dot{y}_i(t)| = 1$. Note that at the moment 0, $x_i = 0$ and $y_i = 1$. The result follows.

Let $\boldsymbol{x}^*$ be the optimal solution of $\max\limits_{\boldsymbol{x} \in [0,1]^n} f(\boldsymbol{x})$ and $\boldsymbol{p} = (\boldsymbol{x}^* \vee \boldsymbol{x}) \wedge \boldsymbol{y}$. By the DR-submodularity of f, we have the following observation.

Lemma 7. *Consider an arbitrary moment of Algorithm 5.*

(1) If there is some i such that $\nabla_i f(\boldsymbol{x}) \leq 0$, set $\boldsymbol{y}' = (x_i, \boldsymbol{y}_{-i})$ and $\boldsymbol{p}' = (\boldsymbol{x}^* \vee \boldsymbol{x}) \wedge \boldsymbol{y}'$. Then

$$f(\boldsymbol{y}') - f(\boldsymbol{y}) \geq 0 \text{ and } f(\boldsymbol{p}') - f(\boldsymbol{p}) \geq 0.$$

(2) If there is some i such that $\nabla_i f(\boldsymbol{y}) \geq 0$, set $\boldsymbol{x}' = (y_i, \boldsymbol{x}_{-i})$ and $\bar{\boldsymbol{p}} = (\boldsymbol{x}^* \vee \boldsymbol{x}') \wedge \boldsymbol{y}$. Then

$$f(\boldsymbol{x}') - f(\boldsymbol{x}) \geq 0 \text{ and } f(\bar{\boldsymbol{p}}) - f(\boldsymbol{p}) \geq 0.$$

(3) If each i satisfies $\nabla_i f(\boldsymbol{x}) > 0$ and $\nabla_i f(\boldsymbol{y}) < 0$,

$$\frac{d[f(\boldsymbol{x}) + f(\boldsymbol{y})]}{dt} + 2\frac{df(\boldsymbol{p})}{dt} \geq 0.$$

Proof. (1) It is easy to see that $\boldsymbol{x} \leq \boldsymbol{p}' \leq \boldsymbol{y}'$. From the DR-submodularity of f and the assumption that $\nabla_i f(\boldsymbol{x}) \leq 0$, we have $\nabla_i f(\boldsymbol{z}) \leq 0$ for any $\boldsymbol{z} \geq \boldsymbol{x}$. Specially, $\nabla_i f(s_i, \boldsymbol{z}_{-i}) \leq 0$ for any $s_i \in [x_i, y_i]$ and for any $\boldsymbol{z}_{-i}$ such that $\boldsymbol{x}_{-i} \leq \boldsymbol{z}_{-i} \leq \boldsymbol{y}_{-i}$. Thus $f(s_i, \boldsymbol{z}_{-i})$ is decreasing in the interval $[x_i, y_i]$ along coordinate i. Since $\boldsymbol{y}' = (x_i, \boldsymbol{y}_{-i})$ and $\boldsymbol{p}' = (x_i, \boldsymbol{p}_{-i})$, we have

$$f(\boldsymbol{y}') - f(\boldsymbol{y}) \geq 0 \text{ and } f(\boldsymbol{p}') - f(\boldsymbol{p}) \geq 0.$$

(2) Take a similar discussion as that in (1), one can conclude the inequalities.

(3) We first analyze $\frac{df(\boldsymbol{p})}{dt}$. One can see that: if $x_i^* \in [0, x_i)$, $\dot{p}_i(t) = \dot{x}_i(t)$; if $x_i^* \in [x_i, y_i]$, $\dot{p}_i(t) = 0$; if $x_i^* \in (y_i, 0]$, $\dot{p}_i(t) = \dot{y}_i(t)$. Thus,

$$\frac{df(\boldsymbol{p})}{dt} = \sum_{i=1}^{n} \nabla_i f(\boldsymbol{p})\dot{p}_i(t)$$

$$= \sum_{i:x_i^* \in [0,x_i)} \nabla_i f(\boldsymbol{p})\dot{x}_i(t) + \sum_{i:x_i^* \in (y_i,0]} \nabla_i f(\boldsymbol{p})\dot{y}_i(t)$$

$$\geq \sum_{i:x_i^* \in [0,x_i)} \nabla_i f(\boldsymbol{y})\dot{x}_i(t) + \sum_{i:x_i^* \in (y_i,0]} \nabla_i f(\boldsymbol{x})\dot{y}_i(t)$$

$$= \sum_{i:x_i^* \in [0,x_i)} \frac{\nabla_i f(\boldsymbol{x})\nabla_i f(\boldsymbol{y})}{\nabla_i f(\boldsymbol{x}) - \nabla_i f(\boldsymbol{y})} + \sum_{i:x_i^* \in (y_i,0]} \frac{\nabla_i f(\boldsymbol{x})\nabla_i f(\boldsymbol{y})}{\nabla_i f(\boldsymbol{x}) - \nabla_i f(\boldsymbol{y})},$$

where the first inequality holds from the submodularity of f, the fact that $\boldsymbol{x} \leq \boldsymbol{p} \leq \boldsymbol{y}$ and the assumption that $\nabla_i f(\boldsymbol{x}) > 0$ and $\nabla_i f(\boldsymbol{y}) < 0$. On the other hand,

$$\frac{\mathrm{d}[f(\boldsymbol{x}) + f(\boldsymbol{y})]}{\mathrm{d}t} = \sum_{i=1}^{n}[\nabla_i f(\boldsymbol{x})\dot{x}_i(t) + \nabla_i f(\boldsymbol{y})\dot{y}_i(t)]$$

$$= \sum_{i=1}^{n} \frac{\nabla_i f(\boldsymbol{x})^2 + \nabla_i f(\boldsymbol{y})^2}{\nabla_i f(\boldsymbol{x}) - \nabla_i f(\boldsymbol{y})}$$

$$\geq \sum_{i:x_i^* \in [0,x_i)} \frac{\nabla_i f(\boldsymbol{x})^2 + \nabla_i f(\boldsymbol{y})^2}{\nabla_i f(\boldsymbol{x}) - \nabla_i f(\boldsymbol{y})} + \sum_{i:x_i^* \in (y_i,0]} \frac{\nabla_i f(\boldsymbol{x})^2 + \nabla_i f(\boldsymbol{y})^2}{\nabla_i f(\boldsymbol{x}) - \nabla_i f(\boldsymbol{y})}.$$

Therefore,

$$\frac{\mathrm{d}[f(\boldsymbol{x}) + f(\boldsymbol{y})]}{\mathrm{d}t} + 2\frac{\mathrm{d}f(\boldsymbol{p})}{\mathrm{d}t}$$

$$\geq \sum_{i:x_i^* \in [0,x_i)} \frac{(\nabla_i f(\boldsymbol{x}) + \nabla_i f(\boldsymbol{y}))^2}{\nabla_i f(\boldsymbol{x}) - \nabla_i f(\boldsymbol{y})} + \sum_{i:x_i^* \in (y_i,0]} \frac{(\nabla_i f(\boldsymbol{x}) + \nabla_i f(\boldsymbol{y}))^2}{\nabla_i f(\boldsymbol{x}) - \nabla_i f(\boldsymbol{y})}$$

$$\geq 0.$$

Theorem 4. *Algorithm 5 returns a solution $\boldsymbol{x}$ with $f(\boldsymbol{x}) \geq \frac{1}{2}f(\boldsymbol{x}^*)$.*

Proof. Suppose that at the moment t of Algorithm 5, $\boldsymbol{x} = \boldsymbol{x}^{(t)}$, $\boldsymbol{y} = \boldsymbol{y}^{(t)}$. Let $\boldsymbol{p}^{(t)} = \boldsymbol{x}^* \vee \boldsymbol{x}^{(t)} \wedge \boldsymbol{y}^{(t)}$. Partition the time interval $[0,1]$ into k small intervals

$$[0,t_1),[t_1,t_2),\ldots,[t_k,1],$$

where t_j $(j = 1,\ldots,k)$ is a moment that Algorithm 5 executes line 3 or 4. Let $\boldsymbol{x}^{(t_j^-)}$ $(\boldsymbol{x}^{(t_j^+)})$ be the state of $\boldsymbol{x}$ exactly before (after) t_j and $\boldsymbol{y}^{(t_j^-)}$ $(\boldsymbol{y}^{(t_j^+)})$ be the state of $\boldsymbol{y}$ exactly before (after) t_j. For convenience, let $t_0 = 0$, $f(\boldsymbol{x}^{(t_0^-)}) = f(\boldsymbol{x}^{(t_0^+)}) = f(\boldsymbol{x}^{(0)})$, and $t_{k+1} = 1$, $f(\boldsymbol{x}^{(t_{k+1}^-)}) = f(\boldsymbol{x}^{(t_{k+1}^+)}) = f(\boldsymbol{x}^{(1)})$. Note that if line 3 or line 4 happens, the value of the i-th coordinate is fixed and the function considered in the sequel has one less variable. Thus in each $[t_{j-1}^+, t_j^-]$ $(j = 1,\ldots,k+1)$ it satisfies $\nabla_i f(\boldsymbol{x}) > 0$ and $\nabla_i f(\boldsymbol{y}) < 0$ for every coordinate i of a variable left. Consider an arbitrary time interval $[t_{j-1}^+, t_j^-]$. From (3) of Lemma 7, for each $j = 1,\ldots,k+1$, we have

$$[f(\boldsymbol{x}^{(t_j^-)}) - f(\boldsymbol{x}^{(t_{j-1}^+)})] + [f(\boldsymbol{y}^{(t_j^-)}) - f(\boldsymbol{y}^{(t_{j-1}^+)})]$$

$$= \int_{t_{j-1}^+}^{t_j^-} \frac{\mathrm{d}[f(\boldsymbol{x}^{(t)}) + f(\boldsymbol{y}^{(t)})]}{\mathrm{d}t}$$

$$\geq -2\int_{t_{j-1}^+}^{t_j^-} \frac{\mathrm{d}f(\boldsymbol{p})}{\mathrm{d}t}$$

$$= 2[f(\boldsymbol{p}^{(t_{j-1}^+)}) - f(\boldsymbol{p}^{(t_j^-)})].$$

Consider an arbitrary moment that Algorithm 5 executes line 3 or 4. From (1) and (2) of Lemma 4, we have

$$[f(\boldsymbol{x}^{(t_j^+)}) - f(\boldsymbol{x}^{(t_j^-)})] + [f(\boldsymbol{y}^{(t_j^+)}) - f(\boldsymbol{y}^{(t_j^-)})] \geq 2[f(\boldsymbol{p}^{(t_j^-)}) - f(\boldsymbol{p}^{(t_j^+)})].$$

Thus,

$$[f(\boldsymbol{x}^{(1)}) - f(\boldsymbol{x}^{(0)})] + [f(\boldsymbol{y}^{(1)}) - f(\boldsymbol{y}^{(0)})] \geq f(\boldsymbol{p}^{(0)}) - f(\boldsymbol{p}^{(1)}).$$

Note that $f(\boldsymbol{x}^{(0)}) \geq 0$, $f(\boldsymbol{y}^{(0)}) \geq 0$, $\boldsymbol{p}^{(0)} = \boldsymbol{x}^*$ and $\boldsymbol{p}^{(1)} = \boldsymbol{x}^{(1)} = \boldsymbol{y}^{(1)}$. The result follows.

Researchers have also studied the trade-off between the number of adaptive sequential rounds of parallel computations (adaptive complexity), the total number of objective function evaluations (query complexity) and the resulting solution quality. Chen et al. [8] considered the USM problem. Using the idea of double-greedy algorithm, they proposed the first algorithm for this problem that achieves a tight $(\frac{1}{2}-\epsilon)$-approximation guarantee using $\tilde{O}(\frac{1}{\epsilon})$ adaptive rounds and a linear number of function evaluations. The algorithm can be extend to the maximization of a non-negative continuous DR-submodular function subject to a box constraint, and achieves a tight $(\frac{1}{2} - \epsilon)$-approximation guarantee for this problem while keeping the same adaptive and query complexities.

References

1. Ahmed, S., Atamtürk, A.: Maximizing a class of submodular utility functions. Math. Program. **128**(1–2), 149–169 (2011)
2. Bach, F.: Submodular functions: from discrete to continuous domains. Math. Program. **175**(1–2), 419–459 (2019)
3. Balcan, M.-F., Harvey, N.-J.-A.: Learning submodular functions. In: Proceedings of the 43rd ACM Symposium on Theory of Computing, pp. 793–802. ACM. San Jose, CA, USA (2011)
4. Bian, A., Mirzasoleiman, B., Buhmann, J., Krause, A.: Guaranteed nonconvex optimization: submodular maximization over continuous domains. In: Proceedings of the 20th International Conference on Artificial Intelligence and Statistics, pp. 111–120. JMLR. Fort Lauderdale, Florida, USA (2017)
5. Buchbinder, N., Feldman, M.: Deterministic algorithms for submodular maximization problems. ACM Trans. Algorithms **14**(3), 1–20 (2018). Article 32
6. Buchbinder, N., Feldman, M., Naor, J.-S., Schwartz, R.: Submodular maximization with cardinality constraints. In: Proceedings of the 25th Annual ACM-SIAM Symposium on Discrete Algorithms, SODA, Oregon, Portland, pp. 1433–1452 (2014)
7. Buchbinder, N., Feldman, M., Seffi, J., Schwartz, R.: A tight linear time (1/2)-approximation for unconstrained submodular maximization. SIAM J. Comput. **44**(5), 1384–1402 (2015)
8. Chen, L., Feldman, M., Karbasi, A.: Unconstrained submodular maximization with constant adaptive complexity. In: Proceedings of the 51st Annual ACM SIGACT Symposium on Theory of Computing, STOC, Phoenix, AZ, USA, pp. 102–113 (2019)
9. Ene, A., Nguyễn, H.-L.: Constrained submodular maximization: beyond $1/e$. In: 2016 IEEE 57th Annual Symposium on Foundations of Computer Science, FOCS, New Brunswick, NJ, USA, pp. 248–258 (2016)
10. Ene, A., Nguyễn, H.-L., Vladu, A.: A parallel double greedy algorithm for submodular maximization (2018). https://arxiv.org/abs/1812.01591

11. Feige, U., Mirrokni, V.-S., Vondrák, J.: Maximizing non-monotone submodular functions. SIAM J. Comput. **40**(4), 1133–1153 (2011)

12. Feldman, M., Naor, J., Schwartz, R.: A unified continuous greedy algorithm for submodular maximization. In: 2011 IEEE 52nd Annual Symposium on Foundations of Computer Science, FOCS, pp. 570–579. Palm Springs, CA, USA (2011)

13. Gharan, S.-O., Vondrák J.: Submodular maximization by simulated annealing. In: Proceedings of the Twenty-Second Annual ACM-SIAM Symposium on Discrete Algorithms, pp. 1098–1117. Society for Industrial and Applied Mathematics, San Francisco, California, USA (2011)

14. Gottschalk, C., Peis, B.: Submodular function maximization on the bounded integer lattice. In: Sanità, L., Skutella, M. (eds.) WAOA 2015. LNCS, vol. 9499, pp. 133–144. Springer, Cham (2015). https://doi.org/10.1007/978-3-319-28684-6_12

15. Hartline, J., Mirrokni, V., Sundararajan, M.: Optimal marketing strategies over social networks. In: Proceedings of the 17th International Conference on World Wide Web, pp. 189–198. ACM. Beijing, China (2008)

16. Hochbaum, D.S.: An efficient algorithm for image segmentation. Markov random fields and related problems. J. ACM **48**(4), 686–701 (2001)

17. Hochbaum, D.S., Hong, S.P.: About strongly polynomial time algorithms for quadratic optimization over submodular constraints. Math. Program. **69**(1), 269–309 (1995)

18. Kapralov, M., Post, I., Vondrák, J.: Online submodular welfare maximization: greedy is optimal. In: Proceedings of the 24th Annual ACM-SIAM Symposium on Discrete Algorithms, pp. 1216–1225. SIAM, New Orleans, Louisiana, USA (2013)

19. Niazadeh, R., Roughgarden, T.: Optimal algorithms for continuous non-monotone submodular and DR-submodular maximization. In: the 32nd Conference on Neural Information Processing Systems, NIPS, Montréal, Canada, pp. 9617–9627 (2018)

20. Soma, T., Yoshida, Y.: A generalization of submodular cover via the diminishing return property on the integer lattice. In: Advances in Neural Information Processing Systems, pp. 847–855 (2015)

21. Soma, T., Yoshida, Y.: Non-monotone DR-submodular function maximization. In: Proceedings of the 31st AAAI Conference on Artificial Intelligence, pp. 898–904. AAAI, San Francisco, California, USA (2017)

Sequential Location Game on Continuous Directional Star Networks

Xujin Chen[1,2], Xiaodong Hu[1,2], and Mengqi Zhang[1,2(✉)]

[1] Academy of Mathematics and Systems Science, Chinese Academy of Sciences,
Beijing 100190, China
{xchen,xdhu,mqzhang}@amss.ac.cn
[2] School of Mathematical Sciences, University of Chinese Academy of Sciences,
Beijing 100049, China

Abstract. We consider a sequential location game on a continuous directional star network, where a finite number of players (facilities) sequentially choose their locations to serve their consumers who are uniformly and continuously distributed in the network. Each consumer patronizes all the closest locations that have been chosen, bringing them equal shares of payoff. In turn, each location distributes the total payoff it receives evenly to every player choosing it. We study hierarchical Stackelberg equilibria (HSE), a.k.a, subgame perfect equilibria of the game, under which every player chooses a location to maximize its payoff. We establish a universal lower bound for payoff to a player under any HSE outcome. The lower bound is then strengthened with better estimations, and some HSE outcomes are explicitly presented, provided that the number of players and the network parameters satisfy certain relations.

Keywords: Sequential location game · Directional network · Hierarchical Stackelberg Equilibrium · Subgame perfect equilibrium

1 Introduction

Competitive location model was first introduced by Hotelling [12]. He considered the competitive location of two facilities in a finite nondirectional interval with uniformly spread consumers. The market (network) of consumers could be nondirectional or directional. The directional constraint that consumers can only move in one direction (instead of the reverse) play an important role in some applications, like TV broadcasting time, air schedules, among other things [4].

Following Prescott and Visscher [2] and Yates [13], we are concerned with the n-player sequential location game in continuous directional markets. In this setting, consumers are uniformly distributed on an directional network and each will buy one unit of goods from the closest facilities. Facilities (also called firms or players) $1, 2, \ldots, n$ enter the directional network (market) sequentially according

Research supported in part by NNSF of China under Grant No. 11531014.

to an exogenously given order, say $1, 2, \ldots, n$, and each will choose a location on the network when knowing the locations of the preceding facilities. Assume that facilities sell a heterogeneous product with a fixed price, i.e., they only compete on location and the payoff of each facility is the number of consumers served by it. We consider sequential location with foresight, where each facility locates to maximize its payoff, given the locations of facilities that have located and the information that all succeeding facilities will try to maximize their payoffs. Facility n chooses an optimal location given the locations selected by the preceding $n - 1$ facilities. Its strategy is a response function on the $n - 1$ known locations. For facility $n - 1$, given the locations selected by the preceding $n - 2$ facilities, it assumes that facility n is a payoff maximizer, i.e., facility n will always make a best response to each of its decisions; so facility $n - 1$ will make decision to maximize payoff considering the decision rule of facility n. Continuing in this way, facility 1 will choose an optimal location considering the decision rule of the succeeding $n - 1$ facilities. Prescott and Visscher [2] presented the optimal solutions for the case of sequential locations of 2 players and the case of 3 players in an unit interval $[0, 1]$, which models the linear market.

The solution of the game we investigate is the hierarchical Stackelberg equilibrium (HSE) [1], which is an extension of Stackelberg equilibrium of two players into n-player sequential game. It is a subgame perfect equilibrium in which every facility knows how to make a best response (i.e., choose an optimal location to maximize its payoff in response) to every location profile of preceding players, if all succeeding facilities will make best response to every location profile they are facing. When every player makes a best response to preceding players, the resulting profile of locations is called an HSE outcome.

Our Results. We study the sequential location game of n players on a continuous directional star network S which consists of s incoming arcs to and t outgoing arcs from a common center. Each player can choose its location to be any endpoint or interior point of an arc in S, under the constraint that no two players are allowed to choose the same interior point. The chosen locations along with endpoints of arcs divide S into a set of internally disjoint closed directional intervals. All the points in each of the intervals have the same set of closest locations; these locations split the length of the interval equally as their market shares obtained from it. In turn, each location will distribute its total market share evenly to all players choosing it as their payoffs. We call S a *normal star* if all of its arcs have equal length, an *out-star* if $s = 0$, and an *in-star* if $t = 0$.

We contribute to the study on sequential location on directional stars by estimating player payoffs and locations in any HSE outcome, as well as constructing typical HSE outcomes.

- We establish a universal lower bound $l_{\min}/(\lfloor \frac{n}{s+t} \rfloor + 1))$ for player payoffs under any HSE outcome, where $l_{\min}$ is the shortest length of the $s + t$ arcs in S. The lower bound is shown to be tight when S is a normal out-star (i.e., $s = 0$) and t does not divide n.

– For in-stars and normal starts with $n \equiv 0 \pmod{s+t}$, we show that under any HSE outcome, the total arc length of S is evenly divided into n shares to be player payoffs.

The equal-payoff results (see Corollary 1 and Theorem 2) generalize the counterpart for a directional interval proved by Yates [13], and show that there is no first-mover advantage or last-mover advantage in the corresponding cases. This differs from to the results for the nondirectional model in Prescott and Visscher [2], where the preceding players might get larger market shares than the succeeding ones.

The most technical part of our work is the proof that in some cases a kind of *weak monotonicity* holds, which means that some succeeding players can always find a location to obtain a market share at least as good as those of some preceding players.

– For normal stars with $n \equiv 1 \pmod{s+1}$ and $t \geq 1$, we show that under any HSE, the payoffs of the first $f := (s+1)\lfloor \frac{n}{s+t} \rfloor + 1$ players must equal to the lower bound $\rho := 1/(\lfloor \frac{n}{s+t} \rfloor + \frac{1}{s+1})$. A common pattern of locations in all HSE outcomes in this case is also presented.
– For normal out-stars with $n \not\equiv 0 \pmod{t}$, we show that under any HSE, the payoffs of the first $f := (n \bmod t)(\lfloor \frac{n}{t} \rfloor + 1)$ players must equal to the lower bound $\rho := 1/(\lfloor \frac{n}{t} \rfloor + 1)$.

Using these properties (see Theorems 4 and 5), we construct HSE outcomes where the first f players adopt a conservative (and actually optimal) strategy to obtain the bottom payoff ρ, while the remaining players obtain an equally larger payoff $1/\lfloor \frac{n}{s+t} \rfloor$.

In deriving the above results, we do not directly use the backward induction procedure as in Prescott and Visscher [2] to find an HSE since it is almost impossible for the sequential location game with multiple players. Our basic method uses the lower bound of the payoff that each player can guarantee to rule out most strategies that cannot lead to an HSE outcome, which simplifies the process of computing an equilibrium outcome.

Related Work. Our work is an extension of the sequential location model introduced by Prescott and Visscher [2], who introduced the sequential location game in a continuous nondirectional linear market. They considered the sequential entry of a fixed number facilities freely and found the optimal locations of for the games with 2 facilities and 3 facilities, respectively. In the early of 1980s, Drezner [6] and Hakimi [11] presented a series of results on the sequential location game in the plane and networks, respectively. The basic setting is that the leader who plays first locates p facilities and the follower who plays second locates q facilities. Most works on sequential location since then are the extensions of this setting. Recently, Gentile *et al.* [9] used integer programming to deal with three sequential location problems under the framework of leader-follower. Reviews of sequential location models can be found in [7, 10, 14, 16].

Location model with directional constraints (called directional location model or one-sided location model) was originally proposed by Cancian [3]. He modeled the television news scheduling as a Hotelling location game [12] with directional constraints and showed that there exists no pure Nash equilibrium. After that, many researchers have developed this model in different aspects. Yates [13] regarded the stock exchange listings as a one-sided location competition and first analyzed the HSE outcomes of the n-player sequential location game on a directional line $[0,1]$, which extends the sequential game in Prescott and Visscher [2] to a multi-player setting. He showed that all players obtain equal payoffs under the equilibrium. Lai [15] and Sun [17] considered discrete location choices and characterized the subgame perfect equilibria of 2-player and 3-player sequential location games in directional linear markets, respectively. Colombo [4,5] considered the spatial price discrimination and spatial Cournot competition in directional Hotelling model.

This paper is organized as follows. In Sect. 2, we introduce our formal model of the sequential location game on directional networks and present formal definitions of best responses and HSE (Definition 1). In Sect. 3, we discuss general properties of HSE for games on directional stars, among which Lemma 1 is the major technical tool for deriving our results. As its quick corollaries, we obtain the universal payoff lower bound $l_{\min}/(\lfloor \frac{n}{s+t} \rfloor + 1))$ and equal-payoff result on in-stars. In Sect. 4, we focus on normal stars whose number of arcs have some special relations with n players. First, we extend the equal-payoff result to the case with $n \equiv 0 \pmod{s+t}$ in Sect. 4.1. Then we prove the aforementioned weak monotonicity for the case with $n \equiv 1 \pmod{s+t}$ and $t \geq 1$ in Sect. 4.2, and for normal out-stars with $n \not\equiv 0 \pmod{t}$ in Sect. 4.3. We conclude the paper in Sect. 5 with remarks on future research.

2   Model

We are concerned with the situation where n facilities enter a directional market sequentially, and each facility chooses a location in the market based on the choices of preceding facilities, under the assumption that the consumers are uniformly distributed on the market and each of them will buy one unit of goods from the closest facility. This situation is modeled as a game $(\mathcal{H}, [n], \pi)$, where $\mathcal{H}$ is a directional network that represents the market, $[n] = \{1, 2, \ldots, n\}$ is the set of facilities that are often called *players*, and $\pi = (\pi_\kappa)_{\kappa=1}^n$ consists of the payoff functions π_κ of players $\kappa \in [n]$.

Directional Network. The directional network $\mathcal{H}$ is spanned by its skeleton, which is a digraph $H = (V, A, \ell)$ with vertex set V, arc (directed edge) set A, and length function $\ell \in \mathbb{R}_{\geq 0}^A$. For any arc in A with tail vertex u and head vertex v, we write it as (u, v), and use $\ell(u, v)$ to represent its *length*. For every real number $r \in [0, 1]$, let (u, v, r) denote the *point* on (u, v) at a *distance* $r \cdot \ell(u, v)$ from u. In particular, $u = (u, v, 0)$ and $v = (u, v, 1)$. For any two points $w_1 = (u, v, r_1)$ and $w_2 = (u, v, r_2)$ on arc (u, v) with $r_1 \leq r_2$, we call $[w_1, w_2] = \{(u, v, r) | r_1 \le r \le r_2\}$ a *directional interval*, whose left limit is w_1, right limit is w_2, and *interior* points are points in $[w_1, w_2] - \{w_1, w_2\}$.[1] The *length* of $[w_1, w_2]$ is $\ell(w_1, w_2) = \ell(u, v) \times (r_2 - r_1)$, which is also considered the *distance* from w_1 to w_2. The *directional network* $\mathcal{H} = \{(u, v, r) | (u, v) \in A, r \in [0, 1]\}$ consists of (all the points in) all directional intervals $[u, v]$ with $(u, v) \in A$, where points (vertices) of V are *skeleton* points and the remaining are *non-skeleton* points. Abusing notation, we identify each arc $(u, v) \in A$ with its corresponding direction interval $[u, v]$. Given two points $p, q \in \mathcal{H}$ that are not contained by the same arc, if there exist distinct points $p_1, p_2, \ldots, p_h \in \mathcal{H}$, where $l \ge 1$, such that $[p, p_1], [p_1, p_2], \ldots, [p_i, p_{i+1}], \ldots, [p_h, q]$ are internally disjoint directional intervals in $\mathcal{H}$, let such $p_1, p_2, \ldots, p_h \in \mathcal{H}$ be taken to minimize the number $\sum_{i=0}^{h} \ell(p_i, p_{i+1})$, where $p_0 = p$ and $p_{l+1} = q$. We consider this minimum number as the *distance* from p to q, and the concatenation of $[p, p_1], [p_1, p_2], \ldots, [p_h, q]$ as a *shortest p-q path* in $\mathcal{H}$. If no such $p_1, p_2, \ldots, p_h \in \mathcal{H}$ exist, the distance from p to q is infinite.

Market Share. The participants of the game $(\mathcal{H}, [n], \pi)$ are facilities (players) and consumers. Players $1, 2, \ldots, n$ enter the directional network $\mathcal{H}$ sequentially in this order, where player κ *precedes* player μ, or equivalently, μ *succeeds* κ if and only if $\kappa < \mu$. Each player $\kappa \in [n]$, knowing the locations of the $\kappa - 1$ preceding players, chooses a point $x_\kappa \in \mathcal{H}$ as its location, where a point is called a *location* only if it has been chosen by some player(s). The following *co-location rule* is applied:

(R1) No two players are allowed to co-locate at the same non-skeleton point of $\mathcal{H}$, while any number of players could choose the same skeleton point of $\mathcal{H}$.

Consumers are uniformly distributed on $\mathcal{H}$ in the sense that the number of consumers on a directional interval is proportional to the interval length, Each consumer will buy one unit of goods from their closest players, i.e., the ones (among all n candidates) at a shortest distance to the player. Our model adopts the same *tie-breaking rule* as that in [8]:

(R2) A consumer with exactly d closest locations contributes a market share of $\frac{1}{d}$ to each of them. If exactly h players choose the same location, then each of them obtains a payoff equal to a fraction $\frac{1}{h}$ of the total market share the location receives.

Formally, given a location profile $x = (x_1, \ldots, x_n)$, for every $\kappa \in [n]$, we say that location x_κ as well as all players choosing x_κ is *favored* by directional interval $[a, b]$ if x_κ is a closest location of all the points (consumers) in $[a, b]$. It is easy to see that the set $\mathcal{I}$ of maximal directional intervals that favor x_κ is unique and finite. In turn, each $I \in \mathcal{I}$ is an internally disjoint union of finitely many directional intervals $J_{I,1}, \ldots, J_{I,\nu(I)}$ such that each $J_{I,i}$, $i \in \nu(I)$ favors a common set $X_{I,i}$ of locations (no points inside $J_{I,i}$ has a closest location

---

[1] In our discussion, all directional intervals are closed.

outside $X_{I,i}$). The tie-breaking rule (R2) yields that every player ζ who chooses $x_\zeta = x_\kappa$ obtains a *payoff*

$$\pi_\zeta(x) = \frac{\sum_{I \in \mathcal{I}} \sum_{i=1}^{\nu(I)} \frac{\ell(J_{,I,i})}{|X_{I,i}|}}{|\{\eta \in [n] | x_\eta = x_\kappa\}|},$$

where the numerator is the market share that location x_κ receives.

Hierarchical Stackelberg Equilibrium. For each $\zeta \in [n]$, the (partial) location profile $(x_1, \ldots, x_\zeta)$ is often abbreviated to $(x_\kappa)_{\kappa=1}^\zeta$ or $x_{[\zeta]}$. In a mild abuse of notation, we also use $x_{[\zeta]}$ to denote the corresponding *multiset* $\{x_1, \ldots, x_\zeta\}$ of locations. For convenience, let $x_{[0]}$ denote the *null* location profile. The *strategy* of player $\kappa \in [n]$ is a function λ_κ which maps every partial location profile $x_{[\kappa-1]}$ to a point $\lambda_\kappa(x_{[\kappa-1]}) \in \mathcal{H}$, and let Λ_κ denote the set of such functions. Every strategy profile $(\lambda_k)_{\kappa=1}^n$ *induces* a location profile $(x_\kappa)_{\kappa=1}^n$ such that $x_\kappa = \lambda_\kappa(x_{[\kappa-1]})$ for every $\kappa \in [n]$. This location profile is often referred to as the *outcome* of $(\lambda_k)_{\kappa=1}^n$.

Naturally, every player κ would choose a location that maximizes its payoff based on the observed choices of the preceding $\kappa - 1$ players and optimal decision rules of the succeeding $n - \kappa$ players. This class of player strategies leads to solutions of the sequential location game, which is an extension of widely-studied Stackelberg equilibria. We call a strategy profile $(\lambda_\kappa^*)_{\kappa=1}^n$ a *hierarchical Stackelberg equilibrium* (HSE) of the game $(\mathcal{H}, [n], \pi)$, if for all $\kappa = n, n-1, \ldots, 1$, given any partial location profile $x_{[\kappa-1]}$, point $\lambda_\kappa^*(x_{[\kappa-1]})$ is an optimal location that player κ can choose to maximize its payoff, provided that all the succeeding players make the best responses. Formally, for any $x_{[\kappa-1]}$, function λ_κ^* is an optimal solution $\lambda_\kappa \in \Lambda_\kappa$ of the following optimization problem:

$$\max_{\lambda_\kappa \in \Lambda_\kappa} \pi_\kappa(x_{[\kappa-1]}, x_\kappa, x_{\kappa+1}^*, \ldots, x_n^*) \tag{2.1}$$

$$\text{s.t. } x_\kappa = \lambda_\kappa(x_{[\kappa-1]})$$
$$x_{\kappa+1}^* = \lambda_{\kappa+1}^*(x_{[\kappa-1]}, x_\kappa)$$
$$x_\mu^* = \lambda_\mu^*(x_{[\kappa-1]}, x_\kappa, x_{\kappa+1}^*, \ldots, x_{\mu-1}^*), \ \mu = \kappa+2, \ldots, n$$

where the optimal functions $\lambda_\mu^*, \mu = \kappa+1, \ldots, n$, has been derived before seeking for λ_κ^*.

Before proceeding, a technical issue has to be addressed. In the process of maximization (2.1), it may happen that function λ_κ wants to take location x_κ to be a point as close to some point p as possible but cannot be p itself. The approaching to p can be assumed to happen along an arc $e \in A$ containing p. Firstly, we define such a location choice x_κ selected by λ_κ as a *pseudo point* p^e on e, whose distance to p is an infinitesimal amount $\varepsilon > 0$, such that

(R3) p^e "touches" p along (the direction of) e, and no location choices made afterwards can be "inserted" between p^e and p.
(R4) The payoffs of players in a location profile which contains some pseudo points are computed by taking limit with $\varepsilon \to 0^+$.

Henceforth, by a single word "point" without any modifier we mean a real or pseudo one, and by a "(partial) location profile" we mean a general one which may contain some pseudo locations (unless otherwise noted). Given a profile $x_{[n]}$ of real or pseudo locations and player $\kappa \in [n]$, we define

$$\tilde{\pi}_\kappa(x) = \lim_{\varepsilon \to 0^+} \pi(x).$$

as the *payoff* of player κ under x. When x contains no pseudo locations, $\tilde{\pi}_k(x)$ is noting but $\pi_\kappa(x)$. Replacing (2.1) with the following more "general" maximization which allows $\lambda_\kappa, \ldots, \lambda_n$ to take pseudo points:

$$\max_{\lambda_\kappa \in \Lambda_\kappa} \tilde{\pi}_\kappa(x_{[\kappa-1]}, x_\kappa, x^*_{\kappa+1}, \ldots, x^*_n) \qquad (2.2)$$

$$\text{s.t. } x_\kappa = \lambda_\kappa(x_{[\kappa-1]})$$
$$x^*_{\kappa+1} = \lambda^*_{\kappa+1}(x_{[\kappa-1]}, x_\kappa)$$
$$x^*_\mu = \lambda^*_\mu(x_{[\kappa-1]}, x_\kappa, x^*_{\kappa+1}, \ldots, x^*_{\mu-1}), \ \mu = \kappa + 2, \ldots, n$$

the optimal (real or pseudo) locations of n players can be obtained by solving a n-level optimization problem sequentially.

Given any $\zeta \in [n]$, any partial profile $x'_{[\zeta-1]} = (x'_1, \ldots, x'_{\zeta-1})$ of real or pseudo locations of players $1, \ldots, \zeta - 1$, and any partial strategy profile $(\lambda_\zeta, \lambda_{\zeta+1}, \ldots, \lambda_n) \in \prod^n_{\kappa=\zeta} \Lambda_\kappa$ of players $\zeta, \zeta+1, \ldots, n$, if partial profile $(x_\zeta, x_{\zeta+1}, \ldots, x_n)$ of real or pseudo locations satisfies $x_\kappa = \lambda_\kappa(x'_1, \ldots, x'_{\zeta-1}, x_\zeta, \ldots, x_{\kappa-1})$ for $\kappa = \zeta, \zeta+1, \ldots, n$, we say that it is *induced* by $x'_{\zeta-1}$ and $(\lambda_\zeta, \lambda_{\zeta+1}, \ldots, \lambda_n)$.

Definition 1 (Best Response, HSE). $(\lambda_\zeta, \lambda_{\zeta+1}, \ldots, \lambda_n)$ *is called a best response to* $x'_{[\zeta-1]}$, *if for every* $\kappa = n, n-1, \ldots, \zeta$ *and any partial profile* $x_{[\kappa-1]}$ *of real or pseudo locations in which* $(x_1, \ldots, x_{\zeta-1}) = x'_{[\zeta-1]}$, *function* λ_κ *is an optimal solution of (2.2) and subject to the optimality,* $\lambda_\kappa(x_{[\kappa-1]})$ *is taken to be a real point (instead of a pseudo one) whenever possible. A strategy profile is called a* HSE *if and only if it is a best response to the null location profile.*

3   Sequential Location on Directional Stars

Figure 1 depicts the location game $(\mathcal{S}, [n], \tilde{\pi})$ in a *directional star market* $\mathcal{S}$ spanned by a directed star $S = (V, A, \ell)$, where $V = \{c_i \mid i \in [s]\} \cup \{o\} \cup \{g_j \mid j \in [t]\}$, and A consists of s incoming arcs (c_i, o), $i \in [s]$ and t outgoing arcs (o, g_j), $j \in [t]$. When $s = 0$ (resp. $t = 0$), we call both S and $\mathcal{S}$ out-stars (resp. in-stars). S and $\mathcal{S}$ are *normal* if $\ell(e) = 1$ for every arc $e \in A$.

For any partial location profile $x_{[\zeta]}$, the points in $x_{[\zeta]} \cup V$ divide $\mathcal{S}$ into a set $\mathcal{I}(x_{[\zeta]})$ of internally disjoint closed directional intervals of positive lengths, whose limits are points in $x_{[\zeta]} \cup V$ and which are internally disjoint from $x_{[\zeta]} \cup V$.

– For every $i \in [s]$, we use $C_{i,\zeta} = x_{[\zeta]} \cap \{(c_i, o, r) \mid r \in [0,1)\}$ to denote the multiset of locations of $x_{[\zeta]}$ in arc (c_i, o) with o excluded. We order the points in $C_{i,\zeta}$ as $c_{i,1}, \ldots, c_{i,|C_{i,\zeta}|}$ in nondecreasing distances to o; if $x_\eta, x_\kappa \in C_{i,\zeta}$ with $\eta < \kappa$ are at the same distance to o, then we put x_η in front of x_κ.

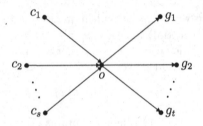

Fig. 1. A directed star S

- For every $j \in [t]$, we use $G_{j,\zeta} = x_{[\zeta]} \cap \{(o, g_j, \mathsf{r}) \mid \mathsf{r} \in (0, 1]\}$ to denote the multiset of locations of $x_{[\zeta]}$ in arc (o, g_j) with o excluded. We order the points in $G_{j,\zeta}$ as $g_{j,1}, \ldots, g_{j,|G_{j,\zeta}|}$ in nondecreasing distances from o; if $x_\eta, x_\kappa \in G_{j,\zeta}$ with $\eta < \kappa$ are at the same distance from o, then we put x_κ in front of x_η.

Let $O_\zeta = \{x \in x_{[\zeta]} \mid x = o\}$ represent the set of players in $[h]$ who choose vertex o as their locations under $x_{[\zeta]}$.

Definition 2 (Happiness). *Given any location profile $x = (x_\kappa)_{\kappa=1}^n$ and real number Π, we say that a player $\kappa \in [n]$ is Π-happy (under x) if either $\tilde{\pi}(x) > \Pi$, or $\tilde{\pi}(x) = \Pi$ and x_κ is a real point in S.*

Lemma 1. *In game $(S, [n], \tilde{\pi})$, let $\tilde{x}_{[f]} = (\tilde{x}_h)_{h=1}^f$ be a fixed partial location profile, where $0 \leq f \leq n - 1$, and let $\Pi > 0$ be a fixed positive number. For every $h \in \{f, f + 1, \ldots, n\}$, let $\mathscr{X}_h$ denote the the set of partial location profiles $x_{[h]}$ with $x_{[f]} = \tilde{x}_{[f]}$. Suppose that for any $\xi \in [n] - [f]$ and any partial location profile $x_{[\xi-1]} \in \mathscr{X}_{\xi-1}$, player ξ can choose some real point $x_\xi^* \in S$ such that one of the following holds:*

(i) $x_\xi^* \neq o$ is the right limit of a unique directional interval in $\mathcal{I}(x_{[\xi-1]}, x_\xi^*)$ whose length equals $\tilde{\pi}_\xi(x_{[\xi-1]}, x_\xi^*) = \Pi$;
(ii) $x_\xi^* = o$, $\tilde{\pi}_\xi(x_{[\xi-1]}, x_\xi^*) \geq \Pi$, and $0 < \ell(c_{i,1}, o) \leq \Pi$ for all $i \in [s]$, where $c_{i,1}$, $i \in [s]$ are defined w.r.t. $x_{[\xi-1]}$, or $\xi = n$ and $\tilde{\pi}_n(x_{[n-1]}, x_n^*) \geq \Pi$.

Given any $\zeta \in [n] - [f]$ and any $\tilde{x}_{[\zeta-1]} \in \mathscr{X}_{\zeta-1}$, let $(\tilde{x}_\zeta, \ldots, \tilde{x}_n)$ be the partial profile induced by $\tilde{x}_{[\zeta-1]}$ and a best response to $\tilde{x}_{[\zeta-1]}$. Then all players in $\{\zeta, \zeta + 1, \ldots, n\}$ are Π-happy under $(\tilde{x}_1, \ldots, \tilde{x}_n)$.

Proof. Suppose that $(\lambda_\zeta, \ldots, \lambda_n)$ is the best response to $\tilde{x}_{[\zeta-1]}$ which together with $\tilde{x}_{[\zeta-1]}$ induces $(\tilde{x}_\zeta, \ldots, \tilde{x}_n)$. By Definition 1 for every $\kappa = [n] - [\zeta - 1]$, the partial strategy profile $(\lambda_\kappa, \ldots, \lambda_n)$ is a best response to all partial location profiles in $\mathscr{X}_{\kappa-1}$.

We apply backward induction on ζ to prove the lemma. The base case where $\zeta = n$ is clear, since by (i) or (ii) player n could choose some real point $\tilde{x}_n^*$ to ensure $\tilde{\pi}_n(\tilde{x}_{[n-1]}, \tilde{x}_n^*) \geq \Pi$. Assuming that the conclusion is true for $n, n-1, \ldots, \zeta+1$, we consider player $\zeta \in [n-1]$. Let $\tilde{x}_\zeta^*$ denote the real point satisfying either (a) $\tilde{x}_\zeta^* \neq o$ is the right limit of a unique directional interval

$[p, \tilde{x}_\zeta^*]$ in $\mathcal{I}(\tilde{x}_{[\zeta-1]}, \tilde{x}_\zeta^*)$ whose length equals $\tilde{\pi}_\zeta(\tilde{x}_{[\zeta-1]}, \tilde{x}_\zeta^*) = \Pi$; or (b) $\tilde{x}_\zeta^* = o$, $0 < \ell(c_{i,1}, o) \leq \Pi$ for all $i \in [s]$, and $\tilde{\pi}_\zeta(\tilde{x}_{[\zeta-1]}, \tilde{x}_\zeta^*) \geq \Pi$, where $c_{i,1}$, $i \in [s]$ are defined w.r.t. $\tilde{x}_{[\zeta-1]}$.

From the induction hypothesis, players $\zeta+1, \ldots, n$ will make a best response $(\lambda_{\zeta+1}, \ldots, \lambda_n)$ to $(\tilde{x}_{[\zeta-1]}, \tilde{x}_\zeta^*)$, inducing partial location profile $(\tilde{x}_{\zeta+1}', \ldots, \tilde{x}_n')$ such that under $(\tilde{x}_{[\zeta-1]}, \tilde{x}_\zeta^*, \tilde{x}_{\zeta+1}', \ldots, \tilde{x}_n')$ each player in $[n] - [\zeta]$ obtains a payoff at least Π and he locates at a real point unless its payoff is larger than Π. Hence, none of $\zeta+1, \ldots, n$ has chosen a point inside $[p, x_\zeta^*] - \{p\}$ in case (a), and none of them has chosen any interior point in $[c_{i,1}, o]$ for any $i \in [s]$ in case (b). So either $\tilde{\pi}_\zeta(\tilde{x}_{[\zeta-1]}, \tilde{x}_\zeta^*, \tilde{x}_{\zeta+1}', \ldots, \tilde{x}_n') = \tilde{\pi}_\zeta(\tilde{x}_{[\zeta-1]}, \tilde{x}_\zeta^*) \geq \Pi$; or $\tilde{x}_\zeta^* = o = \tilde{x}_\eta'$ for some $\eta \geq \zeta+1$, implying

$$\tilde{\pi}_\zeta(\tilde{x}_{[\zeta-1]}, \tilde{x}_\zeta^*, \tilde{x}_{\zeta+1}', \ldots, \tilde{x}_n') = \tilde{\pi}_\eta(\tilde{x}_{[\zeta-1]}, \tilde{x}_\zeta^*, \tilde{x}_{\zeta+1}', \ldots, \tilde{x}_n') \geq \Pi,$$

where the last inequality is guaranteed by the induction hypothesis. Therefore, in any case, choosing real location $\tilde{x}_\zeta^*$ guarantees player ζ to obtain a payoff at least Π provided the succeeding $n - \zeta$ players make a best response to $(\tilde{x}_{[\zeta-1]}, \tilde{x}_\zeta^*)$. It follows from the definition of $(\lambda_\zeta, \ldots, \lambda_n)$ and $(\tilde{x}_\zeta, \ldots, \tilde{x}_n)$ that $\tilde{\pi}_\zeta(\tilde{x}) \geq \Pi$, where $\tilde{x} = (\tilde{x}_1, \ldots, \tilde{x}_n)$, and further that ζ is Π-happy under $\tilde{x}$. Moreover, as $(\lambda_{\zeta+1}, \ldots, \lambda_n)$ is a best response to $(\tilde{x}_1, \ldots, \tilde{x}_\zeta)$, the induction hypothesis guarantees that every $k = \zeta+1, \ldots, n$ is Π-happy under $\tilde{x}$, proving the lemma. $\qquad\square$

Theorem 1. *In game* $(\mathcal{S}, [n], \tilde{\pi})$, *all players are* $\frac{\min_{e \in A} \ell(e)}{\lfloor n/(s+t) \rfloor + 1}$-*happy under every HSE outcome.*

Proof. Given any $\xi \in [n]$ and any partial location profile $x_{[\xi-1]}$, since there is an arc in $\mathcal{S}$ containing at most $\lfloor \frac{n-1}{s+t} \rfloor$ locations in $x_{[\xi-1]}$, we see that this arc contains an directional interval in $\mathcal{I}(x_{[\zeta-1]})$ whose length at least $(\min_{e \in A} \ell(e))/(\lfloor \frac{n-1}{s+t} \rfloor + 1) > (\min_{e \in A} \ell(e))/(\lfloor \frac{n}{s+t} \rfloor + 1)$. So player ξ can find an real interior point x_ξ^* inside this directional interval such that condition (i) in Lemma 1 holds with $\Pi = (\min_{e \in A} \ell(e))/(\lfloor \frac{n}{s+t} \rfloor + 1)$. The result is instant from Lemma 1 by taking $\tilde{x}_{[f]}$ and ζ over there as the null location profile and 1, respectively. $\qquad\square$

For in-stars, the following result shows that players can obtain equal payoff under every HSE, which is a generalization of the result in directional linear markets.

Corollary 1. *Suppose that* $\mathcal{S}$ *is an in-star with* $\Omega = \frac{1}{n} \sum_{i \in [s]} \ell(c_i, o)$. *The following hold for game* $(\mathcal{S}, [n], \tilde{\pi})$.

(i) *If* $x = (x_\kappa)_{\kappa=1}^n$ *is an HSE outcome, then* $\tilde{\pi}_\kappa(x) = \Omega$ *and* x_κ *ia real point for all* $\kappa \in [n]$.

(ii) *If* $n_0 = 0$, $n_i = \sum_{h=1}^i \lfloor \frac{\ell(c_i, o)}{\Omega} \rfloor$ *for* $i \in [s]$, *and for each* $\kappa \leq n_s$, *the positive integer* κ' *is the largest such that* $n_{\kappa'-1} < \kappa$, *then* $(x_\kappa)_{\kappa=1}^n$ *with* $x_\kappa = (c_{k'}, o, (\kappa - n_{\kappa'-1})\Omega)$ *for all* $\kappa \leq n_s$ *and* $x_\kappa = o$ *for all* $\kappa \in [n] - [n_s]$ *is a HSE outcome.*

Proof. Taking $\tilde{x}_{[f]}$ in Lemma 1 as the null location profile with $f = 0$, it is easy to see that condition (i) or (ii) is satisfied with $\Pi = \Omega$. Therefore Lemma 1 (with $\zeta = 1$ over there) implies that κ is Ω-happy under x and particularly $\tilde{\pi}_\kappa(x) \geq \Omega$ for all $\kappa \in [n]$. Now (i) follows from $\sum_{\kappa=1}^n \tilde{\pi}_\kappa(x) \leq \sum_{i \in [s]} \ell(c_i.o)$ and Definition 2. Statement (ii) is clear from (i) and Definition 1.   □

4   Sequential Location in Normal Directional Stars

In this section, we focus on *normal* directional star market $\mathcal{S}$, for which we write the sequential location game $(\mathcal{S}, [n], \tilde{\pi})$ as $\mathfrak{N}$. For the easy case with $n \leq s + t$, the following warmup observation is straightforward.

Observation 1. *Suppose that x is an HSE outcome of game $\mathfrak{N}$.*

(i) *If $n = 1$, then $\tilde{\pi}_1(x) = s + 1$ when $t > 0$, and $\tilde{\pi}_1(x) = s$ when $t = 0$.*

(ii) *If $2 \leq n \leq s - 1$, then $\tilde{\pi}_\kappa(x) = \frac{s}{n}$ for all $\kappa \in [n]$, and $(o, \ldots, o)$ is an HSE outcome.*

(iii) *If $s \leq n \leq s + t$, then $\tilde{\pi}_\kappa(x) = 1$ for all $\kappa \in [n]$, and $(o, \ldots, o, g_1, g_2, \ldots, g_{n-s})$, in which players in $[s]$ chooses o, and player $\kappa \in [n] - [s]$ chooses $g_{\kappa-s}$, is an HSE outcome.*

We next consider $n > s + t$. We divide our discussion into three cases: $n \equiv 0 \pmod{s+t}$ in Sect. 4.1, $n \equiv 1 \pmod{s+t}$ with $t \geq 1$ in Sect. 4.2, and out-stars with $n \not\equiv 0 \pmod{t}$ in Sect. 4.3. For these three cases, we strengthen the general lower bound on payoffs in Theorem 1 to a kind of "characterization" of the HSE outcome.

4.1   Case 1: $n \equiv 0 \pmod{s+t}$

In this subsection, we first prove properties of game $\mathfrak{N}$ with $n = m(s + t)$ (see Lemma 2 and Corollary 2), which helps to find typical HSE outcomes in Theorem 2.

Lemma 2. *In game $\mathfrak{N}$ with $n = m(s + t)$, given any $\zeta \in [n]$ and any partial location profile $x_{[\zeta-1]}$, player ζ can choose some real point $x_\zeta^* \in \mathcal{S}$ such that exactly one of the following holds:*

(i) *$x_\zeta^* \neq o$ is the right limit of a unique directional interval in $\mathcal{I}(x_{[\zeta-1]}, x_\zeta^*)$ whose length equals $\tilde{\pi}_\zeta(x_{[\zeta-1]}, x_\zeta^*) = \frac{1}{m}$;*

(ii) *$x_\zeta^* = o$, $\tilde{\pi}_\zeta(x_{[\zeta-1]}, x_\zeta^*) \geq \frac{1}{m}$, and $0 < \ell(c_{i,1}, o) \leq \frac{1}{m}$ for all $i \in [s] \neq \emptyset$, where $c_{i,1}$, $i \in [s]$ are defined w.r.t. $x_{[\zeta-1]}$.*

Proof. If $\mathcal{I}(x_{[\zeta-1]})$ contains an directional interval $[a, b]$ such that either $\ell(a, b) > \frac{1}{m}$, or $\ell(a, b) = \frac{1}{m}$, $b \neq o$ and $b \notin x_{[\zeta-1]}$, we may take x_ζ^* to be the real point in $[a, b]$ at a distance $\frac{1}{m}$ from a, and $\tilde{\pi}_\zeta(x_{[\zeta-1]}, x_\zeta^*) = \ell(a, x_\zeta^*) = \frac{1}{m}$ implies (i).

So we assume that all directional intervals in $\mathcal{I}(x_{[\zeta-1]})$ have length no more than $\frac{1}{m}$, and their right limits belong to $x_{[\zeta-1]} \cup \{o\}$ whenever their length are

$\frac{1}{m}$. It follows that $\ell(c_{i,1}, o) \in (0, \frac{1}{m}]$ for all $i \in [s]$, and $|G_{j,\zeta-1}| \geq m$ for all $j \in [t]$. These t inequalities along with $(\sum_{i=1}^{s} |C_{i,\zeta-1}|) + |O_{\zeta-1}| + \sum_{j=1}^{t} |G_{j,\zeta-1}| = \zeta - 1 \leq n - 1 = m(s+t) - 1$ enforce

$$|O_{\zeta-1}| + \sum_{i=1}^{s} |C_{i,\zeta-1}| \leq ms - 1. \tag{4.1}$$

Let L denote the total length of directional intervals $[c_{i,1}, o]$, $i \in [s]$. Since every directional interval in $\mathcal{I}(x_{[\zeta-1]})$ has length at most $\frac{1}{m}$, it is easy to see that $\frac{1}{m} \sum_{i=1}^{s} |C_{i,\zeta-1}| + L \geq s$, which along with (4.1) gives $\frac{L}{|O_{\zeta-1}|+1} \geq \frac{1}{m}$. Therefore, choosing $x_\zeta^* = o$ provides $\tilde{\pi}_\zeta(x_{[\zeta-1]}, x_\zeta^*) = \frac{L}{|O_{\zeta-1}|+1} \geq \frac{1}{m}$, implying (ii). $\qquad \square$

The following is an immediate corollary of Lemmas 1 and 2, where we take $f = 0$ (for which the fixed partial location profile is the null one) and $\Pi = \frac{1}{m}$.

Corollary 2. *In game $\mathfrak{N}$ with $n = m(s+t)$, given any $\zeta \in [n]$ and any partial location profile $x_{[\zeta-1]}$, supposed that $(\lambda_\zeta, \ldots, \lambda_n)$ is a best response to $x_{[\zeta-1]}$, and $(x_\zeta, \ldots, x_n)$ is the partial location profile induced by $x_{[\zeta-1]}$ and $(\lambda_\zeta, \ldots, \lambda_n)$. Then all players in $\{\zeta, \zeta+1, \ldots, n\}$ are $\frac{1}{m}$-happy under $(x_1, \ldots, x_n)$.*

For any two positive integers a and b, let $a \oslash b$ denote $(a \mod b)$ if $a \neq 0 \pmod{b}$, and denote b otherwise.

Theorem 2. *If $n = m(s+t)$, then the following hold for game $\mathfrak{N}$.*

(i) *If $x = (x_\kappa)_{\kappa=1}^{n}$ is an HSE outcome, then $\tilde{\pi}_\kappa(x) = \frac{1}{m}$ and x_κ is a real point for all $\kappa \in [n]$.*

(ii) *If $x = (x_\kappa)_{\kappa=1}^{n}$ satisfies $x_\kappa = (c_{\lceil \kappa/m \rceil}, o, \frac{\kappa \oslash m}{m})$ for $\kappa \in [ms]$ and $x_\kappa = (o, g_{\lceil (\kappa-ms)/m \rceil}, \frac{(\kappa-ms) \oslash m}{m})$ for $\kappa \in [n] - [ms]$, then x is an HSE outcome.*

Proof. Since the total length of arcs in $\mathcal{S}$ is $s + t$, Corollary 2 along with Definition 2 implies (i) instantly. By (i) and Definition 1, it is routine to check the correctness of (ii). $\qquad \square$

4.2   Case 2: $n \equiv 1 \pmod{s+t}$ and $t \geq 1$

In this subsection, we assume that $n = m(s+t) + 1$ and $t \geq 1$. Different from the case discussed in Sect. 4.1, the "extra" player makes it impossible to split market share equally, and invalidates the analysis method of the preceding subsection. Fortunately, we still have a tight lower bound

$$\Theta = \frac{s+1}{(s+1)m+1}$$

for the payoff that each player can guarantee himself to obtain. For each $j \in [t]$, we decompose $\mathcal{S}$ into two normal stars: $\mathcal{S}_j$ consisting of $s+1$ arcs (directional intervals) $[c_i, o]$, $i \in [s]$ and $[o, g_j]$; and $\bar{\mathcal{S}}_j$ consisting of the remaining $t-1$ arcs (directional intervals) $[o, g_h]$, $h \in [t] - \{j\}$.

Lemma 3. *In game $\mathfrak{N}$ with $n = m(s+t)+1$ and $t \geq 1$, given any $\zeta \in [n]$ and any partial location profile $x_{[\zeta-1]}$, player ζ can choose some real point $x_\zeta^* \in \mathcal{S}$ such that one of the following holds:*

(i) $x_\zeta^* \neq o$ is the right limit of a unique directional interval in $\mathcal{I}(x_{[\zeta-1]}, x_\zeta^*)$ whose length equals $\tilde{\pi}_\zeta(x_{[\zeta-1]}, x_\zeta^*) = \Theta$;
(ii) $\zeta = n$, and $\tilde{\pi}_n(x_{[n-1]}, x_n^*) \geq \Theta$.

Proof. If $\mathcal{I}(x_{[\zeta-1]})$ contains an directional interval $[a, b]$ such that either $\ell(a, b) > \Theta$, or $\ell(a, b) = \Theta$, $b \neq o$ and $b \notin x_{[\zeta-1]}$, then we may take x_ζ^* to be the real point in $[a, b]$ at a distance Θ from a, giving $\tilde{\pi}_\zeta(x_{[\zeta-1]}, x_\zeta^*) = \ell(a, x_\zeta^*) = \Theta$, as stated in (i).

So we assume that all directional intervals in $\mathcal{I}(x_{[\zeta-1]})$ have length no more than Θ, and their right limits belong to $x_{[\zeta-1]} \cup \{o\}$ whenever their lengths are Θ. It follows that $|C_{i,\zeta-1}| \geq m$ for all $i \in [s]$ and $|G_{j,\zeta-1}| \geq m$ for all $j \in [t]$. These $s+t$ inequalities along with $(\sum_{i=1}^s |C_{i,\zeta-1}|) + |O_{\zeta-1}| + \sum_{j=1}^t |G_{j,\zeta-1}| = \zeta - 1 = m(s+t) - (n-\zeta)$ enforce $|O_{\zeta-1}| \leq \zeta - n$. Therefore, we have $|O_{\zeta-1}| = \zeta - n = 0$, i.e., $\zeta = n$ and $|C_{i,\zeta-1}| = m$ for all $i \in [s]$ and $|G_{j,\zeta-1}| = m$ for all $j \in [t]$. Then we only need to prove (ii) holds.

Renaming the indices $j \in [t]$ if necessary, we may assume that $\ell(o, g_{1,1}) = \min_{j \in [t]} \ell(o, g_{j,1})$. Note that $\mathcal{S}_1$ consists of $(s+1)$ arcs, and contains $(s+1)m$ locations in $x_{[n-1]}$. It follows that $\mathcal{S}_1$ consists of at most $(s+1)(m+1)$ directional intervals from $\mathcal{I}(x_{[n-1]})$. We partition these directional intervals into two sets $\mathcal{I}_1$ and $\mathcal{I}_2$ with $\mathcal{I}_1 = \{[o, g_{1,1}] \cup \{[c_{i,1}, o] \mid i \in [s]\}$ consisting of the $s+1$ directional intervals that contain a common skeleton point o, and $\mathcal{I}_2$ consisting of the remaining at most $m(s+1)$ directional intervals.

For $j = 1$, either $g_{j,|G_{j,n-1}|} = g_j$ which implies $|\mathcal{I}_2| < m(s+1)$, or $g_{j,|G_{j,n-1}|} \neq g_j$, i.e., $g_j \notin x_{[n-1]} \cup \{o\}$, which enforces $\ell(g_{j,|G_{j,n-1}|}, g_j) < \Theta$. Thus $\sum_{I \in \mathcal{I}_2} \ell(I) < \Theta \cdot m(s+1)$. In turn, from $\sum_{I \in \mathcal{I}_1} \ell(I) + \sum_{I \in \mathcal{I}_2} \ell(I) = s+1$ we deduce that $\sum_{I \in \mathcal{I}_1} \ell(I) > \Theta$, and there is an interior point x_n^* in $[o, g_{1,1}]$ such that $L = (\sum_{I \in \mathcal{I}_1 - \{[o,g_{1,1}]\}} \ell(I)) + \ell(o, x_n^*) \geq \Theta$. Since $O_{n-1} = \emptyset$, the minimality of $\ell(o, g_{1,1})$ among all $\ell(o, g_{j,1})$, $j \in [t]$ guarantees that $\tilde{\pi}_n(x_{[n-1]}, x_n^*) = L \geq \Theta$, as stated in (ii). $\qquad \square$

The following result follows instantly from Lemmas 1 and 3, where we take $f = 0$ (for which the fixed partial location profile is the null one) and $\Pi = \Theta$.

Theorem 3. *In game $\mathfrak{N}$ with $n = m(s+t)+1$ and $t \geq 1$, given any $\zeta \in [n]$ and any partial location profile $x_{[\zeta-1]}$, supposed that $(\lambda_\zeta, \ldots, \lambda_n)$ is a best response to $x_{[\zeta-1]}$, and $(x_\zeta, \ldots, x_n)$ is the partial location profile induced by $x_{[\zeta-1]}$ and $(\lambda_\zeta, \ldots, \lambda_n)$. Then all players $\{\zeta, \zeta+1, \ldots, n\}$ are Θ-happy under $(x_1, \ldots, x_n)$.*

Corollary 3. *In game $\mathfrak{N}$ with $n = m(s+t)+1$ and $t \geq 1$, no player locates at the skeleton point o under any HSE outcome.*

Proof. Let $x = x_{[n]}$ be an HSE outcome. In view of Theorem 1, the result is trivial when $s = 0$. Suppose for a contradiction that $s \geq 1$, and the set O_n of players who located at o under x is nonempty. Since, by Theorem 3, $\tilde{\pi}_\kappa(x) \geq \Theta$, for all $k \in [n]$. It follows that $|G_{j,n}| \leq m$ for all $j \in [t]$. Therefore, for every player $\kappa \in O_n$, we have

$$\tilde{\pi}_\kappa(x) \leq \frac{1}{|O_n|} \left(s - (n - mt - |O_n|) \cdot \Theta \right).$$

Substituting $m(s+t)+1$ for n in the above inequality, easy computation give $\tilde{\pi}_\kappa(x) < \Theta$, which is a contradiction to Theorem 3.   □

For convenience, given any direction interval I, we use $\alpha(I)$ and $\beta(I)$ to denote the left and right limits of I, respectively. We next show that certain weak monotonicity holds for HSE.

Theorem 4. *In game $\mathfrak{N}$ with $n = m(s+t)+1$ and $t \geq 1$, let $x = (x_\kappa)_{\kappa=1}^n$ be an HSE outcome. There exists $j^* \in [t]$ such that the following hold with $\vartheta = \frac{1}{(s+1)m+1}$ and $\Theta = \frac{s+1}{(s+1)m+1}$.*

(i) *$|C_{i,n}| = m$, and $c_{i,h} = (c_i, o, (m+1-h)\Theta)$ for $i \in [s]$ and all $h \in [m]$.*
(ii) *$|G_{j^*,n}| = m+1$, and $g_{j^*,h} = (o, g_{j^*}, \vartheta + (h-1)\Theta)$ for all $h \in [m+1]$.*
(iii) *There are exactly $(s+1)m+1$ players who choose locations inside S_{j^*}; the payoff of each of them is Θ.*
(iv) *$|G_{j,n}| = m$ and $\ell(o, g_{j,1}) \geq \Theta$ for all $j \in [t] - \{j^*\}$.*
(v) *$\tilde{\pi}_\kappa(x) = \Theta$ for all $\kappa \in [(s+1)m+1]$.*

Proof. We observe that $\vartheta + m\Theta = 1$, and recall from Theorem 3 that

$$\tilde{\pi}_\kappa(x) \geq \Theta \text{ for every } k \in [n]. \tag{4.2}$$

It follows that

Fact 1. $|C_{i,n}| \leq m$ for all $i \in [s]$ and $|G_{j,n}| \leq m+1$ for all $i \in [t]$.

Let J denote the set of indices $j \in [t]$ such that $\ell(o, g_{j,1}) < \Theta$. Since $n = m(s+t)+1$, there exists $j^* \in [t]$ such that $|G_{j^*,n}| = m+1$. By (4.2), we see that $\ell(o, g_{j^*,1}) \leq 1 - m \cdot \Theta = \vartheta < \Theta$. Hence $j^* \in J$. It is instant from (4.2) that

$$\ell(o, g_{j,1}) = \ell(o, g_{j^*,1}) \leq \vartheta \text{ for all } j \in J.$$

Let μ denote the player locating at $x_\mu = g_{j^*,1}$. Let $(\lambda_\kappa)_{\kappa=1}^n$ be the HSE that induces x. By Definition 1, for every $\zeta \in [n]$, $(\lambda_\kappa)_{\kappa=\zeta}^n$ is a best response to all partial location profiles of players $1, \ldots, \zeta - 1$,

Fact 2. $J = \{j^*\}$.

If $|J| \geq 2$, then $\tilde{\pi}_\mu(x) = \ell(o, g_{j^*,1}) + \frac{1}{|J|} \sum_{i=1}^s \ell(c_{i,1}, o) < \Delta := \ell(o, g_{j^*,1}) + \sum_{i=1}^s \ell(c_{i,1}, o)$. Let $\psi \in \{\mu, \ldots, n\}$ be the *largest* index such that $\tilde{\pi}_\psi(x) < \Delta$. The maximality implies

Claim 1. $\tilde{\pi}(x_\kappa) \geq \Delta$ for all (if any) $\kappa \in \{\psi+1, \ldots, n\}$.

Moreover, we may find an interior point x'_ψ of $[o, g_{j^*,1}]$ (which is very close to $x_\mu = g_{j^*,1}$) such that

Claim 2. $\tilde{\pi}_\psi(x) < \ell(o, x'_\psi) + \sum_{i=1}^{s} \ell(c_{i,1}, o) < \Delta$.

Let $\Psi_1, \ldots, \Psi_f$ be all the directional intervals in $\mathcal{I}(x_{[\psi-1]})$ that (with the left limits excluded) contain some (at least one) location(s) from $\{x_{\psi+1}, \ldots, x_n\}$. Notice that for each $\kappa \in \{\psi + 1, \ldots, n\}$, there is a unique $h \in [f]$ such that either x_κ is an interior point of Ψ_h, or $x_\kappa = \beta(\Psi_h) \neq o$ is a skeleton point, where $x_\kappa \neq o$ is guaranteed by Corollary 3. For each $h \in [f]$, let h' denote the number of locations in $\{x_{\psi+1}, \ldots, x_n\}$ that are contained $\Psi_h - \{\alpha(\Psi_h)\}$. Then

Claim 3. $\sum_{h=1}^{f} h' = n - \psi$.

Since $\ell(o, g_{j^*,1}) \leq \vartheta < \Delta$, we see that $[o, x_\mu] = [o, g_{j^*,1}] \notin \{\Psi_1, \ldots, \Psi_f\}$ contains no locations inside $\{x_{\psi+1}, \ldots, x_n\}$. Suppose w.l.o.g. that x_ψ is contained by one of $\Psi_1 - \{\alpha(\Psi_1)\}, \ldots, \Psi_f - \{\alpha(\Psi_f)\}$ only if $x_\psi \in \Psi_f - \{\alpha(\Psi_f)\}$. Let us examine the intervals in $\mathcal{I}(x_{[\psi-1]}, x'_\psi)$ that (with the left limits excluded) contain some location(s) from $\{x_{\psi+1}, \ldots, x_n\}$. There are $\Psi'_1, \ldots, \Psi'_{f-1}$ such that $\Psi'_h = \Psi_h$ for all $h \in [f-1]$, and either (when $\psi \neq \mu$ or $x_\psi \notin \Psi_f$) $\Psi'_f := \Psi_f$, or (when $\psi = \mu$ and $x_\mu \in \Psi_f$) $\Psi'_f := [x'_\mu, \beta(\Psi_f)]$. In either case, every $\Psi'_h - \{\alpha(\Psi'_h)\}$, $h \in [f]$ contains exactly h' locations from $\{x_{\psi+1}, \ldots, x_n\}$. It is immediate from Claim 1 and Theorem 1 that

Claim 4. For all all $h \in [f]$, either $\ell(\Psi'_h) > h'\Delta$, or $\ell(\Psi'_h) = h'\Delta$ and $\beta(\Psi_h) \notin (x_{[\psi-1]}, x'_\psi)$ is a skeleton point.

For any $\zeta \in [n] - [\psi]$ and any partial location profile $x'_{[\zeta-1]}$ with $x'_{[\psi]} = (x_{[\psi-1]}, x'_\psi)$, by $\sum_{h=1}^{r} h' = n - \psi$ in Claim 3, we deduce from $|\{x'_{\psi+1}, \ldots, x'_{\zeta-1}\}| \leq n - 1 - \psi$ that there exists $d \in [f]$ such that $\Psi'_d - \{\alpha(\Psi'_d)\}$ contains at most $d' - 1$ vertices from $\{x'_{\psi+1}, \ldots, x'_{\zeta-1}\}$. It follows from Claim 4 that

Claim 5. Player ζ can choose some real point $x^*_\zeta \in \Psi'_d$ such that $x^*_\zeta \neq o$ is the right limit of a unique directional interval in $\mathcal{I}(x'_{[\zeta-1]}, x^*_\zeta)$ whose length equals $\tilde{\pi}_\zeta(x'_{[\zeta-1]}, x^*_\zeta) = \Delta$.

Let partial location profile $(x'_{\psi+1}, \ldots, x'_n)$ be induced by $(x_{[\psi-1]}, x'_\psi)$ and its best response $(\lambda_\kappa)_{\kappa=\psi+1}^{n}$. It is instant from Claim 5 and Lemma 1 that $\tilde{\pi}_\kappa(x_{[\psi-1]}, x'_\psi, x'_{\psi+1}, \ldots, x'_n) \geq \Delta$ for every $\kappa = \psi + 1, \ldots, n$. Recalling Claim 2, $\ell(o, x'_\psi) + \sum_{i=1}^{s} \ell(c_{i,1}, o) < \Delta$ implies that none of $x'_\psi, x'_{\psi+1}, \ldots, x'_n$ is inside any of the directional interval (o, x'_ψ) and $(c_{i,1}, o)$, $i \in [s]$. It follows that $\tilde{\pi}_\psi(x_{[\psi-1]}, x'_\psi, x'_{\psi+1}, \ldots, x'_n) = \tilde{\pi}(x_{[\psi-1]}, x'_\psi) = \ell(o, x'_\psi) + \sum_{i=1}^{s} \ell(c_{i,1}, o) > \tilde{\pi}_\psi(x)$, a contradiction to the hypothesis that $(\lambda_\kappa)_{\kappa=1}^{n}$ is an HSE, which justifies Fact 2.

Recall from Corollary 3 that $O_n = \emptyset$. Facts 1 and 2 along with $n = (s+t)m+1$ enforces

Fact 3. $|C_{i,n}| = m$ for all $i \in [s]$, and $|G_{j,n}| = m$ for all $j \in [t] - \{j^*\}$.

It follows from (4.2) and Fact 3 that $\ell(c_{i,1}, o) \leq 1 - |C_{i,n}| \cdot \Theta = \vartheta$ for all $i \in [s]$, and therefore $\frac{s}{(s+1)m+1} = \sum_{i=1}^{s} \ell(c_{i,1}, o) = \tilde{\pi}_\mu(x) - \ell(o, g_{j^*,1}) \geq \Theta - \ell(o, g_{j^*,1}) \geq \frac{s}{(s+1)m+1}$ enforces that

$$\ell(c_{i,1}, o) = \vartheta = \ell(o, g_{j^*,1}) \text{ for all } i \in [s],$$

and the payoff of each of the ms players in $\cup_{i \in [s]} C_{i,n}$ is Θ. Again, from (4.2) we deduce that the payoff of each of the $m + 1$ players in $G_{j^*,n}$ touches the lower bound Θ. Note that the total length $s + 1$ of arcs in $\mathcal{S}_{j^*}$ has to be distributed without any loss, g_{j^*} must be chosen by a unique player under x. This gives

Fact 4. Under x, exactly $(s + 1)m + 1$ players locate inside $\mathcal{S}_{j^*}$; each of them obtains a payoff Θ; one of them locates at the skeleton point g_{j^*}.

Combining Facts 1–4, we have proved (i)–(iv). To prove (v), we suppose on the contrary that some player $\varsigma \in [(s+1)m + 1]$ obtains a payoff $\tilde{\pi}_\varsigma(x) > \Theta$. It follows from Fact 4 that there exists a *largest* $\phi \in [n] - [(s+1)m + 1]$ such that $x_\phi \in \mathcal{S}_{j^*}$. Furthermore, notice from Fact 4 that exactly $(t-1)m$ players locate inside the out-star $\bar{\mathcal{S}}_{j^*}$ of $t - 1$ arcs. By the maximality of ϕ, it is clear that the locations in $x_{\phi+1}, \ldots, x_n$ are exactly those inside $\bar{\mathcal{S}}_{j^*}$ and outside $x_{[\phi]}$. Let $x_{[\phi]}|_{\bar{\mathcal{S}}_{j^*}}$ denote the partial location profile, which is the restriction of $x_{[\phi]}$ to $\bar{\mathcal{S}}_{j^*}$. Applying Lemma 2 to game $(\bar{\mathcal{S}}_{j^*}, \{\kappa \in [n] \mid x_\kappa \in \bar{\mathcal{S}}_{j^*}\}, \tilde{\pi})$, we deduce that for any $\zeta \in \{\phi+1, \ldots, n\}$ and any partial location profile $(x_{[\phi]}|_{\bar{\mathcal{S}}_{j^*}}, z_{\phi+1}, \ldots, z_{\zeta-1}) \subseteq \bar{\mathcal{S}}_{j^*}$, player ζ can choose some real point $z_\zeta^* \in \bar{\mathcal{S}}_{j^*}$ such that $z_\zeta^* \neq o$ is the right limit of a unique directional interval in $\mathcal{I}(x_{[\phi]}|_{\bar{\mathcal{S}}_{j^*}}, z_{\phi+1}, \ldots, z_{\zeta-1}, z_\zeta^*)$ whose length equals $\tilde{\pi}_\zeta(x_{[\phi]}|_{\bar{\mathcal{S}}_{j^*}}, z_{\phi+1}, \ldots, z_{\zeta-1}, z_\zeta^*) = \frac{1}{m}$. Now, Lemma 1 applies, giving

Claim 6. $\tilde{\pi}_\kappa(x) \geq \frac{1}{m} > \Theta$ for all $\kappa \in [n] - [\phi]$.

We conduct an analysis similar to that in the proof of Fact 2. Let $\Phi_1, \ldots, \Phi_{f-1}, [a, x_\varsigma]$ be all the directional intervals in $\mathcal{I}(x_{[\phi-1]})$. Let h' with $h \in [f]$ (resp. f') denote the number of players $\{\phi + 1, \ldots, n\}$ whose locations are inside $\Phi_h - \{\alpha(\Phi_h)\}$ (resp. $[a, x_\varsigma] - \{a\}$). It is straight forward that

Claim 7. $\sum_{h=1}^{f} h' = n - \phi$, and (by Claim 6 and Fact 4) for all $h \in [f - 1]$, either $\ell(\Phi_h) > \frac{h'}{m}$, or $\ell(\Phi_h) = \frac{h'}{m}$ and $\beta(\Phi_h) \notin x_{[\phi]} \cup \{o\}$.

Clearly by Claim 6, $\ell(a, x_\varsigma) \geq \tilde{\pi}_\varsigma(x) + \frac{f'}{m} > \Theta + \frac{f'}{m}$. Hence we can take an interior point x'_ϕ of $[a, x_\varsigma]$ such that

Claim 8. $\frac{1}{m} > \ell(a, x'_\phi) > \Theta$, and $\ell(\Phi_f) \geq \frac{f'}{m}$, where $\Phi_f := [x'_\phi, x_\varsigma]$.

For any $\zeta \in [n] - [\phi]$ and any partial location profile $x'_{[\zeta-1]}$ with $x'_{[\phi]} = (x_{[\phi-1]}, x'_\phi)$, by Claim 7, we see that $|\{x'_{\phi+1}, \ldots, x'_{\zeta-1}\} \cap (\Phi_d - \{\alpha(\Phi_d)\})| < d'$ for some $d \in [f]$. Now, combining the second statement of Claim 7 and $\ell(\Phi_f) > \frac{f'}{m}$ in Claim 8, we deduce that player ζ can choose some real point $x_\zeta^* \in \Phi_d$ such that $x_\zeta^* \neq o$ is the right limit of a unique directional interval in $\mathcal{I}(x'_{[\zeta-1]}, x_\zeta^*)$ whose length equals $\tilde{\pi}_\zeta(x'_{[\zeta-1]}, x_\zeta^*) = \frac{1}{m}$.

Let partial location profile $(x'_{\phi+1}, \ldots, x'_n)$ be induced by $(x_{[\phi-1]}, x'_\phi)$ and best response $(\lambda_\kappa)^n_{\kappa=\phi+1}$. It is from Lemma 1 that $\tilde{\pi}_\kappa(x_{[\phi-1]}, x'_\phi, x'_{\phi+1}, \ldots, x'_n) \geq \frac{1}{m}$ for every $\kappa = \phi + 1, \ldots, n$. Recalling Claim 8, $\ell(a, x'_\phi) < \frac{1}{m}$ implies that none of $x'_\phi, x'_{\phi+1}, \ldots, x'_n$ is inside the directional interval $[a, x'_\phi]$. It follows that $\tilde{\pi}_\phi(x_{[\phi-1]}, x'_\phi, x'_{\phi+1}, \ldots, x'_n) = \ell(a, x'_\phi) > \Theta$, which is a contradiction to the hypothesis that $(\lambda_\kappa)^n_{\kappa=1}$ is an HSE. The contradiction establishes (v), completing the proof.   □

According to Theorem 4 we can construct an HSE outcome of sequential location game with $m(s + t) + 1$ players.

Corollary 4. *If $n = m(s+t)+1$ and $t \geq 1$, then game $\mathfrak{N}$ has an HSE outcome $x = (x_\kappa)^n_{\kappa=1}$ with*

$$x_\kappa = (c_{\lceil \kappa/m \rceil}, o, (\kappa \oslash m)\Theta) \text{ for } \kappa \in [sm],$$
$$x_\kappa = (o, g_1, \vartheta + (\kappa - s - 1)\Theta) \text{ for } \kappa = [(s+1)m + 1] - [sm], \text{ and}$$
$$x_\kappa = (o, g_{\lceil(\kappa-(s+1)m-1)/m\rceil}, \tfrac{(\kappa-(s+1)m-1)\oslash m}{m}) \text{ for } \kappa = [n] - [(s+1)m + 1]$$

such that $\tilde{\pi}_\kappa(x) = \Theta$ for $\kappa \in [(s+1)m + 1]$, and $\tilde{\pi}_\kappa(x) = \frac{1}{m}$ for $\kappa \in [n] - [(s+1)m + 1]$.

Proof. By Theorem 4, the first $(s+1)m+1$ players have already made their best choices. In response to these choices, the last $(t - 1)m$ players are actually play a game on the out-star consisting of $(t - 1)$ arcs (o, g_j), $j = 2, \ldots, t$. It follows from Theorem 2 that x is indeed an HSE outcome.   □

4.3   Case 3: Out-Stars

Different from the game on in-stars (recalling Corollary 1), when players play the game on a normal out-star, under the HSE their payoffs are not necessarily equal, as seen from the following theorem.

Theorem 5. *Suppose that $\mathcal{S}$ is a normal out-star. The following hold for game $(\mathcal{S}, [n], \tilde{\pi})$ with $m = \lfloor \frac{n}{t} \rfloor$ and $1 \leq r = n \pmod{t}$.*

(i) *If $x = (x_\kappa)^n_{\kappa=1}$ is an HSE outcome, then $\tilde{\pi}_\kappa(x) = \frac{1}{m+1}$ for all $\kappa \in [r(m+1)]$ and $\tilde{\pi}_\kappa(x) \geq \frac{1}{m+1}$ for all $\kappa \in [n] - [r(m+1)]$*

(ii) *If $x = (x_\kappa)^n_{i=1}$ satisfies $x_\kappa = (o, g_{\lceil \kappa/(m+1)\rceil}, \frac{\kappa \oslash(m+1)}{m+1})$ for all $\kappa \in [r(m+1)]$ and $x_\kappa = (c_{\lceil(\kappa-r(m+1))/m\rceil}, \frac{(\kappa-r(m+1))\oslash m}{m})$ for all $\kappa \in [n] - [r(m+1)]$, then x is an HSE outcome such that $\tilde{\pi}_\kappa(x) = \frac{1}{m+1}$ for all $\kappa \leq r(m + 1)$ and $\tilde{\pi}_\kappa(x) = \frac{1}{m}$ for all $\kappa \in [n] - [r(m+1)]$.*

Proof. Theorem 1 has guaranteed that $\tilde{\pi}_\kappa(x) \geq \frac{1}{m+1}$ for all $\kappa \in [n]$. It follows that under x at most $(t - r)m = n - r(m + 1)$ players can obtain a payoff larger then $\frac{1}{m+1}$. Suppose that x is the outcome of HSE $(\lambda_\kappa)^n_{\kappa=1}$. To prove (i), we suppose on the contrary that some player $\varsigma \in [r(m + 1)]$ obtains a payoff $\tilde{\pi}_\varsigma(x) > \frac{1}{m+1}$. Then we may take a largest $\phi \in [n] - [r(m + 1)]$ such that $\tilde{\pi}_\phi(x) \leq \frac{1}{m+1}$, and a point x'_ϕ the directional interval $[a, x_\varsigma] \in \mathcal{I}(x_{[\phi-1]})$ such that

Claim 9. $\tilde{\pi}_\kappa(x) > \frac{1}{m+1}$ for all $\kappa \in [n] - [\phi]$, and $\tilde{\pi}_\phi(x) < \ell(a, x'_\phi) < \min_{\kappa=\phi+1}^{n} \tilde{\pi}_\kappa(x)$.

By arguments similar to (and easier than) those for deriving Claims 7 and 8 in the proof of Theorem 4, we deduce that for any $\zeta \in [n] - [\phi]$ and any partial location profile $x'_{[\zeta-1]}$ with $x'_{[\phi]} = (x_{[\phi-1]}, x'_\phi)$, player ζ can choose some real point x^*_ζ such that $x^*_\zeta \neq o$ is the right limit of a unique directional interval in $\mathcal{I}(x'_{[\zeta-1]}, x^*_\zeta)$ whose length equals $\tilde{\pi}_\zeta(x'_{[\zeta-1]}, x^*_\zeta) = \ell(a, x'_\phi)$. Let partial location profile $(x'_{\phi+1}, \ldots, x'_n)$ be induced by $(x_{[\phi-1]}, x'_\phi)$ and its best response $(\lambda_\kappa)_{\kappa=\phi+1}^{n}$. It is instant from Lemma 1 that $\tilde{\pi}_\kappa(x_{[\phi-1]}, x'_\phi, x'_{\phi+1}, \ldots, x'_n) \geq \ell(a, x'_\phi)$ for every $\kappa = \phi+1, \ldots, n$. Thus none of $x'_\phi, x'_{\phi+1}, \ldots, x'_n$ is inside the directional interval $[a, x'_\phi]$. It follows that $\tilde{\pi}_\phi(x_{[\phi-1]}, x'_\phi, x'_{\phi+1}, \ldots, x'_n) = \ell(a, x'_\phi) > \tilde{\pi}_\phi(x)$, where the last inequality is given by Claim 9, The contradiction to the hypothesis that $(\lambda_\kappa)_{\kappa=1}^{n}$ is an HSE establishes (i).

Considering the location profile x given in (ii), by (i), it is easy to check that every player in $r(m+1)$ has already made its best choices. Given these choices, the game is reduced to a $(t-r)m$-player game on the out-star consisting of $t-r$ arcs (o, g_j), $j \in [t] - [r]$. It follows from Theorem 2 that x is indeed an HSE outcome.   □

5   Conclusion

In this paper, we study the properties of HSE for the sequential location game on directional star networks. Due to the complexity of analyzing the full strategies of each player, we focus on discussing the existence and characterizations of the HSE outcomes. While the solution for the cases under investigation might be somewhat satisfactory, many interesting questions remain open. As seen from Corollary 1 and Theorem 2, in-stars and normal stars with $n \equiv 0 \pmod{s+t}$ each guarantees that all HSE outcomes consist of real locations. We suspect that this is true for all star networks without any additional condition. Moreover, our method might be extended to deal with normal stars for the cases with $n \equiv r \pmod{s+t}$ for $r \geq 2$. The game on non-normal stars also deserves research efforts.

References

1. Anderson, S., Engers, M.: Stackelberg versus cournot oligopoly equilibrium. Int. J. Ind. Organ. **10**, 127–135 (1992)
2. Prescott, E.C., Visscher, M.: Sequential location among firms with foresight. Bell J. Econ. **8**, 378–393 (1977)
3. Cancian, M., Bills, A., Bergstrom, T.: Hotelling location problems with directional constraints: an application to television news scheduling. J. Ind. Econ. **43**, 121–124 (1995)
4. Colombo, S.: Spatial price discrimination in the unidirectional hotelling model with elastic demand. J. Econ. **102**(2), 157–169 (2011)

5. Colombo, S.: Spatial cournot competition with non-extreme directional constraints. Ann. Reg. Sci. **51**, 761–774 (2013)
6. Drezner, Z.: Competitive location strategies for two facilities. Reg. Sci. Urban Econ. **12**, 485–493 (1982)
7. Eiselt, H., Marianov, V.: Foundations of Location Analysis, vol. 155. Springer, New York (2011). https://doi.org/10.1007/978-1-4419-7572-0
8. Fournier, G., Scarsini, M.: Hotelling games on networks: existence and efficiency of equilibria. Math. Oper. Res. **44**, 212–235 (2016)
9. Gentile, J., Pessoa, A., Poss, M., Costa Roboredo, M.: Integer programming formulations for three sequential discrete competitive location problems with foresight. Eur. J. Oper. Res. **265**, 872–881 (2017)
10. Gorji, M.: Competitive location: a state-of-art review. Int. J. Ind. Eng. Comput. **6**, 1–18 (2015)
11. Hakimi, S.: On locating new facilities in a competitive environment. Eur. J. Oper. Res. **12**, 29–35 (1983)
12. Hotelling, H.: Stability in competition. Econ. J. **39**(153), 41–57 (1929)
13. Yates, A.J.: Hotelling and the new york stock exchange. Econ. Lett. **56**, 107–110 (1997)
14. Kress, D.: Sequential Competitive Location on Networks (2013)
15. Lai, F.C.: Sequential locations in directional markets. Reg. Sci. Urban Econ. **31**, 535–546 (2001)
16. San Martin, G., Cordera, R., Alonso, B.: Spatial Interaction Models (2017)
17. Sun, C.H.: Sequential location in a discrete directional market with three or more players. Ann. Reg. Sci. **48**, 101–122 (2012)

Core Decomposition, Maintenance and Applications

Feiteng Zhang, Bin Liu$^{(\boxtimes)}$ (iD), and Qizhi Fang

School of Mathematical Sciences, Ocean University of China, Qingdao 266100,
Shandong, People's Republic of China
`binliu@ouc.edu.cn`

Abstract. Structures of large graphs have attracted much attention in recent years, including k-clique, k-core, k-truss, k-club, to name just a few. These structures can help detect the most cohesive or most influential subgraphs of social networks and other massive graphs. In this survey, we summarize the research on k-core, which is the maximal connected subgraph of a graph and the degree for each vertex is equal to or greater than k. We will address the core decomposition problem, the core maintenance problem, and a few applications of k-core.

Keywords: K-core · Core decomposition · Core maintenance

1 Introduction

In many fields, relationships between entities is ubiquity and graph is a suitable model to depict entities and their relationships. For example, in a telecommunication record, a vertex represents a person and an edge between two vertices represents that the two persons have communicated. In a data science model, a graph may consist of millions of vertices and edges, and may evolve over time. Detecting and analyzing the structures of large graphs have therefore become important.

Capturing the structures, such as k-core, k-clique, k-truss, k-club, and cohesive communities of graphs, has attracted much attention. Such structures have been used widely to find densely connected regions in a graph, analyze topological structures of the internet, and identify the most influential spreaders, among other things. A k-core of a graph plays a significant role in analyzing networks. Determining all k-cores in a static graph, which is called the *core decomposition* problem, can be solved in linear time [2]. In this survey, we summarize some of the major contributions to the *core decomposition* problem and

This research was supported in part by the National Natural Science Foundation of China (11971447, 11871442), the Natural Science Foundation of Shandong Province of China (ZR2017QA010), and the Fundamental Research Funds for the Central Universities (201964006).

D.-Z. Du and J. Wang (Eds.): Ko Festschrift, LNCS 12000, pp. 205–218, 2020.
https://doi.org/10.1007/978-3-030-41672-0_12

the *core maintenance* problem. We also address a few applications of these two problems.

The rest of this paper is organized as follows: In Sect. 2, we present a few basic definitions. In Sect. 3 we describe algorithms for solving the core decomposition problem, including a linear-time algorithm [2], distributed algorithms [6,7,10,20], external-memory algorithms [4], semi-external model [14,15], k-core on uncertain graphs [11,12], and structure detecting algorithms for adding some constrains to k-core (such as adding a radius or dual graph). We then present a few core maintenance algorithms in Sect. 4, including streaming algorithms [16–18], distributed algorithms [19], parallel algorithms [22,23], and order-based algorithms [21]. In Sect. 5, we provide a few applications of the core decomposition problem and the core maintenance problem.

2   Basic Definitions and a Problem Statement

The definition of k-core was first presented in 1983 by Seidman [1], which has played a significant role in describing structures of social networks. For an undirected graph $G = (V, E)$, where V is the set of vertices and E is the set of edges, the degree of a vertex $u \in V$ is the number of incident edges of u in G, denoted by $d_G(u)$. We define $\delta(G) = \min\{d_G(u) : u \in V\}$. Next we introduce some definitions and properties.

Definition 1. *Let H be a connected subgraph of $G = (V, E)$. If $\delta(H) \geq k$, where k is a non-negative integer, then H is called a seed k-core of G. Furthermore, if H satisfies the maximality, i.e., there is no other seed k-core H' contains H, then H is called a k-core of G.*

Suppose that H is a k-core that contains a vertex u. Then H is the unique k-core that contains u, denoted by H_k^u. Otherwise, there must have another k-core Q containing u. Since $H \cup Q$ is also a k-core containing u and $H \cup Q \supset H$, it contradicts to the maximality of H. On the other hand, if u has a k-core H_k^u with $k \geq 1$, then a $(k-1)$-core H_{k-1}^u must exist and $H_{k-1}^u \supseteq H_k^u$.

Definition 2. *For a vertex u in $G = (V, E)$, the K value (core number) of u, denoted by $K(u)$, is the largest k, such that there exists a k-core containing u. The max-k-core of u in G, denoted by H^u, is the k-core with $k = K(u)$.*

An example is presented in Fig. 1, illustrating k-cores of a graph. The vertex u is contained in a 1-core and a 2-core. The max-k-core of u is a 2-core and $K(u) = 2$.

Problem Statement. For a graph $G = (V, E)$, without lose generality, it is assumed that G is connected. We want to calculate the K value of every vertex in V, which is equivalent to finding all k-cores with different values of k in G. This problem is known as the *core decomposition* problem. An ideal $O(m)$-time algorithm was proposed in 2003 [2], where $m = |E|$. Built on this initial success, a number of other algorithms have been devised, including distributed algorithms

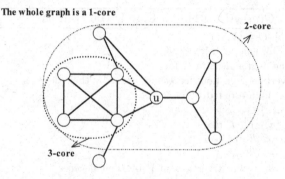

Fig. 1. An example of k-core

[6,7,10,20], external-memory algorithms [4], semi-external model [14,15], and k-core on uncertain graphs [11,12].

The core decomposition problem is formulated on a static graph. However, a lot of graphs in applications are evolving over time, inserting or removing edges or vertices. To obtain k-core structures of the new graphs, it needs to update the K values, which is known as the *core maintenance* problem. Although the $O(m)$ algorithm can be used to the new graph to update the new K values, it will incur too much time. Since only a small portion of vertices would need their K values updated, detecting the subgraph that contains all vertices whose K values need to be updated can help improve the efficiency of this process. This has motivated the work on designing streaming algorithms [16–18], distributed algorithms [19], parallel algorithms [22,23] and order-based algorithms [21].

3 Core Decomposition

3.1 A Linear-Time Algorithm

In 2003, Batagelj and Zaversnik [2] presented a linear-time algorithm for the core decomposition problem. This is the first algorithm for the core decomposition problem to reach be linear time. For a connected graph $G = (V, E)$, if $\delta(G) \geq k$, then G itself is a k-core. Based on the degrees of vertices, all k-cores can be detected. In fact, if we recursively delete all vertices whose degrees are less than k, then the remaining graph is a combination of some k-cores.

Algorithm 1 is the algorithm presented in [2]. The vertices need to be ordered in a nondecreasing order by their degrees after computing them. Since all degrees of vertices are bounded integers, bucket sort can be used to order them and the time complexity is $O(n)$, where $n = |V|$. Testing from vertex u_1 of the smallest degree whether $K(u_1) = d(u_1)$ and subtract 1 from $d(w)$ if $(u_1, w) \in E$ and $d(w) > d(u_1)$. Do the same operation on the remaining vertices recursively to compute the K values of all vertices. In the worst case, we traverse all edges at most once, so the time complexity is $O(m)$. Based on this approach, researchers

---

Algorithm 1. $O(m)$ Algorithm for Core Decomposition

Require: Graph $G = (V, E)$
Ensure: $K(u)$, for each $u \in V$
  1: Compute the degree $d(u)$ in G for each $u \in V$
  2: Order the vertices in a non-decreasing order by their degrees
  3: **for** each $u \in V$ in order **do**
  4:     $K(u) \leftarrow d(u)$
  5:     **for** each $(u, w) \in E$ **do**
  6:         **if** $K(u) < d(w)$ **then**
  7:             $d(w) \leftarrow d(w) - 1$
  8:             Reorder the rest vertices in V by their degrees
  9:         **end if**
 10:     **end for**
 11: **end for**
 12: return $K(u)$

---

have devised algorithms for solving the core decomposition problem and the core maintenance problem.

A *vertex property function* is $p(v, C)$ with real values, where $v \in V$ and $C \subseteq V$. We say $p(v, C)$ is monotone, if for each C_1, C_2 with $C_1 \subseteq C_2$, $p(v, C_1) \leq p(v, C_2)$. Listed below are examples of vertex property functions [3]:

$$
\begin{aligned}
p_1(v, C) &= \deg(v, C), \\
p_2(v, C) &= \mathrm{indeg}(v, C), \\
p_3(v, C) &= \mathrm{outdeg}(v, C), \\
p_4(v, C) &= \mathrm{indeg}(v, C) + \mathrm{outdeg}(v, C).
\end{aligned}
\tag{1}
$$

Batagelj et al. [3] generalized of the notion of core using vertex property functions. Algorithms for the cores of their generalization are similar to determining k-cores by degrees in Algorithm 1, testing from the vertex with smallest $p(v, C)$. They showed that if the vertex property function is monotone, the time complexity of determining the corresponding cores is $O(m \cdot max(\triangle, \log n))$, where $\triangle$ is the maximum degree, $m = |E|$, and $n = |V|$.

3.2 External-Memory Algorithms and I/O Efficient Algorithms

The $O(m)$-time algorithm [2] assumes that the entire graph can be loaded in the main memory and for random access. Thus, this algorithm is not suitable for a very large graph encountered in practice that exceeds the capacity of the underlying memory of a computer. An *External-Memory algorithm* for the core decomposition problem was designed [4] to take care of large graphs efficiently. Furthermore, for the networks that can be kept in the main memory, the external-memory algorithm can obtain comparable results as in-memory algorithms. Compared to the traditional *buttom-up* algorithm, the external-memory algorithm uses a novel *top-down* approach which detects k-cores recursively for

k values from large to small. By removing the vertices in the k-cores we have detected, the I/O cost and search space can be reduced. The entire external-memory algorithm is divided into three parts: (1) Divide the whole graph into several subgraphs so that each subgraph can be loaded in the main memory and an efficient partition algorithm can be devised to scan the graph G only once. (2) Estimate the upper bound on K values of vertices in each subgraph and refine it progressively. (3) Use the *top-down* core decomposition algorithm recursively to determine the K value of every vertex in G. If the graph cannot be stored in the main memory, the algorithm needs $O(k_{max})$ scans of the graph, where $k_{max} = max\{K(u) : u \in G\}$. As a result, the external-memory algorithm determines the K values of all vertices in $O(k_{max}(m+n))$ time, with $O(\frac{k_{max}(m+n)}{B})$ I/O space and $O(\frac{m+n}{B})$ disk block space in the worst case, where $m = |E|$ and $n = |V|$.

To design an I/O efficient core decomposition algorithm, Wen et al. presented a *Semi-External* model in [14] and [15], which only stores the information of vertices to the main memory and the information of edges on the disk of the underlying computer. The semi-external algorithm stores K values in the main memory and updates K values iteratively on the edges that are scanned. Their algorithms can also be used to the core maintenance problem for edge removing. In particular, they first devised I/O efficient core maintenance algorithms for edge inserting, degeneracy order computation and maintenance algorithms, respectively, under the semi-external model. As a result, the space complexity, the time complexity, and the I/O complexity of the I/O efficient core decomposition algorithm are, respectively, $O(n)$, $O(l(m+n))$ and $O(\frac{l(m+n)}{B})$, where l is the number of iterations and is often small in practice.

3.3   Distributed Algorithms

A distributed algorithm is desirable when a graph is too large to store in a single host or the description of the graph is inherently distributed in multiple hosts. Montresor et al. [6] propose distributed core decomposition algorithms to solve it. They considered two computation models:

- *One-to-one* model. One computational unit, namely one host, is associated with one vertex in the graph. Thus the information can be diffused directly through edges between two nodes.
- *One-to-many* model. One computational unit is associated with a set of vertices in the graph. Information diffuses between vertex sets in this model, which is suitable to the situation for the graph inherently distributed in multiple hosts.

At the beginning, use the degrees of vertices as the upper bounds of their K values. Hosts diffuse their information to their neighbors, then neighbors update their upper bounds of K values recursively until the upper bounds of K cannot be updated. The final upper bounds are the K values of vertices and can be reached with at most $O(n)$ iterations, where $n = |V|$. Distributed algorithms are implemented on GraphChi and Webgraph, and can be extended to large datasets on a single PC [8].

Mandal et al. [10] proposed a distributed core decomposition algorithm, called *Spark-kCore*, to run on a Spark cluster computing platform. Using a think-like-a-vertex paradigm and a message passing paradigm for the core decomposition problem, Spark-kCore algorithms can reach the target with reduced I/O cost.

3.4   Core Decomposition on Uncertain Graphs

Definition 3. *A graph $G = (V, E, p)$ is an uncertain graph or a probabilistic graph, where V is the set of vertices, E is the set of edges, and each $(u, v) \in E$ is assigned a probability $p(u, v)$, $p \in (0, 1]$.*

Uncertain graphs have arisen in many fields. For instance, a vertex represents gene and edge represents interactions among genes. Since the interactions among them are derived through noisy and error operation in experiments, edges are existing in a probability. To solve the problem that can the core decomposition problem of uncertain graphs be solved by an efficient approach, Bonchi et al. [11] propose some algorithms for this problem. They introduce the definition of (k, η)-*core* $H = (C, E|C, p)$. H is a maximal subgraph of $G = (V, E, p)$ satisfying $\delta(H) \geq k$ and every vertex in H has probability no less than η, i.e., $Pr[d_H(v) \geq k] \geq \eta$, where $v \in C$ and $\eta \in [0, 1]$ is a threshold representing the level of certainty of the cores. The algorithms are like the $O(m)$-time algorithm for the core decomposition problem on deterministic graphs [2]: computing the initial η-degrees by a novel efficient dynamic-programming approach, removing the vertex with smallest η-degree and updating η-degrees recursively. However, it may have exponential time complexity when computes and updates η-degrees. Bonchi et al. devise an efficient dynamic-programming method to overcome it. As a result, the complexity of computation (k, η)-core is $O(m\triangle)$, where $m = |E|$ and $\triangle$ is the maximum η-degree.

Peng et al. [12] proposed a different probabilistic k-core model on uncertain graphs, named (k, θ)-*core* where θ is a probability threshold, basing on the well-known possible world semantics. In a fundamental uncertain graph $G = (V, E, p)$, where $p \in (0, 1]$ and $G' = (V, E')$ is a deterministic subgraph with probability

$$Pr(G') = \prod_{e \in E'} p(e) \prod_{e \in E \setminus E'} (1 - p(e)). \tag{2}$$

Then they defined

$$p(u) = \sum_{G' \in G} p(G') I_{G'}(k, u) \tag{3}$$

as the probability of u contained in a k-core of G, where $I_{G'}(k, u)$ is an indicator function. If u is contained in a k-core of G', $I_{G'}(k, u) = 1$, otherwise $I_{G'}(k, u) = 0$. They [12] showed that solve the problem of finding all u with $p(u) \geq \theta$, i.e., finding a (k, θ)-core of G is NP-hard. They proposed a sampling-based method to find a (k, θ)-core, and used pruning techniques to reduce the candidate size and a novel membership check algorithm to speed up the computation.

3.5   Core Decomposition Under Additional Constrains

We discuss two constrains. One is adding a attribute to a graph and detecting the maximal seed k-core in which any two vertices are linked by a relationship [26,27], the other is finding k-connected cores in large dual networks [25].

In the real world, some graphs are given attributes, as explained in [27]. The authors of [27] introduced (k, r)-*core* H which is a maximal subgraph such that $\delta(H) \geq k$ and any two vertices in H should satisfy an attribute about a threshold r. Finding a maximal (k, r)-core and a maximum (k, r)-core are both NP-hard. In [26], Wang et al. added a spatial constrain and asked a new question of finding a maximal subgraph H with $\delta(H) \geq k$ in a radius-bounded area. Then they explored three algorithms to find a (k, r)-core where k is the minimum degree of the core and r is the radius. These algorithms are triple-vertex-based paradigm, binary-vertex-paradigm, and a paradigm based on rotating circles. Finding radius-bounded (k, r)-core can be solved in polynomial time.

A dual graph contains a physical graph and a conceptual graph with the same vertices. Yue et al. [25] formulated a k-*connected core* (k-CCO) model, which is a k-core in the conceptual graph and is connected in the physical graph. For a fixed k, they designed a polynomial-time peeling-style algorithm to detect all k-$CCOs$ in a dual graph in $O(hm)$ time, where m is the number of edges in the conceptual graph and h is a value bounded by the number of vertices of the dual graph. Then they designed *bottom-up* and *top-down* algorithms to detect maximum k-$CCOs$ and a binary search algorithm to speedup these algorithms. Finally, they designed an index structure to detect a k-CCO containing a set of query vertices. Basing on the index structure, they presented an efficient query-processing algorithm and a polynomial-time index construction algorithm. The size of index is bounded by $O(n)$ and the time complexity of their query-processing algorithm is $O(m)$.

4   Core Maintenance

How to find the k-cores and determine K values of all vertices in dynamic graphs that evolve over time is an important problem. This is the core maintenance problem of graphs. Let V^* denote the set of vertices whose K values will change if the graph $G = (V, E)$ changes. The main idea of the core maintenance problem is to find a subgraph $H \subseteq G$ (or a subset V' of V) which contains V^*, prune the vertices with unchanged K values in H, and update the K values for the vertices in V^*.

4.1   Streaming Algorithms

Sarıyüce et al. [16,17] presented three streaming algorithms, called *Subcore Algorithm*, *Purecore Algorithm*, *Traversal Algorithm*, to solve the core maintenance problem. Under the assumption that only one edge is inserted to or removed from a graph $G = (V, E)$ each time, they showed that $|K(w) - K'(w)| \leq$ 1 for each vertex $w \in V$, where $K'(w)$ is the new K value of w after inserting or removing. Furthermore, if there is a vertex whose K value changes after inserting or removing an edge $e = (u, v)$ with $K(u) \leq K(v)$, then $K(u)$ must change and we say that u is the root r. In the insertion case, if a vertex w has $K'(w) - K(w) = 1$, then w is connecting to r via a path in which all the vertices have $K = K(r)$ and $K' = K(r) + 1$ in $G + e$. In the deletion case, if a vertex w has $K(w) - K'(w) = 1$, then w is connecting to r via a path in which all the vertices have $K = K(r)$ and $K' = K(r) - 1$ in G. Using these properties, they showed how to find a set of vertices, *subcore* of r, that contains V^*. Next, they used *current degree* (*cd*) of each vertex as a criterion to judge whether a vertex in the subcore will have its K value changed. Define the *cd* value of a vertex w, denoted by $cd(w)$, as the number of adjacent vertices w' satisfying $K(w') \geq K(w)$. If a vertex w has $cd(w) \leq K(r)$ in $G + e$ in the insertion case, then $K(w)$ will not change and we delete w from the subcore of r. Since w cannot help w' to have a high K value, then $cd(w')$ should decrease by 1, where $(w, w') \in E + e$, $cd(w') > cd(w)$, and w' on the subcore of r. Do the same operation recursively until all remaining vertices on the subcore have $cd > K(r)$ and the set of remaining vertices is V^*. If a vertex w has $cd(w) < K(r)$ in $G - e$ in the deletion, then $K(w)$ will change. Updating the *cd* values of the remaining vertices by the same method in the insertion case recursively until all the remaining vertices on subcore have $cd \geq K(r)$, we can get the set of vertices which have been deleted is V^* in the deletion case. This is the subcore algorithm, and the time complexity and the space complexity are both $O(m)$, where $m = |E|$.

Then, they defined three constrains: the MCD value, the PCD value, and the RCD value of a vertex, which are used to judge whether the vertex will have its K value increased. In other words, if a vertex w will have its K value increased, then the MCD value, the PCD value, and the RCD value are both greater than $K(w)$. Define the MCD value of a vertex w, denoted by $MCD(w)$, as the number of adjacent vertices with $K \geq k(w)$. By using the MCD value as a constrain, the subcore can be downsized to the *purecore* based on which they devised a *Purecore Algorithm*. The process of the purecore algorithm is the same as the subcore algorithm except V' is smaller and purecore algorithm only can be used to the insertion case. The time complexity and the space complexity of the purecore algorithm are both $O(m)$. On the base of knowing MCD values, we can get a smaller V' by the PCD values of vertices. And the PCD value of a vertex w, denoted by $PCD(w)$, is defined as the number of adjacent vertices w' with $K(w') > K(w)$ or $(K(w') = K(w)$ and $MCD(w') > K(w))$. Next, the *Traversal Algorithm* is proposed, which uses the depth-first-search to find a smaller subgraph than the purecore algorithm and the subcore algorithm. Because the algorithm will not continue to search when searches a vertex which cannot have its K value changed. The time complexity and the space complexity are $O(m)$ and $O(n)$ respectively, where $n = |V|$. The RCD value of a vertex w, specifically denoted by $RCD(w, n)$, $n \geq 0$, is a generalization of the MCD value and the PCD value, where $RCD(w, 1) = MCD(w)$ and $RCD(w, 2) = PCD(w)$. Define $RCD(w, n)$, $n \geq 0$, as the number of adjacent vertices w' with

$K(w') > K(w)$ or $(K(w') = K(w)$ and $RCD(w', n - 1) > K(w))$. Thus, using the $RCD(w, n)$ values as a constrain, we can get a smaller V' and use more calculation with the increasing of n.

In [18], Li et al. found the same properties that the vertices whose K values will increased have as that in [16,17], and designed efficient core maintenance algorithms independently. They used $Color, RecolorInsert, UpdateInsert$ to find V', prune vertices, and update K values, respectively. In particular, they presented two pruning techniques to find a smaller V'.

4.2 Distributed Algorithms

Since graphs are growing too fast to disposed on a single server, Aksu et al. presented distributed algorithms for the core maintenance problem [19]. The idea is executing against the partitioned graph in parallel and taking advantage of k-core properties to reduce unnecessary computation. In particular, they defined $G = (V, E, M[V, E], C[V, E])$ and $N_G^k(v) = |\{w|(w, v) \in E, d_G(w) \geq k\}|$, where V is the set of vertices, E is the set of edges, $M[V, E]$ is the structured metadata, and $C[V, E]$ is the unstructured context. First, they developed a distributed k-core construction algorithm by the method that prune the vertices with $N_G^k < k$ recursively. Second, they developed a new k-core maintenance algorithm which intents to update the previous k-core for a certain k after the change of a graph. They used a pruning technique to limit the scope of k-core update after insertion or deletion. For a graph G, there is a k-core $G_k = (V_k, E_k)$ in it. Inserting an edge $e = (u, v)$, if both $u, v \in V_k$, do not change G_k; if u or v or both are not in V_k, then the subgraph consisting of vertices in $\{w|w \in V, d_G(w) \geq k, N_G^k(w) \geq k\}$, where every vertex in it is reachable from u or v, may need to be updated to include additional vertices into G_k. Removing an edge $e = (u, v)$, if (u, v) is not in E_k, do not change G_k; if $(u, v) \in E_k$, then the subgraph consisting of $\{w|w \in V, d_G(w) \geq k, N_G^k(w) \geq k\}$, where every vertex is reachable from u or v, may need to be updated to remove additional vertices from G_k. In the end, they proposed a batch k-core maintenance, which accumulates data updates and refreshes k-core in a batch bundles up expensive graphs traversals and can speed up the time of updating, compared to maintaining each update incrementally.

4.3 Parallel Algorithms

Previous work mainly focuses on inserting or removing one edge or vertex at a time. In [22] and [23], the authors presented parallel algorithms for the core maintenance problem when multiple edges or vertices are inserted or removed. The parallel algorithms can make a better use of make use of computation power and avoid extra overhead when inserting or removing one at a time.

Considering the case of inserting multiple edges, Wang et al. [22] discovered a structure named *superior edge set* that can update K values in parallel. Given a graph $G = (V, E)$, an edge $e = (u, v) \in E$ is a *superior edge* of u if $K(u) \leq K(v)$; define $K(e) = K(u)$. Define the k-*superior edge set* as $E_k = \{e_1, e_2, \dots e_p\}$ such that $K(e_i) = k$ and each vertex in V is incident with at most one superior edge. Then the *superior edge set*

$$\varepsilon_q = E_{k_1} \cup E_{k_2} \cup \ldots \cup E_{k_q} \tag{4}$$

is a set of edges made up of several k-superior edge sets with different k values. Insert ε_q, then $K'(w) - K(w) \leq 1$ for each $w \in V$. Next, the set of vertices whose K values will increase is determined by inserting a superior edge set. Define the *Superior Degree* of a vertex u, $u \in V$, as follows:

$$SD(u) = |\{w|(u, w) \in E \cup \varepsilon_q, K(w) \geq K(u)\}|. \tag{5}$$

Define the *Constraint Superior Degree* of a vertex u by

$$CSD(u) = |\{w|(u, w) \in E \cup \varepsilon_q, K(w) > K(u) \text{ or } K(w) = K(u) \wedge SD(w) > K(u)\}|. \tag{6}$$

For a vertex w, if w satisfies these conditions: (1) $CSD(w) > K(w)$; (2) w is connected to a vertex u with $K(u) = K(w)$ by a path whose vertices have $K = K(u)$ in $G + \varepsilon_q$; (3) a superior edge of u is contained in ε_q, then w may have its K value increased. (The K values of u and w are those in G.)

Jin et al. [23] developed the parallel algorithms in [22]. They showed that if the inserted or removed edges constitute a matching, the core number update with respect to each inserted or removed edge can be handled in parallel. Meanwhile, they added parallel core maintenance algorithms for the deletion case. If a matching is inserted to or removed from graph $G = (V, E)$, then $|K'(w) - K(w)| \leq 1$ for each $w \in V$. Since the parallel algorithms can operate a matching other than one edge in an iteration, the number of iterations needed is only $\triangle_c + 1$ where $\triangle_c$ is maximum number of inserted or removed edges connecting to a vertex, which will substantially reduce the time cost over inserting or removing just one edge at a time.

4.4   Order-Based Algorithms

The $O(m)$ algorithm for the core decomposition problem [2] will produce a vertices' sequence when it removes vertices recursively to determine K values. This vertices' sequence is a k-order. Based on the k-order, Zhang et al. [21] proposed a novel order-based approach for the core maintenance problem for both insertion and deletion. Meanwhile, they pointed the drawbacks of the existing traversal algorithms in [17], which need a large search space to find V^* and high overhead to maintain PCD and MCD.

For any pair of vertices u, v in a graph $G = (V, E)$, let $u \preceq v$ denote $K(u) < K(v)$ or $K(u) = K(v)$ but u is removed before v in the $O(m)$ algorithm for the core decomposition problem. In other words, u is in front of v in a k-order. We can summarize that a k-order, $(v_1, v_2, \ldots, v_n)$, for each vertex in graph G, has transitivity, i.e., if $v_h \preceq v_j$ and $v_j \preceq v_i$, then $v_h \preceq v_i$. After inserting or removing an edge, the k-order need to be updated and the new k-order is also a removing sequence produced from the graph after the changing by using the core decomposition algorithm in [2]. Consider inserting or removing one edge at a time, the K values of all vertices change at most 1. Define the *remaining degree* of a vertex u in G, denoted by $deg^+(u)$,:

$$deg^+(u) = |\{v|(u,v) \in E, u \preceq v\}|. \tag{7}$$

Let O_k denotes the sequence of vertices in k-order with $K = k$, and we get a sequence $O_0 O_1 O_2 \cdots$, where $O_i \preceq O_j$ if $i < j$. Then the order $(\preceq)$ on $O_0 O_1 O_2 \cdots$ is a k-order iff $deg^+(u) \leq k$ for every vertex u in O_k for each k. Insert (u,v), $u \in O_k$ and $u \preceq v$, if a vertex $w \in O_l$, and $l > k$ or $l < k$, then w is not in V^*; if $w \in O_k$ but $w \preceq u$, then w is not in V^*; if $w \in O_k$, $u \preceq w$ and there is a path $w_0, w_1, w_2, \cdots, w_t$ such that $w_0 = u$, $w_t = w$, $(w_i, w_{i+1}) \in E$ and $w_i \preceq w_{i+1}$ for $0 \leq i < t$, then w may in V^*. They used the similar idea in the traversal algorithm in [17] to remove an edge, but they used the maintaining k-order method instead of using the PCD values. Define the *candidate degree* of a vertex w in O_k, denoted by $deg^*(w)$, as follows

$$deg^*(w) = |\{w'|(w,w') \in E, w' \preceq w \wedge w' \text{is a potential candidate of } V^*\}|. \tag{8}$$

Then $deg^*(w) + deg^+(w)$ is a criterion to judge whether a vertex w will be in V^*, where $w \in O_k$. Specifically, if $deg^*(w) + deg^+(w) \leq k$, then w is not in V^*. Otherwise, w is a potential vertex in V^*. Finally, they designed OrderInsert and OrderRmoval algorithms for edge inserting and removing respectively.

5 Applications

K-core is a critical structure of graphs. It is used to depict the properties (e.g., cohesiveness, centrality, sustainability), and has been applied to a variety of fields: including detecting communities, analyzing the structures of large networks, finding the most influential subgraphs or vertices, helping to find other structures (k-clique, etc.), large-scale networks fingerprinting and visualization, dealing with problems in bioinformatics, and analyzing software bugs, to name a few.

Seidman et al. [1] introduced the notion of k-core to measure network cohesion and density, which is the first application of k-core. For the resilience of core, randomly deleting edges or vertices can destroy the core resilience, and then destroy the graph resilience [28]. Identifying the most influential spreaders is a significant issue in understanding the dynamics of information diffusion in large scale networks. Bae et al. [30] proposed a novel measure *coreness centrality* to estimate the spreading influence of a node in a network by using its k-shell indices. Rossi et al. [31] further refined the nodes by k-truss, which have a stronger influential ability locating in cores. To find the most influential part of a graph, Li et al. [29] introduced a novel community model called k-*influential community* based on k-core to capture the influence of a community. When it comes to finding a certain community, Papadopoulos et al. [34] used the method for the core decomposition problem to detect communities in large networks of social media. Nasir et al. [35] used the methods of the core decomposition problem and the core maintenance problem to find the *top-k* densest subgraphs in networks. Alduaiji et al. [32] used k-core to detect communities to find the subgraphs and their impact on users' behavior in twitter communities, then they find that community members intend to share positive tweets than negative increasing over time.

In recently years, k-core has been widely used in the fields of software engineering, and bioinformatics. Qu et al. [36] used the core decomposition problem on class dependency networks to analyze software bugs. Cheng et al. [37] proposed a method to find cluster subgraphs made of k-core and r-clique, which is used to gene networks. Ma and Balasundaram [38] focused on a change-constrained version of the minimum spanning k-core problem under probabilistic edge failures. This can help telecommunication networks design, airline networks configuration and freight distribution planning. Alvarez-Hamelin et al. [39] proposed a general visualization algorithm to compare different networks using the method of the core decomposition problem, and a visualization tool to find specific structures of a network.

References

1. Seidman, S.B.: Network structure and minimum degree. Soc. Netw. **5**(3), 269–287 (1983)
2. Batagelj, V., Zaversnik, M.: An O(m) algorithm for cores decomposition of networks. In: The Computing Research Repository (CoRR). arXiv:cs.DS/0310049 (2003)
3. Batagelj, V., Zaveršnik, M.: Fast algorithms for determining (generalized) core groups in social networks. Adv. Data Anal. Classif. **5**(2), 129–145 (2011)
4. Cheng, J., Ke, Y., Chu, S., Özsu, M.T.: Efficient core decomposition in massive networks. In: 27th International Conference on Data Engineering (ICDE), pp. 51–62. IEEE, Hannover (2011)
5. Garas, A., Schweitzer, F., Havlin, S.: A k-shell decomposition method for weighted networks. New J. Phys. **14**(8), 083030 (2012)
6. Montresor, A., De Pellegrini, F., Miorandi, D.: Distributed k-core decomposition. Trans. Parallel Distrib. Syst. **24**(2), 288–300 (2012)
7. Jakma, P., Orczyk, M., Perkins, C.S., Fayed, M.: Distributed k-core decomposition of dynamic graphs. In: Proceedings of the 2012 ACM Conference on CoNEXT Student Workshop, pp. 39–40. ACM, Nice (2012)
8. Khaouid, W., Barsky, M., Srinivasan, V., Thomo, A.: K-core decomposition of large networks on a single PC. Proc. VLDB Endow. **9**(1), 13–23 (2015)
9. Govindan, P., Wang, C., Xu, C., Duan, H., Soundarajan, S.: The k-peak decomposition: mapping the global structure of graphs. In: Proceedings of the 26th International Conference on World Wide Web, pp. 1441–1450. International World Wide Web Conferences Steering Committee, Perth (2017)
10. Mandal, A., Al Hasan, M.: A distributed k-core decomposition algorithm on spark. In: 2017 IEEE International Conference on Big Data (Big Data), pp. 976–981. IEEE, Boston (2017)

11. Bonchi, F., Gullo, F., Kaltenbrunner, A., Volkovich, Y.: Core decomposition of uncertain graphs. In: Proceedings of the 20th ACM SIGKDD International Conference on Knowledge Discovery and Data Mining, pp. 1316–1325. ACM, New York (2014)

12. Peng, Y., Zhang, Y., Zhang, W., Lin, X., Qin, L.: Efficient probabilistic k-core computation on uncertain graphs. In: 2018 IEEE 34th International Conference on Data Engineering (ICDE), pp. 1192–1203. IEEE, Paris (2018)

13. Tripathy, A., Hohman, F., Chau, D.H., Green, O.: Scalable K-core decomposition for static graphs using a dynamic graph data structure. In: 2018 IEEE International Conference on Big Data (Big Data), pp. 1134–1141. IEEE, Seattle (2018)

14. Wen, D., Qin, L., Zhang, Y., Lin, X., Yu, J.X.: I/o efficient core graph decomposition at web scale. In: 2016 IEEE 32nd International Conference on Data Engineering (ICDE), pp. 133–144. IEEE, Helsinki (2016)

15. Wen, D., Qin, L., Zhang, Y., Lin, X., Yu, J.X.: I/O efficient core graph decomposition: application to degeneracy ordering. IEEE Trans. Knowl. Data Eng. **31**(1), 75–90 (2018)

16. Sarıyüce, A.E., Gedik, B., Jacques-Silva, G., Wu, K.L., Çatalyürek, Ü.V.: Streaming algorithms for k-core decomposition. Proc. VLDB Endow. **6**(6), 433–444 (2013)

17. Sarıyüce, A.E., Gedik, B., Jacques-Silva, G., Wu, K.L., Çatalyürek, Ü.V.: Incremental k-core decomposition: algorithms and evaluation. VLDB J. Int. J. Very Large Data Bases **25**(3), 425–447 (2016)

18. Li, R.H., Yu, J.X., Mao, R.: Efficient core maintenance in large dynamic graphs. IEEE Trans. Knowl. Data Eng. **26**(10), 2453–2465 (2013)

19. Aksu, H., Canim, M., Chang, Y.C., Korpeoglu, I., Ulusoy, Ö.: Distributed k-core view materialization and maintenance for large dynamic graphs. IEEE Trans. Knowl. Data Eng. **26**(10), 2439–2452 (2014)

20. Aridhi, S., Brugnara, M., Montresor, A., Velegrakis, Y.: Distributed k-core decomposition and maintenance in large dynamic graphs. In: Proceedings of the 10th ACM International Conference on Distributed and Event-based Systems, pp. 161–168. ACM, Irvine (2016)

21. Zhang, Y., Yu, J.X., Zhang, Y., Qin, L.: A fast order-based approach for core maintenance. In: 2017 IEEE 33rd International Conference on Data Engineering (ICDE), pp. 337–348. IEEE, San Diego (2017)

22. Wang, N., Yu, D., Jin, H., Qian, C., Xie, X., Hua, Q.S.: Parallel algorithm for core maintenance in dynamic graphs. In: 2017 IEEE 37th International Conference on Distributed Computing Systems (ICDCS), pp. 2366–2371. IEEE, Atlanta (2017)

23. Jin, H., Wang, N., Yu, D., Hua, Q.S., Shi, X., Xie, X.: Core maintenance in dynamic graphs: a parallel approach based on matching. IEEE Trans. Parallel Distrib. Syst. **29**(11), 2416–2428 (2018)

24. Bonchi, F., Gullo, F., Kaltenbrunner, A.: Core Decomposition of Massive, Information-Rich Graphs. In: Alhajj, R., Rokne, J. (eds.) Encyclopedia of Social Network Analysis and Mining. Springer, New York (2018). https://doi.org/10.1007/978-1-4939-7131-2_110176

25. Yue, L., Wen, D., Cui, L., Qin, L., Zheng, Y.: K-connected cores computation in large dual networks. In: Pei, J., Manolopoulos, Y., Sadiq, S., Li, J. (eds.) DASFAA 2018. LNCS, vol. 10827, pp. 169–186. Springer, Cham (2018). https://doi.org/10.1007/978-3-319-91452-7_12

26. Wang, K., Cao, X., Lin, X., Zhang, W., Qin, L.: Efficient computing of radius-bounded k-cores. In: 2018 IEEE 34th International Conference on Data Engineering (ICDE), pp. 233–244. IEEE, Paris (2018)

27. Zhang, F., Zhang, Y., Qin, L., Zhang, W., Lin, X.: When engagement meets similarity: efficient (k, r)-core computation on social networks. Proc. VLDB Endow. **10**(10), 998–1009 (2017)

28. Laishram, R., Sariyüce, A.E., Eliassi-Rad, T., Pinar, A., Soundarajan, S.: Measuring and improving the core resilience of networks. In: Proceedings of the 2018 World Wide Web Conference, pp. 609–618. International World Wide Web Conferences Steering Committee, Lyon (2018)

29. Li, R.H., Qin, L., Yu, J.X., Mao, R.: Finding influential communities in massive networks. VLDB J. Int. J. Very Large Data Bases **26**(6), 751–776 (2017)

30. Bae, J., Kim, S.: Identifying and ranking influential spreaders in complex networks by neighborhood coreness. Phys. A Stat. Mech. Appl. **395**, 549–559 (2014)

31. Rossi, M.E.G., Malliaros, F.D., Vazirgiannis, M.: Spread it good, spread it fast: identification of influential nodes in social networks. In: Proceedings of the 24th International Conference on World Wide Web, pp. 101–102. ACM, Florence (2015)

32. Alduaiji, N., Datta, A.: An empirical study on sentiments in twitter communities. In: 2018 IEEE International Conference on Data Mining Workshops (ICDMW), pp. 1166–1172. IEEE, Singapore (2018)

33. Barbieri, N., Bonchi, F., Galimberti, E., Gullo, F.: Efficient and effective community search. Data Min. Knowl. Disc. **29**(5), 1406–1433 (2015)

34. Papadopoulos, S., Kompatsiaris, Y., Vakali, A., Spyridonos, P.: Community detection in social media. Data Min. Knowl. Disc. **24**(3), 515–554 (2012)

35. Nasir, M.A.U., Gionis, A., Morales, G.D.F., Girdzijauskas, S.: Fully dynamic algorithm for top-k densest subgraphs. In: Proceedings of the 2017 ACM on Conference on Information and Knowledge Management, pp. 1817–1826. ACM, Singapore (2017)

36. Qu, Y., et al.: Using K-core decomposition on class dependency networks to improve bug prediction model's practical performance. IEEE Trans. Softw. Eng. 1 (2019). https://doi.org/10.1109/TSE.2019.2892959

37. Cheng, Y., Lu, C., Wang, N.: Local k-core clustering for gene networks. In: 2013 IEEE International Conference on Bioinformatics and Biomedicine, pp. 9–15. IEEE, Shanghai (2013)

38. Ma, J., Balasundaram, B.: On the chance-constrained minimum spanning k-core problem. J. Global Optim. **74**(4), 783–801 (2019)

39. Alvarez-Hamelin, J.I., Dall'Asta, L., Barrat, A., Vespignani, A.: Large scale networks fingerprinting and visualization using the k-core decomposition. In: Advances in Neural Information Processing Systems, pp. 41–50 (2006)

40. Eppstein, D., Löffler, M., Strash, D.: Listing all maximal cliques in sparse graphs in near-optimal time. In: Cheong, O., Chwa, K.-Y., Park, K. (eds.) ISAAC 2010. LNCS, vol. 6506, pp. 403–414. Springer, Heidelberg (2010). https://doi.org/10.1007/978-3-642-17517-6_36

Active and Busy Time Scheduling Problem: A Survey

Vincent Chau[1] and Minming Li[2(✉)]

[1] Shenzhen Institutes of Advanced Technology, Chinese Academy of Sciences, Shenzhen, China
vincentchau@siat.ac.cn
[2] City University of Hong Kong, Hong Kong, China
minming.li@cityu.edu.hk

Abstract. We present an overview of recent research on the busy time and active time scheduling model, which has its applications in energy efficient scheduling for cloud computing systems, optical network design and computer memories. The major feature of this type of scheduling problems is to aggregate job execution into as few time slots as possible to save energy. The difference between busy time and active time is that the former refers to multiple machines while the latter refers to a single machine. After summarizing the previous results on this topic, we propose a few potential future directions for each model.

Keywords: Busy time · Active time · Approximation

1 Introduction

With the rise of the use of data centers, energy consumption has become a major concern. Saving energy is crucial for both economic and ecological reasons.

One of the significant energy consumptions comes from the use of memory which can be turned on and off depending on the underlying usage [2], especially during low utilization periods. The *min-gap* strategy is one of the approaches in dynamic resource sleep management [4,12]. We have the possibility of transiting machines into the sleep state without any cost. However, a small amount of energy will be consumed for transiting the machines back into the active state to process jobs. For more details on the *min-gap* scheduling problems, the interested reader may consult the recent surveys [3,11].

In this survey, we focus on the case when machines are *active*. At any moment, the scheduler can work on a group of at most g jobs by machines that are at the working state. Unlike the *min-gap* strategy, there is no cost for turning on a device. Thus, the aim is to minimize the total time that a machine is on.

This measure was introduced in the cloud computing context [19,23]. Commercial cloud computing systems possess a set of computing resources and allocate them to clients, depending on their requests. They usually charge the clients

D.-Z. Du and J. Wang (Eds.): Ko Festschrift, LNCS 12000, pp. 219–229, 2020.
https://doi.org/10.1007/978-3-030-41672-0_13

according to the time usage independent of their workload. Therefore, it is natural for the clients to maximize the number of tasks they can compute according to the given period, or to minimize the time used to complete all their tasks.

The *active time* is also motivated by problems in optical network design [7, 8, 21]. In fiber-optic communications, optical wavelength-division multiplexing (WDM) is a technology that deals with the growth of traffic in communication networks. The optical carrier signals are multiplexed onto a single optical fiber by using different wavelengths (i.e., colors) of laser light. The communications between nodes are realized by lightpaths. This technique enables bidirectional communications over one strand of fiber, as well as multiplication of capacity. As the energy of the signal decreases along a lightpath, we need to place regenerators to reinforce the signal. Thus the associated hardware cost is proportional to the length of the lightpaths. When a regenerator is set at some node and is operating at some specific color, it can be shared by at most g connections. This is known as *traffic grooming*. The regenerator optimization problem on the path topology can be seen as a scheduling problem where the regenerator cost is measured by the length of lightpaths (the busy time) while grooming corresponds to the machine capacity.

2   Preliminaries

In the scheduling problem that we are interested in, we want to schedule a set of n jobs on m identical machines. Each job J_j is characterized by its release time r_j, its deadline d_j and its processing time p_j. A job can only be scheduled inside the interval $[r_j, d_j)$ and needs to be executed during p_j time-unit. Moreover, we may schedule up to g different jobs at any moment on the same machine. We can see this environment as having m machines, and each of them has g processors.

Definition 1. *A machine is* active *whenever there is at least one job running at a given time.*

The goal is to schedule all the jobs while minimizing the total duration that a machine is on. In this survey, we consider different types of instances which are defined below.

Proper Instance. A job set is *proper* if no job interval is properly contained in another interval. Formally, for any two jobs $J_i, J_j \in \mathcal{J}$, we have $r_i \leq r_j \Leftrightarrow d_i \leq d_j$.

Laminar Instance. A job set forms a *laminar* family if for any two jobs $J_i, J_j \in \mathcal{J}$, it holds that $[r_i, d_i) \subseteq [r_j, d_j)$ or $[r_j, d_j) \subseteq [r_i, d_i)$ or $[r_i, d_i) \cap [r_j, d_j) = \emptyset$.

Clique Instance. A job set is a *clique instance* if any two job intervals intersect, i.e., $\exists t$ such that $t \in [r_j, d_j) \ \forall J_j \in \mathcal{J}$.

One-sided Clique Instance. It is a special case of the *clique instance.* In the *one-sided clique instance*, all jobs have the same release time or the same completion time.

Proper Clique. The *proper clique instance* is a combination of the *proper instance* and the *clique instance.* In this kind of instance, jobs not only share a common time, but also follow the same ordering of release time and deadline.

Note that the *one-sided clique instance* is also a *proper clique instance.*

3   Busy Time

In the busy time model, jobs cannot be preempted, i.e., a job cannot be interrupted once it is started and the goal is to schedule all the jobs while minimizing the total duration that the machine is on.

3.1   Minimizing the Busy Time Length

We first investigate the case when jobs are rigid (interval jobs), i.e., there is no flexibility for scheduling a job, which must start at its release time and complete at its deadline. Then we consider the complementary case when jobs are flexible. Finally, we focus on the online case.

Interval Jobs In The Offline Setting. The problem is NP-Hard for interval jobs even when $g = 2$ [24] by a reduction from the CIRCULAR ARC GRAPH COLORING problem. In the following, we assume that we are given an unlimited number of machines to schedule jobs, and the number of machines is part of the output of the scheduling algorithms. Alicherry and Bhatia [1], and independently Kumar and Rudra [17], developed a 2-approximation algorithm for interval jobs. Flammini et al. [9] proposed later on a simple greedy algorithm with an approximation ratio of 4. The idea is as follows: First, we sort the jobs in non-increasing order of their processing time. Then we assign the next job, J_j, to the first machine that can process it, i.e., find the minimum value of $i \geq 1$ such that, at any time $t \in J_j$, M_i is processing at most $g - 1$ jobs. If no such a machine exists, then open a new machine for J_j. They showed that this algorithm could not have a ratio better than 3. When jobs have agreeable deadline, a 2-approximation greedy algorithm has been independently proposed by two groups: Flammini et al. [9] and Khandekar et al. [13]. The idea is similar to the algorithm for the general case, but instead of considering jobs according to their processing times, they considered the jobs according to their release times.

Flexible Jobs In The Offline Setting. Khandekar et al. [13] were the first to give a constant approximation algorithm when jobs are flexible, i.e., jobs can be scheduled at any time between their release time and their deadline. Their algorithm achieves an approximation ratio of 5. In fact, their algorithm can solve a more general case where jobs have an additional demand on the number of resource $R \in [1, g]$. Therefore, a machine can schedule at any time a set of jobs so that their total demand is no more than g. Jobs are separated into two subsets according to their demand size. After assigning the wide jobs arbitrarily, we apply a greedy algorithm on the remaining jobs.

Finally, Chang et al. [6] improved the approximation ratio to 3 using a rounding technique of linear programming. The above results all assume that there is an unbounded number of machines and the number of machines required by the schedules they generate can be as large as $\Omega(n)$.

Khoeller and Khuller [15] focused on optimizing both objectives, i.e., the busy time length as well as the number of machines used. When optimizing them separately, we know that the busy time can be approximated with a ratio of 3 [6], while the best-known approximation algorithm for minimizing the number of machines has an approximation ratio of $O(\sqrt{\log n/\log\log n})$. When jobs have the same processing time, but not unit, i.e., when $\forall j\ p_j = p$, the minimum number of machines can be computed in polynomial time. The idea of their algorithm is to first compute the minimum number of machines, then minimize the busy time length. This algorithm is a 6-approximation on the busy time using the minimum number of machines. Finally, for the general case, they provided an algorithm whose approximation ratio is $(\alpha + 1)$ on busy time, and using $\lceil \log_\alpha p_{\max}/p_{\min}\rceil(2\lceil\alpha\rceil m_{opt} + 8)$ machines where m_{opt} is the minimum number of machines to get a feasible solution.

A summary of the results are listed can be found in Table 1.

Table 1. Results of offline mininimization of busy time

Jobs	g	Assumption	Results
Rigid	2		NP-hard [24]
		Clique	Polynomial [13, 18]
	Arbitrary		2-approximation [1, 17]
			4-approximation [9]
		Proper	2-approximation [9, 13]
		Laminar	Polynomial [13]
		Clique	PTAS [13]
			$\left(\frac{gH_g}{H_g+g-1}\right)$-approximation[a] [18]
Flexible	Arbitrary		5-approximation [13]
			3-approximation [6]
	Unbounded		Polynomial [13]
Bounded number of machines			(machines, busy time)
Flexible	Arbitrary	$p_j = p\ \forall j$	$(1, 6)$-approximation [15]
		Arbitrary p_j	$(\log \frac{p_{\max}}{p_{\min}}, 5)$-approximation [15]

[a]H_g is the gth harmonic number.

Interval Jobs In Online Setting. Shalom et al. [22] studied the online version of interval-job scheduling with jobs unknown in advance. Jobs are revealed only when they are released, and we need to decide on which machines to schedule them. Moreover, decisions are made irrevocably. They showed that any deterministic online algorithm cannot have an approximation ratio less than g. They subsequently provided an online algorithm which is $(5 \log p_{\max})$-competitive. For one-sided clique instance, they showed a competitive ratio lower bound of 2 and proposed a $(1 + \varphi)$-competitive algorithm, where $\varphi = (1 + \sqrt{5})/2$ is the golden ratio. By extending this result, a factor of 2 is needed for handling the clique instances.

Flexible Jobs In Online Setting. As opposed to the interval jobs, an additional decision needs to be made on when to schedule the jobs as we have more flexibility. Fong et al. [10], Koehler et al. [15] and Ren et al. [20] independently developed a constant competitive algorithm when g is unbounded. The competitive ratios are, respectively, 4, 5 and $4 + 2\sqrt{2}$, while a lower bound of $\frac{1+\sqrt{5}}{2}$ for any deterministic online algorithm was shown in [15,20].

Koehler and Khuller [15] further studied the case when g is bounded and provided a $\left(\log \frac{p_{\max}}{p_{\min}}\right)$-competitive algorithm. Moreover, if we can see the future, in particular a *lookahead* of $2p_{\max}$, then it is possible to get an algorithm with a constant competitive ratio of 12.

On the other hand, Ren and Tang [20] studied the *non-clairvoyant* case where the processing length of each job is not known until it completes execution. A lower bound (resp. upper bound) of $\frac{1+\sqrt{5}}{2}$ (resp. $1 + \frac{1+\sqrt{5}}{2}$) is given in [20].

A summary of the results can be found in the following Table 2.

Table 2. Results of online mininimization of busy time

Jobs	g	Assumption	Results
Rigid	Arbitrary		Upper bound $\min\{g, 5 \log p_{\max}\}$ [22]
			Lower bound g [22]
		One-sided clique	Upper bound $1 + \frac{1+\sqrt{5}}{2}$ [22]
			Lower bound 2 [22]
		Clique	Upper bound $2(1 + \frac{1+\sqrt{5}}{2})$ [22]
Flexible	Unbounded	Clairvoyant	Upper bound 4 [10]
			5 [15]
			$4 + 2\sqrt{2}$ [20]
			Lower bound $\frac{1+\sqrt{5}}{2}$ [15,20]
		Non-clairvoyant	Upper bound $1 + \frac{1+\sqrt{5}}{2}$ [20]
			Lower bound $\frac{1+\sqrt{5}}{2}$ [20]
	Bounded		Upper bound $\left(\log \frac{p_{\max}}{p_{\min}}\right)$ [15]
		Lookahead of $2p_{\max}$	Upper bound 12 [15]
		Lookahead of $p_{\max}$	Lower bound $\sqrt{2}$ [15]

3.2   Maximizing the Throughput

In this variant, we are given a budget T and a machine can be busy for T time-units. The objective is to maximize the number of jobs completed.

It is worth noticing that the throughput maximization problem is at least as hard as the busy time minimization problem. Indeed, if one can efficiently solve the throughput maximization problem, then we can perform a binary search on the budget of the busy time until we get the minimum busy time such that all jobs are scheduled.

Offline Algorithm. Mertzios et al. [18] showed that when jobs have common release time or common deadline (one-sided clique instance), it can be solved optimally in polynomial time by sorting the jobs in non-increasing order of their processing times, and then selecting them within the budget on the busy time and with less than g selected jobs. However, for the general clique instance, there is only a 4-approximation algorithm. Finally, for the proper clique instance, they proposed a dynamic programming algorithm whose running time is $O(n^3 g)$ (Table 3).

Table 3. Results of offline throughput maximization of busy time

Assumption	Results
Clique instance	4-approximation [18]
One-sided clique instance	$O(n \log n)$ [18]
Proper clique instance	$O(n^3 g)$ [18]

Online Algorithm. In the online setting, Shalom et al. [22] showed that any online deterministic algorithm could not have a competitive ratio better than gT, where T is the budget and subsequently proposed an algorithm with the same competitive ratio. They considered feasible instances for which there exist offline schedules that schedule all jobs. In particular, they investigated the one-sided clique instances, showed a lower bound of $2 - 2/(g+1)$ for any deterministic online algorithm and gave a constant competitive online algorithm with the ratio depending on g, but at most $9/2$ in general (Table 4).

Table 4. Results of online throughput maximization of busy time

Assumption	Results
General case	Upper bound gT [22]
	Lower bound gT [22]
Common release time	Upper bound $9/2$ [22]
	Lower bound $2 - 2/(g + 1)$ [22]

3.3   Open Questions

As mentioned in [18], the authors considered the case where scheduling a job incur a unit profit. A natural extension is to consider the weighted case where each job is associated with a weight and the goal is to schedule a set of jobs with the highest profit.

4   Active Time

The *active time* model is similar to the *busy time* model. The main difference is that we only have a single machine and preemption of jobs is allowed. Moreover, we consider that each job J_j is characterized by a set of feasible intervals $T_j = \bigcup_k [r_{jk}, d_{jk}]$. In the following, we differentiate between the *general interval* case where there are at least two distinct feasible intervals for a same job, and the *single interval* case.

In this model, we consider that the time horizon of a schedule is divided into time-slots. A time-slot is considered *active* as long as at least one job is scheduled on it and there are at most g different jobs. Preemption of jobs occurs at integral point. The goal is to schedule all the jobs while minimizing the number of active time slots.

4.1   Minimizing the Active Time Length

Chang et al. [5] considered that jobs are composed of a set of time intervals in which it can be feasibly scheduled. They showed that the problem is NP-hard when $g \geq 3$ with unit processing time jobs: they made a reduction from the 3-EXACT COVER problem. However, the problem can be solved in polynomial time when $g = 2$. They cleverly transformed the scheduling problem into the degree-constrained subgraph problem. Then, by computing the maximum cardinality matching of this graph, we can get the minimum number of active slots used by any schedule corresponding to the graph. Surprisingly, when jobs have a single interval of availability for arbitrary g, the problem can be solved in linear time.

For the case when jobs have arbitrary processing time, they proposed a 2-approximation algorithm which is based on the LP rounding technique. It is subsequently simplified by Kumar and Khuller in [16] by proposing a combinatorial algorithm although the approximation ratio remains the same.

Fong et al. [10] studied the case where g is unbounded and jobs form a proper instance. They showed that without preemption, they are able to solve it in polynomial time via dynamic programming.

4.2 Batch Scheduling

The batch scheduling problems are extensively studied in the literature. However, most of the results are about optimizing classical objective functions such as minimizing makespan, minimizing total completion time etc. In our case, we aim to minimize the number of batches such that all jobs are scheduled. A batch is a set of at most g jobs which all start and finish at the same time. Note that this variant is slightly more restrictive than the active time model, because it is required to schedule all jobs at the same time (when they are in the same batch). See Fig. 1 for an illustration of the difference of the two models.

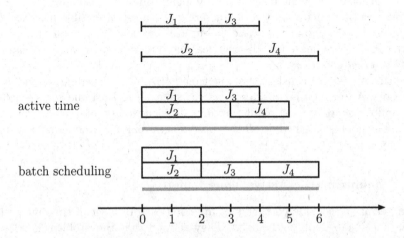

Fig. 1. Illustration of the difference between the active time model and the batch scheduling problem. We have 4 jobs with their respective availability windows $[0, 2)$, $[0, 3)$, $[2, 4)$ and $[3, 6)$. The processing time of the three jobs is 2 each. The optimal solution of the active time model is using 5 time-units while the batch scheduling needs to use 6. Moreover, a solution for the batch scheduling model is also feasible for the active time model, but the reverse is not true.

Chang et al. [5] studied this variant and gave a polynomial-time algorithm when jobs have the same processing time, with a further improvement that reduces the running time from $O(n^8)$ to $O(n^3)$ [14]. Moreover, Koehler and Khuller [14] also considered the case when g is unbounded and gave a faster algorithm (Table 5).

Table 5. Results of offline minimization of active time

Interval	g	Assumption	Results
General	2	Arbitrary p_j	$O(\sqrt{L}n)$ [5]
	≥ 3	$p_j = 1 \ \forall j$	NP-complete [5]
	Arbitrary	Time not slotted	Polynomial (LP) [5]
Single	Arbitrary	$p_j = 1 \ \forall j$	$O(n)$ [5]
		Arbitrary p_j	2-approximation [6,16]
	Unbounded	Proper and non-preemption	$O(n^3)$ [10]
Minimize number of batches (of active time)			
Single	Arbitrary	$p_j = p \ \forall j$	$O(n^8)$ [5]
			$O(n^3)$ [14]
	Unbounded	$p_j = p \ \forall j$	$O(n^2)$ [14]

4.3 Open Questions

Complexity of the Problem. An extensive literature exists for the busy time model while there are only a few papers on the active time model. Although the active time model can be seen as a special case of the busy time model, it is not known whether this problem can be solved optimally in polynomial time.

Throughput Maximization. Chang et al. [5] studied this criterion, but only in the context of batch scheduling. Exploring the general case of the active time model is an interesting direction. The throughput maximization problem is at least as hard as minimizing the active time length. By considering this variant, a polynomial-time algorithm will settle the complexity status of the problem.

5 Concluding Remarks

We present an overview of recent research on the busy time and active time scheduling model. While minimizing the busy time length has been extensively studied; other criteria would also be worth exploring. We propose a few potential future directions for each model.

References

1. Alicherry, M., Bhatia, R.: Line system design and a generalized coloring problem. In: Di Battista, G., Zwick, U. (eds.) ESA 2003. LNCS, vol. 2832, pp. 19–30. Springer, Heidelberg (2003). https://doi.org/10.1007/978-3-540-39658-1_5
2. Amur, H., Cipar, J., Gupta, V., Ganger, G.R., Kozuch, M.A., Schwan, K.: Robust and flexible power-proportional storage. In: Proceedings of the 1st ACM Symposium on Cloud Computing, pp. 217–228. ACM (2010)

3. Bampis, E.: Algorithmic issues in energy-efficient computation. In: Kochetov, Y., Khachay, M., Beresnev, V., Nurminski, E., Pardalos, P. (eds.) DOOR 2016. LNCS, vol. 9869, pp. 3–14. Springer, Cham (2016). https://doi.org/10.1007/978-3-319-44914-2_1

4. Baptiste, P.: Scheduling unit tasks to minimize the number of idle periods: a polynomial time algorithm for offline dynamic power management. In: Proceedings of the Seventeenth Annual ACM-SIAM Symposium on Discrete Algorithm, pp. 364–367. Society for Industrial and Applied Mathematics (2006)

5. Chang, J., Gabow, H.N., Khuller, S.: A model for minimizing active processor time. Algorithmica **70**(3), 368–405 (2014). https://doi.org/10.1007/s00453-013-9807-y

6. Chang, J., Khuller, S., Mukherjee, K.: Lp rounding and combinatorial algorithms for minimizing active and busy time. J. Sched. **20**(6), 657–680 (2017)

7. Chen, S., Ljubić, I., Raghavan, S.: The regenerator location problem. Netw.: Int. J. **55**(3), 205–220 (2010)

8. Flammini, M., Marchetti-Spaccamela, A., Monaco, G., Moscardelli, L., Zaks, S.: On the complexity of the regenerator placement problem in optical networks. IEEE/ACM Trans. Netw. (TON) **19**(2), 498–511 (2011)

9. Flammini, M., et al.: Minimizing total busy time in parallel scheduling with application to optical networks. Theor. Comput. Sci. **411**(40–42), 3553–3562 (2010)

10. Fong, K.C.K., Li, M., Li, Y., Poon, S.-H., Wu, W., Zhao, Y.: Scheduling tasks to minimize active time on a processor with unlimited capacity. In: Gopal, T.V., Jäger, G., Steila, S. (eds.) TAMC 2017. LNCS, vol. 10185, pp. 247–259. Springer, Cham (2017). https://doi.org/10.1007/978-3-319-55911-7_18

11. Gerards, M.E., Hurink, J.L., Hölzenspies, P.K.: A survey of offline algorithms for energy minimization under deadline constraints. J. sched. **19**(1), 3–19 (2016)

12. Irani, S., Pruhs, K.R.: Algorithmic problems in power management. ACM Sigact News **36**(2), 63–76 (2005)

13. Khandekar, R., Schieber, B., Shachnai, H., Tamir, T.: Minimizing busy time in multiple machine real-time scheduling. In: IARCS Annual Conference on Foundations of Software Technology and Theoretical Computer Science (FSTTCS 2010). Schloss Dagstuhl-Leibniz-Zentrum fuer Informatik (2010)

14. Koehler, F., Khuller, S.: Optimal batch schedules for parallel machines. In: Dehne, F., Solis-Oba, R., Sack, J.-R. (eds.) WADS 2013. LNCS, vol. 8037, pp. 475–486. Springer, Heidelberg (2013). https://doi.org/10.1007/978-3-642-40104-6_41

15. Koehler, F., Khuller, S.: Busy Time scheduling on a bounded number of machines (Extended Abstract). Algorithms and Data Structures. LNCS, vol. 10389, pp. 521–532. Springer, Cham (2017). https://doi.org/10.1007/978-3-319-62127-2_44

16. Kumar, S., Khuller, S.: Brief announcement: a greedy 2 approximation for the active time problem. In: SPAA, pp. 347–349 (2018)

17. Kumar, V., Rudra, A.: Approximation algorithms for wavelength assignment. In: Sarukkai, S., Sen, S. (eds.) FSTTCS 2005. LNCS, vol. 3821, pp. 152–163. Springer, Heidelberg (2005). https://doi.org/10.1007/11590156_12

18. Mertzios, G.B., Shalom, M., Voloshin, A., Wong, P.W., Zaks, S.: Optimizing busy time on parallel machines. Theor. Comput. Sci. **562**, 524–541 (2015)

19. Oprescu, A.M., Kielmann, T.: Bag-of-tasks scheduling under budget constraints. In: 2010 IEEE Second International Conference on Cloud Computing Technology and Science, pp. 351–359. IEEE (2010)

20. Ren, R., Tang, X.: Online flexible job scheduling for minimum span. In: Proceedings of the 29th ACM Symposium on Parallelism in Algorithms and Architectures, pp. 55–66. ACM (2017)

21. Saradhi, C.V., et al.: A framework for regenerator site selection based on multiple paths. In: 2010 Conference on Optical Fiber Communication (OFC/NFOEC), Collocated National Fiber Optic Engineers Conference, pp. 1–3. IEEE (2010)
22. Shalom, M., Voloshin, A., Wong, P.W., Yung, F.C., Zaks, S.: Online optimization of busy time on parallel machines. Theor. Comput. Sci. **560**, 190–206 (2014)
23. Shi, W., Hong, B.: Resource allocation with a budget constraint for computing independent tasks in the cloud. In: 2010 IEEE Second International Conference on Cloud Computing Technology and Science, pp. 327–334. IEEE (2010)
24. Winkler, P., Zhang, L.: Wavelength assignment and generalized interval graph coloring. In: Proceedings of the Fourteenth Annual ACM-SIAM Symposium on Discrete Algorithms, pp. 830–831. Society for Industrial and Applied Mathematics (2003)

A Note on the Position Value for Hypergraph Communication Situations

Erfang Shan[1,2(✉)], Jilei Shi[1,3], and Wenrong Lv[1]

[1] School of Management, Shanghai University,
Shanghai 200444, People's Republic of China
efshan@shu.edu.cn, shijilei1987@163.com,
wenrongga@163.com
[2] Department of Mathematics, Shanghai University,
Shanghai 200444, People's Republic of China
[3] Ningbo University of Finance and Economics,
Ningbo 315175, People's Republic of China

Abstract. The position value Meessen [3] is an important allocation rule for communication situations. The axiomatic characterization of the position value for arbitrary graph communication situations were given by Slikker [8]. However, the characterization of the position value for arbitrary hypergraph communication situations remains an open problem. This note provides an axiomatic characterization of the position value for arbitrary hypergraph communication situations by employing component efficiency and partial balanced hyperlink contributions given in Shan et al. [6].

Keywords: TU-game · Position value · Hypergraph communication situations

1 Introduction

A communication situation is one in which participants with an economic or social problem obtain a payoff through cooperation, and their cooperation is restricted to the given network structure. The seminal work on games in which restrictions in the cooperation are given by a graph is due to Myerson [4]. The nodes in the graph represent the players and the links the communication links between the players. One of the most famous allocation rules for graph communication situations is the *Myerson value* [4], which is defined as the Shapley value [7] of the so-called Myerson restricted game. Myerson [5], Borm et al. [1], and Slikker and van den Nouweland [9] provided various characterizations of this allocation rule.

Meessen [3] proposed an alternative allocation rule for the class of graph communication situations, called the *position value*. Borm et al. [1] provided a

Research was partially supported by NSFC (grant number 11971298).

characterization of the position value for graph communication situations with trees. An elegant characterization of this rule for arbitrary graph communication situations was given by Slikker [8]. van den Nouweland et al. [10] extended the position value to hypergraph communication situations. They also gave an axiomatic characterization of the position value for cycle-free hypergraph communication situations. However, an axiomatic characterization of the position value for arbitrary hypergraph communication situations has not yet been found and remains an open problem.

In this note we provide an axiomatic characterization for arbitrary hypergraph communication situations in terms of *component efficiency* and partial balanced hyperlink contributions. Component efficiency states that for each component of the hypergraph the total payoff to its players equals the worth of that component. The partial balanced hyperlink contributions deals with the payoff difference a player experiences if another player breaks one of his hyperlinks. The property is proposed in [6] for the degree value on hypergraph communication situations. It is different from the balanced link contributions due to Slikker [8], the intrinsical difference between the two balanced properties is whether the payoff difference of a player experiences is totally or partially attributing to another player.

2   Preliminaries

A *n-person cooperative game with transferable utility*, or simply a TU-game, is a pair (N, v) where $N = \{1, 2, \ldots, n\}$ is a finite set of players and $v : 2^N \to \mathbb{R}$ a *characteristic function* on the power set 2^N of N with $v(\emptyset) = 0$. For any coalition $S \subseteq N$, the real number $v(S)$ represents the *worth* of coalition S and $|S|$ denotes the cardinality of S. For nonempty $T \subseteq N$, the *subgame* of v with respect to T is $v_T(S) = v(S)$, for all $S \subseteq T$. A game (N, v) is *zero-normalized* if for any $i \in N$, $v(\{i\}) = 0$. Throughout this paper, we consider only zero-normalized game.

For a game (N, v), a *payoff vector* $x = (x_1, x_2, \ldots, x_n) \in \mathbb{R}^n$ assigns a payoff x_i to each player $i \in N$. A *single-valued solution*, also called *value* or *allocation rule*, is a mapping f that assigns to every (N, v) a payoff $f(N, v) \in \mathbb{R}^n$.

For any $T \subseteq N$, the *unanimity game* (N, u_T) corresponding to T is the game defined by $u_T(S) = 1$ if $T \subseteq S$ and $u_T(S) = 0$ otherwise (see [7]). It was proved that each (N, v) can be decomposed into a linear combination of unanimity games; formally,

$$v = \sum_{T \subseteq N : T \neq \emptyset} \lambda_T(v) u_T,$$

where $\lambda_T(v) = \sum_{S \subseteq T} (-1)^{|T|-|S|} v(S)$, called *Harsanyi dividends* [2]. The well-known single-valued solution is the *Shapley value* [7], which is given by

$$Sh_i(N, v) = \sum_{T \subseteq N : i \in T} \frac{1}{|T|} \lambda_T(v). \tag{1}$$

A *hypergraph* on N is a pair (N, H), where $H \subseteq H^N := \{e \subseteq N \mid |e| \geq 2\}$ i.e., H is a family of non-singleton subsets of N, called *hyperlinks*. If N is clear from the context, we write H instead of (N, H). We call each hyperlink a *conference*, the communication is only possible within a conference. In particular, a hypergraph (N, H) is a *graph* if $|e| = 2$ for all $e \in H$.

For each player $i \in N$, let $H_i = \{e \in H \mid i \in e\}$ is the set of hyperlinks containing i in (N, H). For any $S \subseteq N$, the *subhypergraph* $(S, H(S))$ is the hypergraph induced by S where $H(S) = \{e \in H \mid e \subseteq S\}$. We say that (N, A) is a subhypergraph of (N, H) if $A \subseteq H$. We say that nodes i and j are *connected* in H if there exists a vertex-hyperlink alternative sequence $i = i_1, e_1, i_2, e_2, \ldots, i_k, e_k, i_{k+1} = j$ such that $i_l, i_{l+1} \in e_l$ for $l = 1, 2, \ldots, k$. A hypergraph is *connected* if every pair of nodes are connected. A set S is connected in H if $H(S)$ is a connected subhypergraph. Connectedness in H induces a partition of N into components. A *component* is a maximal set of nodes of N in which every pair of nodes are connected. Let N/H be the set of components of (N, H) and S/H instead of $S/H(S)$ the set of components of $(S, H(S))$.

A *hypergraph communication situation* is a triple (N, v, H), where (N, v) is a TU-game and (N, H) is a hypergraph describing a communication possibility. The family of all hypergraph communication situations with fixed player set N is denoted by HCS^N. An allocation rule or value $f(N, v, H)$ on hypergraph communication situations is a n-dimensional vector function f defined on HCS^N.

The Myerson value [4] and the *position value* [1,3] are two important allocation rules widely used in communication situations. [10] generalized the Myerson value and the position value towards hypergraph communication situations. The *Myerson value* for hypergraph communication situations is defined as follows.

$$\mu_i(N, v, H) = Sh_i(N, v^H), \text{ for any } i \in N,$$

where $v^H(S) = \sum_{T \in S/H} v(T)$ for any $S \subseteq N$. The game (N, v^H) is called the *hypergraph-restricted game*.

The *position value* for hypergraph communication situations is given by

$$\pi_i(N, v, H) = \sum_{e \in H_i} \frac{1}{|e|} Sh_e(H, v^N), \text{ for any } i \in N, \tag{2}$$

where $v^N(A) = \sum_{T \in N/A} v(T)$ for any $A \subseteq H$. The game (H, v^N) is called a *hyperlink game*.

3   A Characterization of the Position Value

In this section we shall provide an axiomatic characterization of the position value π for hypergraph communication situations by two important axioms.

Consider the following properties for an allocation rule f defined on a class of hypergraph communication situations HCS^N. The first axiom is the classical axiom–*component efficiency* in [4] below.

Component efficiency: An value f on HCS^N is component efficient if for any hypergraph communication situation $(N, v, H) \in HCS^N$, $T \in N/H$, it holds that

$$\sum_{i \in T} f_i(N, v, H) = v(T).$$

It states that the members of a component ought to allocate to themselves the total worth available to them.

The second property in [6] deals with the gains players contribute to each other. In the property, we consider the influence of asymmetry among hyperlinks, since hyperlinks may have the different numbers of players.

Partial balanced hyperlink contributions: An value f on HCS^N satisfies partial balanced hyperlink contributions if for any $(N, v, H) \in HCS^N$ and any $i, j \in N$, it holds that

$$\sum_{e' \in H_j} \frac{1}{|e'|} [f_i(N, v, H) - f_i(N, v, H \setminus \{e'\})] = \sum_{e \in H_i} \frac{1}{|e|} [f_j(N, v, H) - f_j(N, v, H \setminus \{e\})].$$

The partial balanced hyperlink contributions states that the contribution or threat from a player towards another player equals the reverse contribution or threat, where the contribution or threat of a player towards another player is the sum of a portion payoff differences a player can inflict on another player by building or breaking one of his hyperlinks.

Note that if H is a graph, then the property reduces to balanced link contributions. But, in general, this property is obviously different from the balanced link contributions.

In order to show that the position value π satisfies component efficiency and partial balanced hyperlink contributions, we first give a key lemma below.

Lemma 1. *For any $(N, v, H) \in HCS^N$, $i \in N$,*

$$\pi_i(N, v, H) = \sum_{K \subseteq H, e \in K_i} \left(\sum_{e \in K_i} \frac{1}{|e|} \right) \frac{\lambda_K(v^N)}{|K|}, \tag{3}$$

where $K_i = K \cap H_i$

Proof. Since every TU-game can be written as a linear combination of unanimity games. Hence, for any hypergraph communication situation (N, v, H), the hyerplink game (N, v^N) corresponding to v can be expressed as

$$v^N = \sum_{K \subseteq H} \lambda_K(v^N) u_K.$$

According to the Eqs. (1) and (2), for any $i \in N$, we have

$$\pi_i(N, v, H) = \sum_{e \in H_i} \frac{1}{|e|} Sh_e(H, v^N)$$

$$= \sum_{e \in H_i} \frac{1}{|e|} \sum_{K \subseteq H : e \in K} \frac{\lambda_K(v^N)}{|K|}$$

$$= \sum_{K \subseteq H : e \in K} \sum_{e \in K_i} \frac{1}{|e|} \frac{\lambda_K(v^N)}{|K|}$$

$$= \sum_{K \subseteq H, e \in K_i} \left(\sum_{e \in K_i} \frac{1}{|e|} \right) \frac{\lambda_K(v^N)}{|K|}.$$

By Lemma 1, we show that π satisfies the two properties above.

Lemma 2. *The position value π for any hypergraph communication situations $(N, v, H) \in HCS^N$ satisfies component efficiency and partial balanced hyperlink contributions.*

Proof. We first show that π satisfies component efficiency. By Eq. (2), for any $(N, v, H) \in HCS^N$, $C \in N/H$, we have

$$\sum_{i \in C} \pi_i(N, v, H) = \sum_{i \in C} \sum_{e \in H_i} \frac{1}{|e|} Sh_e(H, v^N) = \sum_{e \in H(C)} |e| \frac{1}{|e|} Sh_e(H, v^N)$$

$$= \sum_{e \in H(C))} Sh_e(H(C), v^N),$$

where the third equality follows the definition of the Shapley value and the fact that, for all $e \in H(C)$ and $K \subseteq H \setminus \{e\}$,

$$v^N(K \cup \{e\}) - v^N(K) = v^N\big((K \cap H(C)) \cup \{e\}\big) - v^N\big(K \cap H(C)\big).$$

Hence, by efficiency of the Shapley value,

$$\sum_{i \in C} \pi_i(N, v, H) = \sum_{e \in H(C))} Sh_e(H(C), v^N) = v^N(H(C)) = v(C).$$

Next we show that the position value π satisfies partial balanced hyperlink contributions. For any $e \in H$ and $K \subseteq H \setminus \{e\} \subseteq H$, let $K_i = K \cap H_i$, $K_j = K \cap H_j$. For any $i, j \in N$, $i \neq j$, by Lemma 1, we have

$$\sum_{e' \in H_j} \frac{1}{|e'|} [\pi_i(N, v, H) - \pi_i(N, v, H \setminus \{e'\})]$$

$$= \sum_{e' \in H_j} \frac{1}{|e'|} \left[\sum_{K \subseteq H : e \in K_i} \left(\sum_{e \in K_i} \frac{1}{|e|} \right) \frac{\lambda_K(v^N)}{|K|} \right.$$

$$\left. - \sum_{K \subseteq H \setminus \{e'\} : e \in K_i} \left(\sum_{e \in K_i} \frac{1}{|e|} \right) \frac{\lambda_K(v^N)}{|K|} \right]$$

$$= \sum_{e' \in H_j} \frac{1}{|e'|} \left[\sum_{K \subseteq H : e \in K_i, e' \in K} \left(\sum_{e \in K_i} \frac{1}{|e|} \right) \frac{\lambda_K(v^N)}{|K|} \right]$$

$$= \sum_{K \subseteq H : e \in K_i, e' \in K_j} \left(\sum_{e' \in K_j} \frac{1}{|e'|} \right) \left(\sum_{e \in K_i} \frac{1}{|e|} \right) \frac{\lambda_K(v^N)}{|K|}.$$

Similarly, we have

$$\sum_{e \in H_i} \frac{1}{|e|} [\pi_j(N, v, H) - \pi_j(N, v, H \setminus \{e\})]$$

$$= \sum_{K \subseteq H : e \in K_i, e' \in K_j} \left(\sum_{e' \in K_j} \frac{1}{|e'|} \right) \left(\sum_{e \in K_i} \frac{1}{|e|} \right) \frac{\lambda_K(v^N)}{|K|}.$$

Therefore,

$$\sum_{e' \in H_j} \frac{1}{|e'|} [\pi_i(N, v, H) - \pi_i(N, v, H \setminus \{e'\})]$$

$$= \sum_{e \in H_i} \frac{1}{|e|} [\pi_j(N, v, H) - \pi_j(N, v, H \setminus \{e\})].$$

This complete the proof of Lemma 2.

The following result shows that the two properties can completely characterize the position value for hypergraph communication situations. Its proof is similar to the proof of Theorem 3.1 in [8] which characterizes the position value for arbitrary (graph) communication situations.

Theorem 1. *The position value for hypergraph communication situations is the unique allocation rule that satisfies component efficiency and partial balanced hyperlink contributions.*

Proof. By Lemma 2, it is proved that the position value for hypergraph games satisfies component efficiency and partial balanced hyperlink contributions. It remains to show that the position value is the unique value that satisfies the two properties. Suppose f is an allocation rule satisfies the two properties, we show that $f = \pi$. We proceed by induction on $|H|$. For $|H| = 0$, the assertion immediately follows from component efficiency. Next we may assume that f coincides with the position value π if $|H| \leq k - 1$. We consider the case when $|H| = k$. For any component $C \in N/H$, let $C = \{1, 2, \ldots, c\}$. By the two properties and the induction hypothesis, we immediately obtain the following system of linearly independent equations,

$$\sum_{e \in H_2} \frac{1}{|e|} f_1(v, H) - \sum_{e \in H_1} \frac{1}{|e|} f_2(v, H)$$

$$= \sum_{e \in H_2} \frac{1}{|e|} \pi_1(v, H \setminus \{e\}) - \sum_{e \in H_1} \frac{1}{|e|} \pi_2(v, H \setminus \{e\}),$$

$$\ldots$$

$$\sum_{e \in H_c} \frac{1}{|e|} f_1(v, H) - \sum_{e \in H_1} \frac{1}{|e|} f_c(v, H)$$

$$= \sum_{e \in H_c} \frac{1}{|e|} \pi_1(v, H \setminus \{e\}) - \sum_{e \in H_1} \frac{1}{|e|} \pi_c(v, H \setminus \{e\}),$$

$$\sum_{i \in T} f(v, H) = v(T),$$

where write $f_i(v, H)$ and $f_i(v, H \setminus \{e\})$ instead of $f_i(N, v, H)$ and $f_i(N, v, H \setminus \{e\})$, respectively, for $i = 1, 2, \ldots, c$. It is easily verified that the above system has a unique solution. Since the position value satisfies component efficiency and partial balanced hyperlink contributions, the position value is a solution of the above system. Consequently, we conclude that $f = \pi$ for any hypergraph communication situations with $|H| = k$.

Note that the axiomatic characterization of the position value for graph communication situations is an immediate consequence of Theorem 1.

Theorem 2 [8]. *The position value for graph communication situations is the unique allocation rule that satisfies component efficiency and balanced link contributions.*

References

1. Borm, P., Owen, G., Tijs, S.: On the position value for communication situations. SIAM J. Disc. Math. **5**(3), 305–320 (1992)
2. Harsanyi, J.C.: A bargaining model for cooperative n-person games. In: Tucker, A.W., Luce, R.D. (eds.) Contributions to the Theory of Games IV, pp. 325–355. Princeton University Press, Princeton (1959)
3. Meessen, R.: Communication games, Masters thesis, Department of Mathematics. University of Nijmegen, The Netherlands (1988). (in Dutch)
4. Myerson, R.B.: Graphs and cooperation in games. Math. Oper. Res. **2**(3), 225–229 (1977)
5. Myerson, R.B.: Conference structures and fair allocation rules. Int. J. Game Theory **9**(3), 169–182 (1980)

6. Shan, E., Zhang, G., Shan, X.: The degree value for games with communication structure. Int. J. Game Theory **47**, 857–871 (2018)
7. Shapley, L.S.: A value for n-person games. In: Kuhn, H., Tucker, A. (eds.) Contributions to the Theory of Games II, pp. 307–317. Princeton University Press, Princeton (1953)
8. Slikker, M.: A characterization of the position value. Int. J. Game Theory **33**(4), 505–514 (2005)
9. Slikker, M., van den Nouweland, A.: Social and Economic Networks in Cooperative Games. Kluwer Academic Publishers (2001)
10. van den Nouweland, A., Borm, P., Tijs, S.: Allocation rules for hypergraph communication situations. Int. J. Game Theory **20**(3), 255–268 (1992)

An Efficient Approximation Algorithm for the Steiner Tree Problem

Chi-Yeh Chen◉ and Sun-Yuan Hsieh$^{(\boxtimes)}$◉

Department of Computer Science and Information Engineering,
National Cheng Kung University, No. 1, University Road, Tainan, Taiwan, ROC
chency@csie.ncku.edu.tw, hsiehsy@mail.ncku.edu.tw
http://www.csie.ncku.edu.tw

Abstract. Given an arbitrary weighted graph, the Steiner tree problem seeks a minimum-cost tree spanning a given subset of the vertices (terminals). Byrka *et al.* proposed an interactive method that achieves an approximation ratio of $1.3863 + \epsilon$. Moreover, Goemans *et al.* shown that it is possible to achieve the same approximation guarantee while only solving hypergraphic LP relaxation once. However, solving hypergraphic LP relaxation is time consuming. This article presents an efficient two-phase heuristic in greedy strategy that achieves an approximation ratio of 1.4295.

Keywords: Steiner trees · Approximation algorithms · Graph Steiner problem · Network design

1 Introduction

The *Steiner tree* problem is one of the classic and most fundamental $\mathcal{NP}$-hard problems. The Steiner tree problem is $\mathcal{NP}$-hard [19] and it is $\mathcal{NP}$-hard to approximate the Steiner tree problem within a factor 96/95 [13]. Numerous variants of the Steiner tree problem include the terminal Steiner tree problem [11,12,15,18,33–35], the partial-terminal Steiner tree problem [25], the Steiner tree problem with distances 1 and 2 [5,7,34], the internal Steiner tree problem [26,27], the prize-collecting Steiner tree problem [2,13], the group Steiner tree problem [14,17,20,22,23], and the Steiner forest problem [1,4,13]. The applications include VLSI routing [29], wireless communications [32,36], transportation [28], wirelength estimation [10], and network routing [31].

In Euclidean and rectilinear minimum cost Steiner trees problem, a near-optimal solution of can be efficiently found [3]. In arbitrary weighted graphs, a sequence of improved approximation algorithms appeared in the literatures [6,9,24,30,37–41] in which the best approximation ratio achievable within polynomial time was improved from 2 to 1.39. Byrka *et al.* developed an LP-based algorithm that achieves approximation ratio of $\ln 4 + \epsilon$ [9]. However, the

---

This article appeared in 2019 the 2nd International Conference on Information Science and Systems (ICISS), under the title "An Efficient Approximation Algorithm for the Steiner Tree Problem".

D.-Z. Du and J. Wang (Eds.): Ko Festschrift, LNCS 12000, pp. 238–251, 2020.
https://doi.org/10.1007/978-3-030-41672-0_15

linear program is solved many times. Goemans *et al.* [21] shown that it is possible to achieve the same approximation guarantee while only solving hypergraphic LP relaxation once. Borchers and Du [8] show that $\rho_k \leq 1 + \lfloor \log_2 k \rfloor^{-1}$ where ρ_k is the worst-case ratio of the cost of optimal k-restricted Steiner tree to the cost of optimal Steiner tree. We can obtain a $1 + \epsilon$ approximation from hypergraphic LP relaxation in which $k = 2^{1/\epsilon}$. The number of variables and constraints will consequently be more than $n^{2^{1/\epsilon}}$ where n is the number of terminals [16]. Therefore, solving hypergraphic LP relaxation is time consuming. To overcome this problem, this article then presents an efficient two-phase heuristic for the general Steiner tree in greedy strategy that achieves an approximation ratio of 1.4295.

2 Notation and Preliminaries

Given an arbitrary weighted graph $G = (V, E, c)$ with nonnegative edge costs $c : E \to \mathbb{R}^+$ and a vertex subset $R \subseteq V$, the Steiner tree problem asks for a minimum-cost subtree spanning the vertex subset R (terminals). Any tree in G spanning R is called a Steiner tree, and any non-terminal vertices contained in a Steiner tree are referred to as Steiner points. The graph G is assumed to be a complete graph and let G_R be a complete graph that induced by R.

For any graph H, let $MST(H)$ be a minimum spanning tree of a graph H and $cost(H)$ be the sum of the costs of all edges in H. We thus abbreviate $mst(H) = cost(MST(H))$, i.e., the cost of a minimum spanning tree of H.

A *terminal-spanning tree* is a Steiner tree that does not contain any Steiner points. Let mst denote the cost of minimum terminal-spanning tree $MST(G_R)$. A *full component* is a minimum-cost Steiner tree spanning subset $R' \subset R$ in which all terminals are leaves. Any Steiner tree can be decomposed into full components by splitting all the non-leaf terminals [38]. The proposed algorithm iteratively chooses full components to improve a minimum-cost terminal spanning tree. The full components do not share Steiner points since it can be assumed to have its own copy of each Steiner point.

Let $\Gamma(K)$ denote the terminal set of a given full component K. Let $E_0(R')$ denote the set of zero-cost edges in which all edges connect all pairs of terminals in R'. For brevity, let $E_0(H) = E_0(\Gamma(H))$. We call a Steiner tree S is a *well solution* if any two full components in this Steiner tree has at most one share terminal. In other words, $|\Gamma(K_i) \cap \Gamma(K_j)| \leq 1$ for any two full components K_i and K_j in S. Let $Loss(K)$ be the minimum-cost sub-forest of K. A simple method of computing $Loss(K)$ is given by the following lemma.

Lemma 1 [38]. *For any full component K, $Loss(K) = MST(K \cup E_0(K)) - E_0(K)$.*

The cost of $Loss(K)$ is denoted by $loss(K)$. Let $\mathcal{C}[K]$ be a loss-contracted full component that can be obtained by collapsing each connected component of $Loss(K)$ into a single node. An optimal k-restricted Steiner tree is denoted by Opt_k. Let opt_k and $loss_k$ denote the cost and loss of Opt_k, respectively. Let opt denote the cost of the optimal Steiner tree. For brevity, this article uses $T/E_0(R')$ to denote the minimum spanning tree of $T \cup E_0(R')$ for $R' \subset R$.

The *gain* of a full component K with respect to T is defined as

$$gain_T(K) = cost(T) - mst(T \cup E_0(K)) - cost(K),$$

and the *load* of of a full component K with respect to T is defined as

$$load_T(K) = cost(K) + mst(T \cup E_0(K)) - cost(T).$$

Let $\Psi_{T_1,T_2}(K) = cost(T_1) - cost(T_2) - mst(T_1 \cup E_0(K)) + mst(T_2 \cup E_0(K))$. The following lemma shows that if no full component can improve a terminal-spanning tree T, then $cost(T) \le opt_k$.

Lemma 2 [38]. *Let T be a terminal-spanning tree; if $gain_T(K) \le 0$ for any k-restricted full component K, then $cost(T) \le opt_k$.*

3   Two-Phase Algorithm

The k-restricted two-phase heuristic (k-TPH) is described in Algorithm 1. Let T^t be the terminal-spanning tree at the end of iteration t and let K_t be the chosen full component at the end of iteration t. The concept of first phase is to find a terminal-spanning tree T_{base} such that no full component can improve it. Then, we can use the terminal-spanning tree T_{base} as based criterion for the second phase. The solution in the first phase is denoted by S_1, and the solution in the second phase is denoted by S_2. The first phase is a loss-contracting algorithm. The criterion function of K with respect to T^{t-1} is defined as

$$r = \frac{gain_{T^{t-1}}(K)}{loss(K)}.$$

In a loss-contracting algorithm, a chosen full component K_i may be modified by other chosen full component. That is, when $C[K_t]$ is added to T^{t-1}, some edges $\{e_1, e_2, \ldots\}$ in T^{t-1} that are corresponding to $C[K_i]$ are deleted. The components are obtained by $K_i - \{e_1, e_2, \ldots\}$ and each component can be replaced by a full component with same terminals. Then, the full component K_i can be replaced by these full components. That is because we want to ensure that $\frac{1}{2} \cdot cost(S_1) \le cost(T_{base})$. If no edge in T^{t-1} is corresponding to $C[K_i]$, we keep a *basic component* from K_i that is a Steiner point directly connect to two terminals in which an edge belongs to $Loss(K_i)$ and another edge belongs to $K_i - Loss(K_i)$ (see Fig. 1). It guarantees that the chosen full components never be chosen again. However, it may bring that some Steiner points are leaves in S_1. Fortunately, these Steiner points can be removed. Therefore, this article assume that no Steiner point is leaf in S_1.

The second phase is the k-restricted enhanced relative greedy heuristic (k-ERGH), which is described in Algorithm 2, to obtain a Steiner tree S_2. The k-ERGH iteratively finds a full component K for modifying the terminal-spanning trees $T_{origin}^0 = MST(G_R)$ and T_{base}^0. When a full component K_t has been chosen, the algorithm contracts the cost of the corresponding edges in

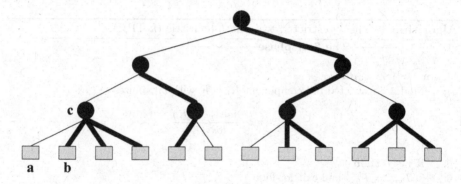

Fig. 1. A full component K: squares denote terminals, circles denote Steiner and bold black edges indicate $K - Loss(K)$. A subgraph $B = (\{a, b, c\}, \{\{a, c\}, \{b, c\}\})$ is a basic component in K where an edge $\{a, c\}$ belongs to $Loss(K)$ and another edge $\{b, c\}$ belongs to $K - Loss(K)$.

T_{origin}^{t-1} to zero, that is, $T_{origin}^{t} = MST(T_{origin}^{t-1} \cup E_0(K_t))$. Similarly, $T_{base}^{t} = MST(T_{base}^{t-1} \cup E_0(K_t))$. The criterion function of K with respect to T_{origin}^{t-1} and T_{base}^{t-1} is defined as

$$ f(K) = \frac{load_{T_{base}^{t-1}}(K)}{\Psi_{T_{origin}^{t-1}, T_{base}^{t-1}}(K)}. $$

The following steps analyze the complexity of k-TPH. Let n be the number of terminals. In the first phase, the number of iterations cannot exceed the number of full Steiner components $O(n^k)$. The gain of a full component K can be found in time $O(k)$ after precomputing the longest edges between any pair of nodes in the current minimum spanning tree, which may be accomplished in time $O(n \log n)$ [38]. Thus, the runtime of all the iterations in the first phase can be bounded by $O(kn^{2k+1} \log n)$. We also can obtain the runtime of all the iterations in the second phase is bounded by $O(kn^{2k+1} \log n)$. Thus, the total runtime is $O(kn^{2k+1} \log n)$.

4   Approximation Ratio of the k-TPH

This section shows the approximation result of the k-TPH. The following lemma shows that the first phase never repeatedly choose the same full component even it has been replaced by some full components.

Lemma 3. *The first phase never choose the chosen full components again.*

Proof. Assume that the first phase choose a full component $K_t = K$. If no chosen full component modifies edges of $\mathcal{C}[K_i]$ in $MST(T \bigcup_{i=1}^{t'} \mathcal{C}[K_i])$, $gain_T(K) \leq 0$ and the first phase never choose the full component K again.

Algorithm 1. The k-restricted two-phase heuristic (k-TPH)

1: ———————The first phase———————
2: $T^0 = MST(G_S)$
3: **for** $t = 1, 2, \ldots$ **do**
4:    Find a k-restricted full component $K_t = K$ with maximizes

$$r = \frac{gain_{T^{t-1}}(K)}{loss(K)}$$

5:    **if** $r \leq 0$ **then**
6:       $T_{base} = T^{t-1}$ and exit for-loop
7:    **end if**
8:    **if** there exist some edges $\{e_1, e_2, \ldots\} \subseteq T^{t-1} - MST(T^{t-1} \cup E_0(K_t))$ and $\{e_1, e_2, \ldots\} \subseteq \mathcal{C}(K_i)$ for $i \neq t$ **then**
9:       Some components are obtained by $K_i - \{e_1, e_2, \ldots\}$ and each components can be replaced by a full component with same terminals.
10:       Replaced the full component K_i by these full components.
11:       (for convenient to describe algorithm, we reuse the notain K_i to represent these full components.)
12:    **end if**
13:    $T^t = MST(T^0 \cup \mathcal{C}[K_1] \cup \cdots \mathcal{C}[K_t])$
14:    $S_1 = MST(T^0 \cup K_1 \cup \cdots \cup K_t)$
15:    **if** no edge in T^t is corresponding to $\mathcal{C}[K_i]$ for $i \neq t$ **then**
16:       Keep a basic component from K_i.
17:       (we also reuse the notain K_i to represent this basic component.)
18:    **end if**
19: **end for**
20: ———————The second phase———————
21: $S_2 = k$-ERGH(T_{base})
22: **return** the minimum-cost tree S between S_1 and S_2.

Algorithm 2. The k-restricted enhanced relative greedy heuristic (k-ERGH)

Require: T_{base}.
1: $T_{base}^0 = T_{base}$ and $T_{origin}^0 = MST(G_S)$
2: **for** $t = 1, 2, \ldots$ **do**
3:    Find a k-restricted full component $K_t = K$ which minimizes

$$f(K) = \frac{load_{T_{base}^{t-1}}(K)}{\Psi_{T_{origin}^{t-1}, T_{base}^{t-1}}(K)}$$

4:    $T_{origin}^t = MST(T_{origin}^{t-1} \cup E_0(K_t))$
5:    $T_{base}^t = MST(T_{base}^{t-1} \cup E_0(K_t))$
6:    **if** $c(T_{origin}^t) = c(T_{base}^t)$ **then**
7:       **return** $MST(T_{origin}^0 \cup K_1 \cup K_2 \cdots \cup K_t)$
8:    **end if**
9: **end for**

If $MST(T \bigcup_{i=1}^{t'} C[K_i])$ does not contain some edge $e \in C[K]$ in the iteration $t' > t$, the full component K is divided into two components by removing the edge e. Let A and B be two connected components of $K - \{e\}$. The full component K_t is replaced by two full components K_A and K_B with terminals sets $\Gamma(A)$ and $\Gamma(B)$, respectively. We have $T^{t'} = MST(T \bigcup_{i=1}^{t-1} C[K_i] \cup K_A \cup K_B \bigcup_{i=t+1}^{t'} C[K^i])$, $gain_{T^{t'}}(K) \leq gain_{T^{t'}}(A \cup B) \leq gain_{T^{t'}}(A) + gain_{T^{t'}}(B)$ and $loss(K) = loss(A) + loss(B)$. Finally,

$$\frac{gain_{T^{t'}}(K)}{loss(K)} \leq \frac{gain_{T^{t'}}(A) + gain_{T^{t'}}(B)}{loss(A) + loss(B)}$$

$$\leq \max\left\{\frac{gain_{T^{t'}}(A)}{loss(A)}, \frac{gain_{T^{t'}}(B)}{loss(B)}\right\}.$$

We knows that $cost(K_A) \leq cost(A)$ and $gain_{T^{t'}}(K_A) \leq 0$. The full component K_A is superior to A. We also can obtain that K_B is superior to B. The first phase never choose the full component K again.

If no edge in $T^{t'}$ is corresponding to $C[K]$, we keep a basic component in K. Then, we can find a full component that superior to K. The chosen full components never be chosen again. □

Lemma 4. $cost(T_{base}^0) \geq \frac{1}{2} \cdot cost(S_1)$.

Proof. The cost of the Steiner tree in the first phase is

$$cost(S_1) = cost(T_{base}^0) + \sum_{K_j \in S_1} loss(K_j).$$

Since $loss(K) \leq \frac{1}{2} \cdot cost(K)$ [38] for any full component K,

$$cost(S_1) \leq cost(T_{base}^0) + \sum_{K_j \in S_1} \frac{1}{2} \cdot cost(K_j)$$

$$\leq cost(T_{base}^0) + \frac{1}{2} \cdot cost(S_1)$$

which yields $cost(T_{base}^0) \geq \frac{1}{2} \cdot cost(S_1)$. □

Lemma 5. *If no full component can improve the terminal-spanning tree T,*

$$load_T\left(\bigcup_{i=1}^{n} K_i\right) \geq \sum_{i=1}^{n} load_T(K_i)$$

for full components $K_1, K_2, \ldots, K_n$.

Proof. The proof can be obtained by the following chain of inequalities:

$$
\begin{aligned}
load_T \left(\bigcup_{i=1}^{n} K_i \right) &= cost \left(\bigcup_{i=1}^{n} K_i \right) + mst \left(T \cup \bigcup_{i=1}^{n} E_0\left(K_i\right) \right) - cost(T) \\
&= \sum_{i=1}^{n} cost(K_i) + mst \left(T \cup \bigcup_{j=1}^{i} E_0\left(K_j\right) \right) - cost \left(T / \bigcup_{j=1}^{i-1} E_0\left(K_j\right) \right) \\
&\geq \sum_{i=1}^{n} cost(K_i) + mst\left(T \cup E_0\left(K_i\right) \right) - cost(T) \\
&= \sum_{i=1}^{n} load_T(K_i).
\end{aligned}
$$

$\square$

The following lemma guarantees that the solution of k-TPH at the second phase is a well solution.

Lemma 6. *For any chosen full components K_i and K_j, $|\Gamma(K_i) \cap \Gamma(K_j)| \leq 1$.*

Proof. Assume that $|\Gamma(K_i) \cap \Gamma(K_j)| = 2$ and $j < i$. The terminal-spanning trees $T_{origin}^{i-1} - MST(T_{origin}^{i-1} \cup E_0(K_i))$ and $T_{base}^{i-1} - MST(T_{base}^{i-1} \cup E_0(K_i))$ contain a zero-cost edge that is from $E_0(K_j)$. Since no full component can improve T_{base}^0, we have $MST(T_{base}^0 \cup K) = T_{base}^0 \cup Loss(K)$ for any full component K. We can find a edge $e \in K_i - Loss(K_i)$ such that $\Psi_{T_{origin}^{i-1}, T_{base}^{i-1}}(K_i) = \Psi_{T_{origin}^{i-1}, T_{base}^{i-1}}(A) + \Psi_{T_{origin}^{i-1}, T_{base}^{i-1}}(B)$ and $load_{T_{base}^{i-1}}(K_i) \geq load_{T_{base}^{i-1}}(A \cup B) \geq load_{T_{base}^{i-1}}(A) + load_{T_{base}^{i-1}}(B)$ (from Lemma 5) where A and B are two connected components of $K_i - \{e\}$. Finally,

$$
\begin{aligned}
\frac{load_{T_{base}^{i-1}}(K_i)}{\Psi_{T_{origin}^{i-1}, T_{base}^{i-1}}(K_i)} &\geq \frac{load_{T_{base}^{i-1}}(A) + load_{T_{base}^{i-1}}(B)}{\Psi_{T_{origin}^{i-1}, T_{base}^{i-1}}(A) + \Psi_{T_{origin}^{i-1}, T_{base}^{i-1}}(B)} \\
&\geq \min \left\{ \frac{load_{T_{base}^{i-1}}(A)}{\Psi_{T_{origin}^{i-1}, T_{base}^{i-1}}(A)}, \frac{load_{T_{base}^{i-1}}(B)}{\Psi_{T_{origin}^{i-1}, T_{base}^{i-1}}(B)} \right\}
\end{aligned}
$$

which contradicts the choice of K_i.

$\square$

Lemma 7. *For any Steiner tree S, $load_{T_{base}^0}(S) \geq load_{T_{base}^{i-1}} \left(S / \bigcup_{j=1}^{i-1} E_0\left(K_j\right) \right)$.*

Proof. Since no full component can improve the terminal-spanning tree T_{base}^0, $cost(S) - cost(T_{base}^0) - mst \left(S \cup \bigcup_{j=1}^{i-1} E_0\left(K_j\right) \right) + mst \left(T_{base}^0 \cup \bigcup_{j=1}^{i-1} E_0\left(K_j\right) \right) \geq 0$. The proof can be obtained by the following chain of inequalities:

$$load_{T^0_{base}}(S) = cost(S) - cost(T^0_{base})$$

$$= mst\left(S \cup \bigcup_{j=1}^{i-1} E_0(K_j)\right) - mst\left(T^0_{base} \cup \bigcup_{j=1}^{i-1} E_0(K_j)\right)$$

$$+ cost(S) - cost(T^0_{base}) - mst\left(S \cup \bigcup_{j=1}^{i-1} E_0(K_j)\right) + mst\left(T^0_{base} \cup \bigcup_{j=1}^{i-1} E_0(K_j)\right)$$

$$\geq mst\left(S \cup \bigcup_{j=1}^{i-1} E_0(K_j)\right) - mst\left(T^0_{base} \cup \bigcup_{j=1}^{i-1} E_0(K_j)\right)$$

$$= load_{T^{i-1}_{base}}\left(S/\bigcup_{j=1}^{i-1} E_0(K_j)\right).$$

$\square$

Lemma 8. *If* $load_{T^{i-1}_{base}/E_0(C)}(K) \leq \Psi_{T^{i-1}_{origin}, T^{i-1}_{base}/E_0(C)}(K)$ *for any full components* C *and* K,

$$\frac{load_{T^{i-1}_{base}/E_0(C)}(K)}{\Psi_{T^{i-1}_{origin}, T^{i-1}_{base}/E_0(C)}(K)} \geq \frac{load_{T^{i-1}_{base}}(K)}{\Psi_{T^{i-1}_{origin}, T^{i-1}_{base}}(K)}.$$

Proof. Since $load_{T^{i-1}_{base}/E_0(C)}(K) \leq \Psi_{T^{i-1}_{origin}, T^{i-1}_{base}/E_0(C)}(K)$ and $cost(T^{i-1}_{base}/E_0(C)) - mst(T^{i-1}_{base} \cup E_0(C) \cup E_0(K)) \leq cost(T^{i-1}_{base}) - mst(T^{i-1}_{base} \cup E_0(K))$, the proof can be obtained by the following chain of inequalities:

$$\frac{load_{T^{i-1}_{base}/E_0(C)}(K)}{\Psi_{T^{i-1}_{origin}, T^{i-1}_{base}/E_0(C)}(K)}$$

$$= \frac{cost(K) + mst(T^{i-1}_{base} \cup E_0(C) \cup E_0(K)) - cost(T^{i-1}_{base}/E_0(C))}{cost(T^{i-1}_{origin}) - cost(T^{i-1}_{base}/E_0(C)) - mst(T^{i-1}_{origin} \cup E_0(K)) + mst(T^{i-1}_{base} \cup E_0(C) \cup E_0(K))}$$

$$\geq \frac{cost(K) + mst(T^{i-1}_{base} \cup E_0(K)) - cost(T^{i-1}_{base})}{cost(T^{i-1}_{origin}) - cost(T^{i-1}_{base}) - mst(T^{i-1}_{origin} \cup E_0(K)) + mst(T^{i-1}_{base} \cup E_0(K))}$$

$$= \frac{load_{T^{i-1}_{base}}(K)}{\Psi_{T^{i-1}_{origin}, T^{i-1}_{base}}(K)}.$$

$\square$

Based on the analysis in [41], the bound on the cost of our solution is as follows.

Theorem 9. *The k-TPH finds a Steiner tree* S *such that*

$$cost(S) \leq \left(\ln \frac{mst - cost(T^0_{base})}{opt_k - cost(T^0_{base})} + 1\right) \cdot (opt_k - cost(T^0_{base})) + cost(T^0_{base}).$$

Proof. Let $M_i = cost(T^i_{origin}) - cost(T^i_{base})$ and $m_i = M_{i-1} - M_i$. Hence, $f(K_i) = \frac{load_{T^{i-1}_{base}}(K_i)}{m_i}$. Let $Opt^{i-1}_k = \left(Opt_k/\bigcup_{l=1}^{i-1} E_0(k_l)\right) - \bigcup_{l=1}^{i-1} E_0(K_l)$. For $i = 1, \ldots, r+1$ and $load_{T^0_{base}}(Opt_k) \leq M_{i-1}$, we have

$$\frac{load_{T_{base}^0}(Opt_k)}{M_{i-1}} = \frac{load_{T_{base}^0}(Opt_k)}{\Psi_{T_{origin}^{i-1},T_{base}^{i-1}}(Opt_k)}$$

$$\overset{Lem\,7}{\geq} \frac{load_{T_{base}^{i-1}}(Opt_k^{i-1})}{\Psi_{T_{origin}^{i-1},T_{base}^{i-1}}(Opt_k^{i-1})}$$

$$= \frac{\sum_{X_j \in Opt_k^{i-1}} load_{T_{base}^{i-1}/\bigcup_{l=1}^{j-1} E_0(X_l)}(X_j)}{\sum_{X_j \in Opt_k^{i-1}} \Psi_{T_{origin}^{i-1}/\bigcup_{l=1}^{j-1} E_0(X_l),T_{base}^{i-1}/\bigcup_{l=1}^{j-1} E_0(X_l)}(X_j)}$$

$$\geq \frac{\sum_{X_j \in Opt_k^{i-1}} load_{T_{base}^{i-1}/\bigcup_{l=1}^{j-1} E_0(X_l)}(X_j)}{\sum_{X_j \in Opt_k^{i-1}} \Psi_{T_{origin}^{i-1},T_{base}^{i-1}/\bigcup_{l=1}^{j-1} E_0(X_l)}(X_j)}$$

$$\overset{Lem\,8}{\geq} \frac{\sum_{X_j \in Opt_k^{i-1}} load_{T_{base}^{i-1}}(X_j)}{\sum_{X_j \in Opt_k^{i-1}} \Psi_{T_{origin}^{i-1},T_{base}^{i-1}}(X_j)}$$

$$\geq \min_{X_j \in Opt_k^{i-1}} \left\{ \frac{load_{T_{base}^{i-1}}(X_j)}{\Psi_{T_{origin}^{i-1},T_{base}^{i-1}}(X_j)} \right\}$$

$$\geq \frac{load_{T_{base}^{i-1}}(K_i)}{m_i}.$$

Replacing $m_i = M_{i-1} - M_i$ into the above inequality yields

$$M_i \leq M_{i-1}\left(1 - \frac{load_{T_{base}^{i-1}}(K_i)}{load_{T_{base}^0}(Opt_k)}\right) \tag{1}$$

for $i = 1, 2, \ldots, t$. From the inequality (1),

$$M_r \leq M_0 \prod_{i=1}^{t}\left(1 - \frac{load_{T_{base}^{i-1}}(K_i)}{load_{T_{base}^0}(Opt_k)}\right).$$

Taking the natural logarithms of both sides and using the inequality $\ln(1 + x) \leq x$,

$$\ln\frac{M_0}{M_r} \geq -\sum_{i=1}^{t} \ln\left(1 - \frac{load_{T_{base}^{i-1}}(K_i)}{load_{T_{base}^0}(Opt_k)}\right)$$

$$\geq \frac{\sum_{i=1}^{t} load_{T_{base}^{i-1}}(K_i)}{load_{T_{base}^0}(Opt_k)}. \tag{2}$$

Since k-TPA interrupts at $M_t = c(T_{origin}^t) - c(T_{base}^t) = 0$, there exists $M_r > load_{T_{base}^0}(Opt_k) \geq M_{r+1}$ for some $r < t$.

The value m_{r+1} can be split into two values m^* and m' such that

$$m^* = M_r - load_{T_{base}^0}(Opt_k), \tag{3}$$

$$m' = load_{T_{base}^0}(Opt_k) - M_{r+1}, \tag{4}$$

According to inequality (3), we have

$$M_{r+1}^* = M_r - m^* = M_r - M_r + load_{T_{base}^0}(Opt_k) = load_{T_{base}^0}(Opt_k). \quad (5)$$

The value $load_{T_{base}^r}(K_{r+1})$ also can be split into w^* and w' such that $\frac{load_{T_{base}^r}(K_{r+1})}{m_{r+1}} = \frac{w^*}{m^*} = \frac{w'}{m'}$. Since $\frac{load_{T_{base}^r}(K_{r+1})}{m_{r+1}} = \frac{w^*}{m^*}$, inequality (2) implies that

$$\ln \frac{M_0}{M_{r+1}^*} \geq \frac{\sum_{i=1}^r load_{T_{base}^{i-1}}(K_i) + w^*}{load_{T_{base}^0}(Opt_k)}. \quad (6)$$

Since $\frac{load_{T_{base}^r}(K_{r+1})}{m_{r+1}} \leq \frac{load_{T_{base}^0}(Opt_k)}{M_r} \leq 1$, we have

$$w' \leq m'. \quad (7)$$

The ratio related to the cost of approximate Steiner tree after $r+1$ iterations is at most

$$\frac{cost(S_2) - cost(T_{base}^0)}{opt_k - cost(T_{base}^0)} = \frac{mst(T_{origin}^0 \cup \bigcup_{i=1}^t K_i) - cost(T_{base}^0)}{load_{T_{base}^0}(Opt_k)}$$

$$\overset{\text{Lem 6}}{\leq} \frac{\sum_{i=1}^{r+1} load_{T_{base}^{i-1}}(K_i) + M_{r+1}}{load_{T_{base}^0}(Opt_k)}$$

$$= \frac{\sum_{i=1}^r load_{T_{base}^{i-1}}(K_i) + w^* + w' + M_{r+1}}{load_{T_{base}^0}(Opt_k)}$$

$$\overset{(6)}{\leq} \ln \frac{M_0}{M_{r+1}^*} + \frac{w' + M_{r+1}}{load_{T_{base}^0}(Opt_k)}$$

$$\overset{(7)}{\leq} \ln \frac{M_0}{M_{r+1}^*} + \frac{m' + M_{r+1}}{load_{T_{base}^0}(Opt_k)}$$

$$\overset{(4)}{=} \ln \frac{M_0}{M_{r+1}^*} + 1$$

$$\overset{(5)}{=} \ln \frac{M_0}{load_{T_{base}^0}(Opt_k)} + 1$$

$$= \ln \frac{cost(T_{origin}^0) - cost(T_{base}^0)}{opt_k - cost(T_{base}^0)} + 1$$

$$= \ln \frac{mst - cost(T_{base}^0)}{opt_k - cost(T_{base}^0)} + 1$$

which yields

$$cost(S) \leq cost(S_2)$$

$$\leq \left(\ln \frac{mst - cost(T_{base}^0)}{opt_k - cost(T_{base}^0)} + 1 \right) \cdot \left(opt_k - cost(T_{base}^0) \right) + cost(T_{base}^0). \quad (8)$$

$\square$

According to Lemmas 2 and 4, we have $cost(T_{base}^0) \leq opt_k$ and $cost(T_{base}^0) \geq \frac{1}{2} \cdot cost(S_1) \geq \frac{1}{2} \cdot opt_k$. Assume that $cost(T_{base}^0) = \alpha \cdot opt_k$ for $\alpha \in \left(\frac{1}{2}, 1\right)$. The following result can be obtained.

Theorem 10. *If* $cost(T_{base}^0) = \alpha \cdot opt_k$ *for* $\alpha \in \left(\frac{1}{2}, 1\right)$, *the* k-*TPH finds a Steiner tree* S *such that*

$$cost(S) \leq \left(\ln \frac{mst - \alpha \cdot opt_k}{opt_k - \alpha \cdot opt_k} + 1 \right) \cdot (opt_k - \alpha \cdot opt_k) + \alpha \cdot opt_k.$$

and

$$cost(S) \leq 2 \cdot \alpha \cdot opt_k.$$

Proof. From Theorem 9, we have

$$cost(S) \leq \left(\ln \frac{mst - \alpha \cdot opt_k}{opt_k - \alpha \cdot opt_k} + 1 \right) \cdot (opt_k - \alpha \cdot opt_k) + \alpha \cdot opt_k.$$

According to Lemma 4, $cost(S) \leq 2 \cdot cost(T_{base}^0) = 2 \cdot \alpha \cdot opt_k$.   □

5   Performance of the k-TPH in General Graphs

The following corollaries gives a bound on the cost of the Steiner tree generated by k-TPH.

Corollary 11. *The* k-*TPH has an approximation ratio of at most* 1.4295.

Proof. Since $mst \leq 2 \cdot opt$ (see [39]), Theorem 10 yield

$$\frac{cost(S)}{opt} \leq \left(\ln \frac{2 \cdot opt - \alpha \cdot opt_k}{opt_k - \alpha \cdot opt_k} + 1 \right) \cdot (1 - \alpha) \frac{opt_k}{opt} + \alpha \cdot \frac{opt_k}{opt}$$

$$= \left(\ln \frac{\frac{2}{\rho_k} - \alpha}{1 - \alpha} + 1 \right) \cdot (1 - \alpha) \rho_k + \alpha \cdot \rho_k$$

and

$$\frac{cost(S)}{opt} \leq 2 \cdot \alpha \cdot \rho_k,$$

where ρ_k is the worst-case ratio of $\frac{opt_k}{opt}$. Borchers and Du [8] show that $\rho_k \leq 1 + \lfloor \log_2 k \rfloor^{-1}$ and $\lim_{k \to \infty} \rho_k = 1$. When $k \to \infty$, the approximation ratio of the k-TPH converges to

$$A(\alpha) = \left(\ln \frac{2 - \alpha}{1 - \alpha} + 1 \right) \cdot (1 - \alpha) + \alpha.$$

and

$$B(\alpha) = 2 \cdot \alpha.$$

Since $A(\alpha)$ is decreasing in α and $B(\alpha)$ is increasing in α, solving $A(\alpha) = B(\alpha)$ yeilds $\alpha^* \approx 0.7147$. The k-TPH has an approximation ratio of at most $A(\alpha^*) \approx 1.4295$.   □

References

1. Agrawal, A., Klein, P., Ravi, R.: When trees collide: an approximation algorithm for the generalized steiner problem on networks. SIAM J. Comput. **24**(3), 440–456 (1995)
2. Archer, A., Bateni, M., Hajiaghayi, M., Karloff, H.: Improved approximation algorithms for prize-collecting steiner tree and TSP. SIAM J. Comput. **40**(2), 309–332 (2011)
3. Arora, S.: Polynomial time approximation schemes for euclidean traveling salesman and other geometric problems. J. ACM **45**(5), 753–782 (1998)
4. Bateni, M., Hajiaghayi, M., Marx, D.: Approximation schemes for steiner forest on planar graphs and graphs of bounded treewidth. J. ACM **58**(5), 21:1–21:37 (2011)
5. Berman, P., Karpinski, M., Zelikovsky, A.: 1.25-approximation algorithm for steiner tree problem with distances 1 and 2. In: Dehne, F., Gavrilova, M., Sack, J.-R., Tóth, C.D. (eds.) WADS 2009. LNCS, vol. 5664, pp. 86–97. Springer, Heidelberg (2009). https://doi.org/10.1007/978-3-642-03367-4_8
6. Berman, P., Ramaiyer, V.: Improved approximations for the steiner tree problem. J. Algorithms **17**(3), 381–408 (1994)
7. Bern, M., Plassmann, P.: The steiner problem with edge lengths 1 and 2. Inf. Process. Lett. **32**(4), 171–176 (1989)
8. Borchers, A., Du, D.Z.: The k-steiner ratio in graphs. SIAM J. Comput. **26**(3), 857–869 (1997)
9. Byrka, J., Grandoni, F., Rothvoss, T., Sanitá, L.: Steiner tree approximation via iterative randomized rounding. J. ACM **60**(1), 6:1–6:33 (2013)
10. Caldwell, A.E., Kahng, A.B., Mantik, S., Markov, I.L., Zelikovsky, A.: On wire-length estimations for row-based placement. In: ISPD 1998: Proceedings of the 1998 International Symposium on Physical Design, pp. 4–11. ACM, New York, NY, USA (1998)
11. Chen, Y.H., Lu, C.L., Tang, C.Y.: On the full and bottleneck full steiner tree problems. In: Warnow, T., Zhu, B. (eds.) COCOON 2003. LNCS, vol. 2697, pp. 122–129. Springer, Heidelberg (2003). https://doi.org/10.1007/3-540-45071-8_14
12. Chen, Y.H.: An improved approximation algorithm for the terminal steiner tree problem. In: Murgante, B., Gervasi, O., Iglesias, A., Taniar, D., Apduhan, B.O. (eds.) ICCSA 2011, Part III. LNCS, vol. 6784, pp. 141–151. Springer, Heidelberg (2011). https://doi.org/10.1007/978-3-642-21931-3_12
13. Chlebík, M., Chlebíková, J.: The steiner tree problem on graphs: Inapproximability results. Theor. Comput. Sci. **406**(3), 207–214 (2008). Algorithmic Aspects of Global Computing
14. Demaine, E.D., Hajiaghayi, M., Klein, P.N.: Node-weighted steiner tree and group steiner tree in planar graphs. ACM Trans. Algorithms **10**(3), 13:1–13:20 (2013)
15. Drake, D.E., Hougardy, S.: On approximation algorithms for the terminal steiner tree problem. Inf. Process. Lett. **89**(1), 15–18 (2004)
16. Feldmann, A.E., Könemann, J., Olver, N., Sanità, L.: On the equivalence of the bidirected and hypergraphic relaxations for Steiner tree. Math. Program. **160**(1), 379–406 (2016)
17. Ferreira, C.E., de Oliveira Filho, F.M.: New reduction techniques for the group steiner tree problem. SIAM J. Optim. **17**(4), 1176–1188 (2006)
18. Fuchs, B.: A note on the terminal steiner tree problem. Inf. Process. Lett. **87**(4), 219–220 (2003)

19. Garey, M.R., Johnson, D.S.: Computers and Intractability, vol. 29. W.H. Freeman, New York (2002)
20. Garg, N., Konjevod, G., Ravi, R.: A polylogarithmic approximation algorithm for the group Steiner tree problem. In: Proceedings of the Ninth Annual ACM-SIAM Symposium on Discrete Algorithms, SODA 1998, pp. 253–259. Society for Industrial and Applied Mathematics, Philadelphia, PA, USA (1998)
21. Goemans, M.X., Olver, N., Rothvoß, T., Zenklusen, R.: Matroids and integrality gaps for hypergraphic steiner tree relaxations. In: Proceedings of the Forty-Fourth Annual ACM Symposium on Theory of Computing, TOC 2012, pp. 1161–1176. ACM, New York, NY, USA (2012)
22. Halperin, E., Kortsarz, G., Krauthgamer, R., Srinivasan, A., Wang, N.: Integrality ratio for group Steiner trees and directed steiner trees. SIAM J. Comput. 36(5), 1494–1511 (2007)
23. Halperin, E., Krauthgamer, R.: Polylogarithmic inapproximability. In: Proceedings of the Thirty-Fifth Annual ACM Symposium on Theory of Computing, STOC 2003, pp. 585–594. ACM, New York, NY, USA (2003)
24. Hougardy, S., Prömel, H.J.: A 1.598 approximation algorithm for the Steiner problem in graphs. In: SODA 1999: Proceedings of the Tenth Annual ACM-SIAM Symposium on Discrete Algorithms, pp. 448–453. Society for Industrial and Applied Mathematics, Philadelphia, ACM, New York (1999)
25. Hsieh, S.Y., Gao, H.M.: On the partial terminal Steiner tree problem. J. Supercomput. 41(1), 41–52 (2007)
26. Hsieh, S.-Y., Gao, H.-M., Yang, S.-C.: On the internal Steiner tree problem. In: Cai, J.-Y., Cooper, S.B., Zhu, H. (eds.) TAMC 2007. LNCS, vol. 4484, pp. 274–283. Springer, Heidelberg (2007). https://doi.org/10.1007/978-3-540-72504-6_25
27. Huang, C.W., Lee, C.W., Gao, H.M., Hsieh, S.Y.: The internal Steiner tree problem: hardness and approximations. J. Complex. 29(1), 27–43 (2013)
28. Hwang, F.K., Richards, D.S., Winter, P.: The Steiner Tree Problem. Annuals of Discrete Mathematics, vol. 53. Elsevier Science Publishers, Amsterdam (1992)
29. Kahng, A.B., Robins, G.: On Optimal Interconnections for VLSI. Kluwer Academic, Boston (1995)
30. Karpinski, M., Zelikovsky, A.: New approximation algorithms for the Steiner tree problems. J. Comb. Optim. 1, 47–65 (1997)
31. Korte, B., Prömel, H.J., Steger, A.: Steiner trees in VLSI-layout. Paths, Flows, and VLSI-Layout, pp. 185–214 (1990)
32. Liang, W.: Constructing minimum-energy broadcast trees in wireless ad hoc networks. In: Proceedings of the 3rd ACM International Symposium on Mobile Ad Hoc Networking and Computing, MobiHoc 2002, pp. 112–122. ACM, New York, NY, USA (2002)
33. Lin, G.H., Xue, G.L.: On the terminal steiner tree problem. Inf. Process. Lett. 84(2), 103–107 (2002)
34. Lu, C.L., Tang, C.Y., Lee, R.C.T.: The full Steiner tree problem. Theor. Comput. Sci. 306(1–3), 55–67 (2003)
35. Martinez, F.V., de Pina, J.C., Soares, J.: Algorithms for terminal Steiner trees. J. Theor. Comput. Sci. 389(1—2), 133–142 (2007)
36. Min, M., Du, H., Jia, X., Huang, C.X., Huang, S.C.H., Wu, W.: Improving construction for connected dominating set with Steiner tree in wireless sensor networks. J. Glob. Optim. 35(1), 111–119 (2006)
37. Prömel, H.J., Steger, A.: RNC-approximation algorithms for the steiner problem. In: Reischuk, R., Morvan, M. (eds.) STACS 1997. LNCS, vol. 1200, pp. 559–570. Springer, Heidelberg (1997). https://doi.org/10.1007/BFb0023489

38. Robins, G., Zelikovsky, A.: Tighter bounds for graph Steiner tree approximation. SIAM J. Discrete Math. **19**(1), 122–134 (2005)
39. Takahashi, H., Matsuyama, A.: An approximate solution for the Steiner problem in graphs. Math. Jpn. **24**, 573–577 (1980)
40. Zelikovsky, A.: An 11/6-approximation algorithm for the network Steiner problem. Algorithmica **9**(5), 463–470 (1993)
41. Zelikovsky, A.: Better approximation bounds for the network and euclidean Steiner tree problems. In: Technical report CS-96-06. University of Virginia. Charlottesville, VA, USA (1996)

A Review for Submodular Optimization on Machine Scheduling Problems

Siwen Liu[1,2(✉)]

[1] School of Management, Hefei University of Technology, Hefei, Anhui, China
liusiwen67@126.com
[2] Department of Computer Science, University of Texas at Dallas,
Richardson, TX, USA

Abstract. This paper provides a review of recent results on machine scheduling problems solved by methods of submodular optimization. We present some basic definitions of submodular functions and their connection to scheduling models. Based on the classification of problem features, we conclude different scheduling models, applications of these scheduling scenarios, approaches of submodular optimization, and the performance of corresponding algorithms. It is shown that the use of these submodular optimization methodologies yields fast and efficient algorithms for specific scheduling models such as controllable processing time, unreliable job processing, and common operation scheduling. By identifying the trends in this field, we discuss some potential directions for future research.

Keywords: Submodular optimization · Machine scheduling · Review

1 Introduction

The submodular functions are set functions characterized by *diminishing return* property. In other words, the marginal gain of adding an element to a smaller subset of S is higher than that of adding it to a larger subset of S. Recently, submodular optimization (SO) has emerged as one of the most popular optimization tools in Computer Science. It appears in a variety of applications includes machine learning problems [18], mobile robotic sensing, door-to-door marketing [45], and image segmentation [17]. Due to the structure and characteristic of submodular functions, in these applications, algorithms developed based on the concepts of SO proved their efficiency and effectiveness (see, e.g., [7,42], and [12]).

Machine scheduling is a classical research field in combinatorial optimization and remains a constant research subject in the last three decades. There are different types of approaches proposed to solve scheduling problems over these years. These approaches can be divided into three categories: exact algorithms (e.g., branch-and-bound algorithm [40,41], dynamic programming algorithm [44,46]), heuristic algorithms [23,24], meta-heuristic algorithms [6,11,20].

D.-Z. Du and J. Wang (Eds.): Ko Festschrift, LNCS 12000, pp. 252–267, 2020.
https://doi.org/10.1007/978-3-030-41672-0_16

Though SO is found applicable in a large number of research topics in computer science, in the past two decades, it is noticed that there exists some specific scheduling scenarios can be effectively handled using the notion of SO. [43] were among the first who solve the scheduling of controllable processing time (SCPT) models by methods of SO. Since then, there is an increasing amount of investigation about how to solve the scheduling problem using SO methods. Among these research work, the main field still exists in SCPT problems while some other scheduling problems are also proposed, such as speed scaling machines scheduling problems, unreliable jobs scheduling problems, common operation scheduling problems. Figure 1 shows the distribution of machine scheduling literature using the approach of SO reviewed in this paper. It can be seen from this figure that there is already some existing research (more than 10) about SCPT using SO. Thus, SCPT scheduling articles are illustrated separately in Sect. 3, while other scheduling articles are reviewed in Sect. 4. Considering the mathematical characteristics of submodular functions, it can somehow combine with these scheduling models, which extends the existing scheduling methods to SO. However, it is worth noticing that there is no existing survey paper about submodular approaches applied in more general scheduling problems.

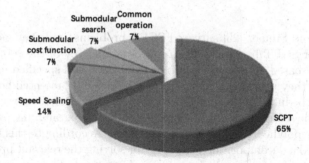

Fig. 1. Distribution of different scheduling publications using SO

In traditional scheduling problems, the job processing time is always assumed to be fixed and known in advance. However, there are different types of uncertain processing time scheduling models, which include machine deterioration, worker's learning effect, and machine maintenance. Another type of model deals with scheduling with controllable processing time (SCPT). In real-life applications, the use of an additional resource such as facilities, energy, human-resource may decrease jobs' original processing times. In a typical SCPT problem, compressing the job's processing time can reduce the planned completion time of a schedule. Meanwhile, it also causes an extra cost due to the usage of resources. Despite the similarity to project planning, this phenomenon is commonly observed in scheduling and sequencing problems in manufacturing enterprises. SCPT models are initially investigated and analyzed by [38]. Though SCPT scheduling problems were primarily studied, the traditional ways to solve SCPT models focus on dynamic programming algorithms [8], assignment problem formulation [9], and heuristic algorithm [25]. Since 2005, [28] started to combine the SCPT models with SO-based approaches, and they extended their research by considering different SCPT models and SO-based approaches (Table 4).

Table 1. List of abbreviations

Abbreviations	Expressions
SCPT	Scheduling with controllable processing times
SO	Submodular optimization
SIC	Scheduling with imprecise computation
GA	Greedy approach
COS	Common operation scheduling
LP	Linear programming
UJP	Unreliable job scheduling problem
SSP	Speed scaling machines scheduling problem

Shabtay and Steiner [26] surveyed SCPT problems, and they gave a unified framework for SCPT by providing an up-to-date survey of the results and performance. SO-based approaches were not mentioned and specified in this paper. Later in [34], they provided a general SCPT model and presented how to handle this typical scheduling problem by methods of SO. They demonstrated that this model could be reformulated as maximization linear programming problems over a submodular polyhedron intersected with a box. According to this formulation, they addressed a decomposition algorithm for solving the relevant problems both on a single machine and parallel machines. After that, [36] gave another survey on the preemptive models of SCPT and scheduling with imprecise computation (SIC). Different from their previous survey paper, they reviewed recent results on SCPT and emphasized on the methodological aspects. Besides, they mentioned SIC models and SCPT models actually studying the same range of problems and established relations between the SCPT and SIC models. Some scheduling problems with other characteristics solved by SO have appeared recently. However, throughout these previous survey papers, only SCPT problems using the methods of SO have been addressed.

We conceive our survey as a convincing example of the connection between more general machine scheduling problems and SO. This paper, which builds upon the classification of different scheduling features, makes another contribution toward the potential research fields of scheduling models, which can be solved by SO. We also highlight some important results and introduce the solution performance by the methods of SO.

This paper is organized as follows: Sect. 2 describes the notations and some preliminaries about submodular concepts applied in the paper. Sect. 3 summarizes the submodular approaches in controllable processing time scheduling problems, single machine, and parallel machine scheduling problems are both discussed. Sect. 4 introduces some other scheduling problems which are also involved in submodular optimization. Finally, concluding remarks are made in Sect. 5.

2 Preliminaries

There are many different classification factors to divide the scheduling problems, such as the number of stages jobs need to be processed, the number of machines at each stage, the jobs release dates/due dates requirements, different job processing times functions, and the number of objectives to optimize or the type of the objectives. Regarding there is a limited number of papers about the submodular method applied in scheduling problems, we aim to find the potential scheduling categories which can be solved using SO. Therefore, this paper is developed with the characteristics of scheduling problems.

Table 2. Notations list

Notations	Description
N	Job set
n	Number of jobs
m	Number of machines
i	Index for machines
j	Index for jobs
M_i	The ith machine, $i \in 1, 2, ..., m$
J_j	The jth job, $j \in 1, 2, ..., n$
l_j	Lower bound of the processing time of job J_j, $j \in 1, 2, ..., n$
u_j	Upper bound of the processing time of job J_j, $j \in 1, 2, ..., n$
p_j	Actual processing time of job J_j, $j \in 1, 2, ..., n$
x_j	Compression amount of job J_j, $j \in 1, 2, ..., n$
w_j	Non-negative cost unit compression cost of job J_j, $j \in 1, 2, ..., n$
r_j	Release dates of job J_j, $j \in 1, 2, ..., n$
d_j	Due dates of job J_j, $j \in 1, 2, ..., n$
ω_j	Weight of job J_j, $j \in 1, 2, ..., n$
T_j	Tardiness of job J_j, $j \in 1, 2, ..., n$
sp_j	Success probability of job J_j, $j \in 1, 2, ..., n$
rw_j	Reward of job J_j, $j \in 1, 2, ..., n$ if it is processed successively
s_j	Processing speed of job J_j, $j \in 1, 2, ..., n$

An adapted version of the notations in this paper is introduced as follows and presented in Table 2. In all the scheduling problems considered the number of jobs and the number of machines are assumed to be finite and denote as n and m, respectively. $N = \{1, 2, ..., n\}$ is used to denote a job set contains n jobs. The subscript j refers to a job, while i refers to a machine. Besides, to explain the scheduling problems addressed in this paper intuitively, the commonly utilized three-field notation $\alpha|\beta|\gamma$ first proposed by [15] is also introduced. The first field α describes the machine(shop) environment. The β field includes one or multiple processing restrictions and constraints for processing while the objectives are specified in the γ field. Table 3 gives a description of the three fields. For example, a parallel machine scheduling problem with different release dates and precedence constraints to minimize the makespan can be expressed as $Pm|r_j, prec|C_{max}$.

Table 3. Description of α, β, γ fields

α		β		γ	
Notation	Description	Notation	Description	Notation	Description
1	Single	r_j	Release dates	C_{max}	Makespan
Pm	Identical	d	Common due dates	$\sum C_j$	Total completion time
Qm	Uniform	d_j	Due dates	E_{max}	Maximum earliness
Rm	Unrelated	$pmtn$	Preemptive schedule	L_{max}	Maximum lateness
		$prec$	Precedence constraints	$\sum w_j C_j$	Total weighted completion time

In the following, we give some basic definitions and concepts related to submodular optimization follow the contents in Krause and Golovin [19]. Let N be a finite set, commonly called the ground set and 2^N denote the family of all subsets of N. Functions $f : 2^N \rightarrow \mathbb{R}$ is a set function that assign each subset $S \subseteq N$ a value $f(S)$.

For a set function $f : 2^N \rightarrow \mathbb{R}$, $S \subseteq N$, and $e \in N$, let $\Delta_f(e|S) := f(S \cup e) - f(S)$ be the *discrete derivative* of f at S with respect to e. A set function is called *submodular* if the following inequality

$$f(A \cup B) + f(A \cap B) \leq f(A) + f(B) \tag{1}$$

holds for every $A \subseteq B \subseteq N$. Or

$$\Delta(e|A) \geq \Delta(e|B) \tag{2}$$

for every $A \subseteq B \subseteq N$ and $e \in N$. For a submodular function f defined on 2^N such that $f(\emptyset) = 0$, the pair $(2^N, f)$ is called submodular system on N, whereas f is referred to as the rank function of the system. Besides, a submodular function is said to be *monotone* iff all its discrete derivatives are non-negative, i.e., iff for every $A \subseteq N$ and $e \in N$ it holds that $\Delta(e|A) \geq 0$. Equivalently, a function $f : 2^N \rightarrow \mathbb{R}$ is *monotone* if for every $A \subseteq B \subseteq N$, $f(A) \leq f(B)$.

For a submodular system $(2^N, f)$, the following two polyhedra are

$$P(f) = \{\mathbf{p} \in \mathbb{R}^N | p(X) \leq f(X), X \in 2^N\} \tag{3}$$

$$B(f) = \{\mathbf{p} \in \mathbb{R}^N | \mathbf{p} \in P(f), p(N) = f(N)\} \tag{4}$$

called a *submodular polyhedron* and *base polyhedron*, respectively. In the case where $|N| = 2$, the submodular polyhedron $P(f)$ and the base polyhedron $B(f)$ are represented in Fig. 2. It can be seen that $B(f)$ is the set of all maximal vectors in $P(f)$.

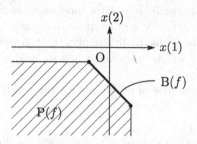

Fig. 2. The *submodular polyhedron* $P(f)$ *and the base polyhedron* $B(f)$

3   Review of SCPT Problems Using so

In this section, we first give the formal description of the SCPT problem and the objectives involved with controllable processing time. Meanwhile, the applications of SCPT problems existing in different manufacturing scenarios are illustrated. Based on the preliminary knowledge, some related definitions and conceptions of the submodular approaches used in the SCPT problem are introduced. Afterward, a detailed review of the submodular approaches applied in different SCPT production scenarios is given.

In scheduling problems with controllable processing times, the actual duration of the jobs are not fixed in advance but have to be chosen from a given interval. For a typical SCPT model, there are n jobs to be scheduled on a single machine or m parallel machines, the actual processing time of a job J_j is within $[l_j, u_j]$. The compression amount of the longest processing time u_j down to p_j is denoted as x_j and calculated by $x_j = u_j - p_j$. On the one hand, the compression decreases the processing time. On the other hand, it results in additional costs. Use w_j to express the non-negative unit compression cost, $w_j x_j$ is the additional compression cost. Besides, preemption is allowed during the processing of any job J_j, which means that the processing can be interrupted on any machine at any time and resumed later, possibly on a different machine or on the same machine. It is assumed that a job can only be processed on one machine at a time, and machine processes at most one job at a time.

As a result of compressing processing time, two types of decisions have to be made in a typical SCPT model: (i) how to assign the actual processing time for each job, and (ii) how to optimize the objective while remaining the restrictions satisfied. Usually, the quality of the resulting schedule is measured with respect to the cost of assigning the actual processing times that guarantee a certain scheduling performance. The wide used objectives in these type of scheduling problems include the following but not limited: C_{max}, $\sum w_j x_j$, L_{max}, $max_{j \in N} x_j / w_j$.

The research into SCPT problems has been active since the 1980s, see surveys by Nowicki and Zdrzalka [22]. After that, SCPT problems are found to be applicable in many practical scenarios include manufacturing, make-or-buy decision making, supply chain management, and imprecise computation. In the survey paper of Shabtay and Steiner [26], they concluded the application of submodular approaches in SCPT problems. It is later mentioned in Shioura et al. [36] that the SCPT and the SIC actually studies the same range of problems, and often apply the same methods. Thus, in this paper, the term SCPT is adopted as the main contents, and SIC will be specified when we only refer to the situation in SIC.

Some application examples of SCPT are introduced from the following three aspects.

Example 1: In manufacturing enterprises, each product J_j has a standard requirement time for its processing, which can be denoted as u_j. Nevertheless, an additional resource (workforce, better equipment condition, new technology) can effectively speed up the original manufacturing process. As a result, the standard processing time can be decreased to p_j, and x_j is the amount of compression time, but still, the processing requirement can never be less than a lower bound l_j. The compression cost $w_j x_j$ in this example indicates the resource cost spent to accelerate the process.

Example 2: In operations management, especially in supply chain logistics, managers always need to make make-or-buy decisions in order to balance the internal production and outsourcing. It may be profitable for a contractor to process only a part of the order internally for p_j time units instead of its full processing requirement u_j using its own facilities and to hire a subcontractor to perform the remaining part of the order for $x_j = u_j - p_j$ time units. A low-risk strategy is achieved by setting the lower bounds l_j of internal production reasonably close to the size of the order u_j. To maximizing the profit, the subcontractor's fee $\sum w_j x_j$ is taken into account at the planning stage.

Example 3: In computing, in systems that support imprecise computations, a task with processing requirement u_j can be decomposed into a mandatory part which takes l_j time, and an optional part which may take up to $u_j - l_j$. If instead of an ideal computation time u_j a task is executed for $p_j = u_j - x_j$ time, then the computation is imprecise, and x_j corresponds to the error of computation. In this application, total compression cost $\sum w_j x_j$ is the total weighted error.

Nemhauser and Wolsey [21] were among the first who noticed that the SCPT models could be handled by methods of Submodular Optimization (SO). Systematic development of a general framework for solving the SCPT problems via submodular methods has been initiated by Shakhlevich and Strusevich [28] and further advanced in their afterward publications. As a result, a powerful toolkit of the SO techniques can be used for designing and justification of algorithms for solving a wide range of SCPT problems.

We give the formal description of the controllable processing times scheduling problems (SCPT) as follows: There is a set of n jobs $N = \{1, 2, ..., n\}$ to be scheduled either on a single machine M_1 or on parallel machines $M_1, M_2, ..., M_m$, where $m \geq 2$. The processing time of each job J_j is not known in advance but has to be chosen from a given interval $[l_j, u_j]$. Let p_j denote the actual processing time, the value $x_j = u_j - p_j$ is called the compression amount of job J_j. Compression may decrease the completion time of each job J_j but incurs additional cost $w_j x_j$, where w_j is a given non-negative unit compression cost.

Moreover, each job $j \in N$ is given a release date r_j, before which it can not be processed, and a deadline d_j, by which its processing must be completed. In the processing of any job, preemption is allowed, which means that the processing can be interrupted on any machine at any time and resumed later, possibly on a different machine. It is assumed that a job can only be processed on one machine at a time, and machine processes at most one job at a time. A schedule is called feasible only if two conditions are both satisfied: (i) job processing time of any job $J_j : j \in N$ is within the interval $[l_j, u_j]$; (ii) the starting time and finishing time of any job $J_j : j \in N$ is within the time interval $[r_j, d_j]$.

Connecting scheduling problems with the basic conceptions of submodular functions, some related definitions are explained. Let $N = 1, 2, ..., n$ be a ground set to denote the job set, where n is a positive integer which indicates the job number, and 2^N denote the family of all subsets of N. For a subset $X \subseteq N$, let $\mathbb{R}^X$ denote the set of all vectors $\mathbf{p}$ with real components p_j, where $j \in X$. For two vectors $\mathbf{p} = (p_1, p_2, ..., p_n) \in \mathbb{R}^N$ and $\mathbf{q} = (q_1, q_2, ..., q_n) \in \mathbb{R}^N$, we write $\mathbf{p} \leq \mathbf{q}$ if $p_j \leq q_j$ for each $j \in N$.

Assuming that the objective is to minimize the total compression cost, the main problem of the controllable processing time scheduling problem can be represented as follows:

$$(LP) : min \sum_{j \in N} w_j x_j$$
$$s.t. p(X) \leq \varphi(X), X \in 2^N, \tag{5}$$
$$l_j \leq p_j \leq u_j, j \in N$$

where $\varphi : 2^N \to \mathbb{R}$ is a submodular function which is also called the rank function with $\varphi(\emptyset) = 0$, $\mathbf{w} \in \mathbb{R}^N_+$ is a nonnegative weight vector, and $\mathbf{u}, \mathbf{l}$ are upper and lower bound vectors, respectively. We refer to Eq. (5) as Problem (LP). This problem can be established as a mathematical model for many SCPT problems and transformed to maximize a linear function over a submodular polyhedron intersected with a box.

In Table 4, we give a detailed summary of SCPT problems using the tool of SO. The machine environment, problem features, the objectives, and the time complexity of its solution are listed. Shakhlevich and Strusevich [28] first combined a range of single machine preemptive SCPT models with SO in their paper. For each model, they studied a single criterion problem to minimize the compression cost of the processing times subject to the non-identical release date r_j and due date d_j constraints. Their main contribution is that they formulated

Table 4. Previous literature about solving SCPT problems using SO

Publications	Problems			Time complexity
	Machine	Features	Objectives	
[28]	1	r_j, d	$\sum w_j x_j$	$O(n \log n)$
	1	r_j	$(C_{max}, \sum w_j x_j)$	$O(n \log n)$
	1	d_j	$\sum w_j x_j$	$O(n^2)$
	1	d_j	$(L_{max}, \sum w_j x_j)$	$O(n^2)$
	Pm	d	$\sum w_j x_j$	$O(n)$
	Pm	-	$(C_{max}, \sum w_j x_j)$	$O(n \log n)$
[29]	Qm	r_j, d_j	$\sum w_j x_j$	$O(mn^4)$
	Qm	d	$\sum w_j x_j$	$O(n \log n + nm)$
	Qm	-	$(C_{max}, \sum w_j x_j)$	$O(n \log n + nm^4)$
[27]	1	-	$(max_{j \in N} f_j(C_j), \sum w_j x_j)$	$O(n^3 L)$
	1	$f_j(C_j) \le U$	$\sum w_j x_j$	$O(n \log n + \sum \log m_j)$
	1	$\sum w_j x_j \le V$	$max f_j(C_j)$	$O(L + n^2 + (\sum \log m_j + n \log n) \log L)$
[30]	Pm	r_j	$(C_{max}, \sum w_j x_j)$	$O(n^2 \log m)$
	Qm	-	$(C_{max}, \sum w_j x_j)$	$O(nm \log m)$
	Qm	r_j	$(C_{max}, \sum w_j x_j)$	$O(n^2 m)$
[31]	Qm	d	$\sum w_j x_j$	$O(n \log n)$ or $O(n + m \log m \log n)$
	Pm	r_j, d	$\sum w_j x_j$	$O(n \log m \log n)$
	Qm	r_j, d	$\sum w_j x_j$	$O(nm \log n)$
[33]	1	r_j, d_j	$\sum w_j x_j$	$O(n \log n)$
[37]	1	r_j, d	x_j / w_j	$O(n \log n)$
	Pm	d	x_j / w_j	$O(n)$
	Pm	r_j, d	x_j / w_j	$O(n^2)$
	Qm	d	x_j / w_j	$O(n \log n + nm)$
	Qm	r_j, d	x_j / w_j	$O(mn^2)$

$pmtn$ and $p_j = u_j - x_j$ are omitted in the Features

each single criterion problem as minimizing a linear function over a polymatroid, and this justified the greedy approach to its solution. Two objectives are applied to measure the quality of a schedule: the makespan $C_{max} = max_{j \in N} C_j$, i.e., the maximum completion time of all jobs, and the maximum lateness $L_{max} = max_{j \in N} \{C_j - d_j\}$. Both single criteria and bicriteria models are considered in their paper.

For the single criteria model $1|r_j, p_j = u_j - x_j, C_j \le d, pmtn| \sum w_j x_j$ and $1|r_j, p_j = u_j - x_j, C_j \le d_j, pmtn| \sum w_j x_j$, they first replaced the processing constraints by a monotone submodular set function $\varphi(X) : 2^N \to \mathbb{R}$, then they proposed a balanced 2–3-tree to implement Greedy Algorithm (GA) in

$O(n \log n)$ time and $O(n^2)$ time for the common due date and arbitrary due date, respectively. Based on the results for single criteria, they further investigated bicriteria problems, typically $1|r_j, p_j = u_j - x_j, C_j \leq d, pmtn|(C_{max}, \sum w_j x_j)$ and $1|r_j, p_j = u_j - x_j, C_j \leq d_j, pmtn|(C_{max}, \sum w_j x_j)$. They are solvable by GA in $O(n^2)$ time. As for the parallel machines scheduling problems with equal release dates, the same procedure is applied and solve the problem $Pm|p_j = u_j - x_j, C_j \leq d, pmtn| \sum w_j x_j$ in $O(n)$ time and $Pm|p_j = u_j - x_j, pmtn|(C_{max}, \sum w_j x_j)$ in $O(n \log n)$ time.

Shakhlevich and Strusevich [29] extended the former mentioned work [28] in a way that they provided a more general approach to solving preemptive scheduling problems with uniform parallel machines and controllable processing times. For the single criterion circumstance with arbitrary release dates and due dates and common due dates, they developed a combinatorial structure of generalized polymatroid and solve these problems in $O(mn^4)$ and $O(n \log n + nm)$ time, respectively. For the bicriteria circumstance with no release date and due date constraints, they addressed an algorithm that constructs the trade-off curve between the compression cost and the makespan.

As a generalized version of [29], Shioura et al. [30] and [31] both investigated the parallel machine SCPT problems. Different from [29], [30] considered all bicriteria problems ($C_{max}, \sum w_j x_j$) and only release dates constraints are taken into account. They first reformulated the SCPT models in terms of optimization over submodular polyhedra, then gave the corresponding efficient frontier in the form of break points. As for [31] in which the objective is to maximize the total compression cost $\sum w_j x_j$, they presented a decomposition recursive algorithm for solving the original submodular optimization for a linear function. Their results in [31] not only contributed to the SO but also extended the toolkit in SCPT scheduling problems. The obtained time complexity of this decomposition recursive algorithm outperforms those previously algorithms in this field of scheduling literature.

In the paper of Shakhlevich et al. [27], they only considered single machine circumstances. Their main model is a bicriteria SCPT model, they solved this model by determining the trade-off between the two objectives ($max f_j(C_j), \sum w_j x_j$) in which $f_j(C_j)$ refers to the makespan C_j or tardiness $max\{C_j - d_j, 0\}$. They represented the feasible region as a submodular polyhedron. Additionally, they considered a pair of associated single criterion problems, in which one of the objective functions is bounded while the other one is to be minimized. Thus, the corresponding problems can be solved by the greedy algorithm that runs two orders of magnitude faster than known previously.

Shioura et al. [37] considered a series of common due date SCPT problems both on a single machine and parallel machines. The main difference between the previous work and [37] is the objective. Their objective in this paper is to minimize the maximum compression cost x_j/w_j while the traditional stream of research on SCPT focus on the total compression cost $\sum w_j x_j$. In their paper, they outlined general principles aimed at solving problems $\alpha|r_j, p_j = u_j - x_j, C_j \leq d, pmtn|\Phi_{max}$ for $\alpha \in \{1, P, Q\}$. It is proved that the running times are effectively reduced compared to the best-known running time in previous literature.

Later in the paper of Shioura et al. [36], they provided a review of solution approaches on SCPT problems. The main aspects they included in this paper are parametric flow techniques and methods for solving mathematical programming problems with submodular constraints. Three different types of methodologies are introduced: Flows in networks, Optimization over submodular polyhedra, and submodular optimization via decomposition algorithm. These reviewed methodologies can generalize fast algorithms that outperform those corresponding fixed processing time SCPT problems on a single or parallel machine for a single criterion or bicriterion. Additionally, the best possible algorithms for a range of SCPT problems on parallel machines are concluded.

4   Review of Other Scheduling Problems Using SO

4.1   Common Operation Scheduling

Common operation scheduling (COS) problems are identified in many real-world applications: movie shooting [10,13], progressive network recovery [39], and pattern sequencing in stock cutting [5]. The specific application in manufacturing exists in material cutting or component dismantling process. The main characteristic of the COS problems is that distinct jobs may share the same operation, and when one operation is finished, all jobs share this operation are finished at the same time. In the paper of [2], they introduced an industrial application in cut sequencing: the manager needs to decide the optimal pattern to cut given large plates (stock items) into objects (small items). Besides, the sequence of jobs influences the final production quality. Thus, a feasible solution generally includes the cutting patterns corresponds to operations and the sequence of the jobs on each operation.

Denote the job tardiness as $T_j = max\{C_j - d_j, 0\}$, the job weight ω_j, this paper is aiming to minimize the weighted number of tardy jobs: $\sum \omega_j t_j$ where $t_j = 1$ if $T_j > 0$ and $t_j = 0$ otherwise. Arbib, Felici, and Servilio [2] formulated the problem they investigated as follows: Schedule a set of o operations O on Qm, give each operation a machine and a starting time so that precedence relations among operations are respected and $\sum \omega_j t_j$ is minimized. They transformed this formulation to SET COVERING with inequalities increasingly exponentially with the number of jobs. Separation/lifting of cover inequalities is realized through the constrained maximization of a submodular set function. A heuristic and a branch-and-cut algorithm are then proposed to solve the model, and a series of computational experiments are carried out. It can be observed in the results that the method outperforms previous method for two cases: $1|cos| \sum \omega_j t_j$ and $P_2|cos, p_j = 1| \sum \omega_j t_j$.

4.2   Unreliable Jobs

In the typical production in an unsupervised automated manufacturing shop, jobs are assigned to a set of parallel machines and automatically loaded on the machine whenever the machine becomes idle. If the production of a job J_j fails, the remaining jobs on that machine cannot be processed and can only be resumed until the machine is cleared at the end of the unsupervised shift. In this model, each job J_j has a success probability sp_j and a reward rw_j if it can be processed successively. In the paper of [1], they considered this model, and their objective is to maximize the expected reward.

Denote the sequence of jobs assigned to a machine as σ, and σ_j is the job in jth position in σ. Thus, the expected reward $ER[\sigma]$ for a machine can be expressed as

$$ER[\sigma] = sp_{\sigma(1)}rw_{\sigma(1)} + sp_{\sigma(1)}sp_{\sigma(2)}rw_{\sigma(2)} + sp_{\sigma(1)} \cdots sp_{\sigma(k)}rw_{\sigma(k)} \quad (6)$$

The objective value is the sum of expected reward on m machines: $ER[\zeta] = ER[\sigma_1] + ER[\sigma_2] + \cdots + ER[\sigma_m]$. A solution to this problem consists of an assignment and a sequence of the n jobs to m machines. Agnetis et al. [1] denotes this problem as $UJP(m)$. For the single machine case, they formulated $UJP(1)$ as maximizing a submodular function over a polymatroid and solved this problem using greedy algorithm. For the parallel machines case, they first proved this problem is NP-hard even for two machines. Then, a round-robin heuristic algorithm is proposed to solve the parallel machines circumstance. Through the experimental results comparing to the performance of a upper bound, the proposed heuristic algorithm proved its superiority in solving $UJP(m)$.

4.3 Speed Scaling Machines

In the scheduling models with speed scaling machines (SSP), machines are able to work at different voltage levels, which enable the manufacturing process to achieve a lower level of energy consumption. SSP is also known as "energy-aware scheduling"[3], "green scheduling"[4]. Since [16] first proposed the concept of SSP, this topic has been well studied due to the increasing demands for energy saving in manufacturing enterprises. Different from most of the previous literature, [32] and [35] provided new insights into solving this sort of SSP models by connecting its link to submodular constraints. Except for different release dates and due dates, they also considered controllable processing times in their paper. They developed a new methodological framework for handling the SSP problem. Based on submodular optimization, algorithms with lower time complexity are proposed for both single- and multiple machines.

Formally, a set of n jobs have to be processed on a set of m parallel machines. Each job has a release date r_j and a due date d_j. Let v_j denote the volume of computation of job J_j, the actual processing time can be rewritten as $p_j = v_j/s_j$. The cost of keeping the processing speed of this job equal to speed s_j for one unit is $f_j(s_j)$. The objective function becomes

$$\hat{F} = \sum_{j=1}^{n} p_j f_j\left(\frac{v_j}{p_j}\right) \quad (7)$$

In the paper of [32] and [35], they formulated this problem as minimizing above $\hat{F}$ in the form of min-cost maxflow problem with a non-linear convex separation objective function. Linking this problem to a non-linear convex minimization problem under submodular constraints, they adapted a decomposition algorithm and implemented $O(n^4)$ and $O(n^2)$ time algorithms for original SSP problem on parallel machines and single machines, respectively.

4.4 Submodular Search in Scheduling

In the paper of [14], they proposed a submodular search model which can be formally described as follows: there is a finite set S of hiding places, a submodular cost function $f : 2^S \to [0, \infty)$, and a supermodular weight function $g : 2^S \to [0, \infty)$. For each search, π of S has to be chosen, and denote $S_j = S_j^\pi$ the union of j and all the locations that precede j in π. Thus, the search cost of j under π is $f(S_j^\pi)$ and the probability that all the objects are in S_j^π is $g(S_j^\pi)$. Their objective is to find an ordering of S that finds all the objects with minimal expected cost, which can be written as

$$c(\pi) = \sum_{j=1}^{n} (g(S_j^\pi) - g(S_j^\pi - j))f(S_j^\pi) \tag{8}$$

This model can be found in many scheduling problems: single machine scheduling with precedence constraints $1|prec| \sum \omega_j C_j$, scheduling with more general costs $1|prec| \sum \omega_j h(C_j)$ in which h are some monotonically increasing functions of the completion times of jobs, and scheduling with subset weights $1|prec| \sum \omega_A C_A$ where A is a subset of all jobs. Considering these problems are NP-hard, [14] developed an efficient combinatorial 2-approximation algorithm using the notion of series-parallel decomposability to solve above-mentioned precedence-constrained scheduling problems.

5   Conclusion

In this paper, we reviewed recent articles on machine scheduling problems solved by methods of submodular optimization. We presented the notations, problem representations, and concepts of submodularity for relevant machine scheduling models. The classification adopted in this paper is the scheduling features. Through the analysis, we find that the SCPT models are the most intensively investigated. For single machine or parallel machines, single criterion or bicriterion, with or without release dates and due dates constraints, a general SO-based framework can be proposed to solve the relevant SCPT scheduling problems. The details of these SCPT models and performances of the algorithms are discussed. Other scheduling problems solved by submodular optimization exist in common operation scheduling, unreliable job scheduling, speed scaling machines scheduling, and submodular search in scheduling. Though there is limited relevant literature on these topics, it is worth noticing that the application of submodular optimization extends traditional scheduling methods and improve the solution qualities.

For potential future research, it is interesting to study constraints such as fuzzy processing, learning/deteriorating effect in SCPT models. As for other scheduling models, different objectives (time-dependent or resource-dependent), multi-objective models, and different machine environments (parallel machines, flow shop, or job shop) can also be taken into account. Moreover, it is challenging to see whether there exist algorithms based on submodular optimization which perform better for these proposed scheduling models.

References

1. Agnetis, A., Detti, P., Pranzo, M., Sodhi, M.S.: Sequencing unreliable jobs on parallel machines. J. Sched. **12**(1), 45 (2009)
2. Arbib, C., Felici, G., Servilio, M.: Common operation scheduling with general processing times: a branch-and-cut algorithm to minimize the weighted number of tardy jobs. Omega **84**, 18–30 (2019)
3. Bambagini, M., Lelli, J., Buttazzo, G., Lipari, G.: On the energy-aware partitioning of real-time tasks on homogeneous multi-processor systems. In: 2013 4th Annual International Conference on Energy Aware Computing Systems and Applications (ICEAC), pp. 69–74. IEEE (2013)
4. Bampis, E., Letsios, D., Lucarelli, G.: Green scheduling, flows and matchings. Theor. Comput. Sci. **579**, 126–136 (2015)
5. Belov, G., Scheithauer, G.: Setup and open-stacks minimization in one-dimensional stock cutting. INFORMS J. Comput. **19**(1), 27–35 (2007)
6. Chaudhry, I.A., Khan, A.A.: A research survey: review of flexible job shop scheduling techniques. Int. Trans. Oper. Res. **23**(3), 551–591 (2016)
7. Chen, Y., Krause, A.: Near-optimal batch mode active learning and adaptive submodular optimization. ICML (1) **28**(160–168), 8–1 (2013)
8. Chen, Z.L., Lu, Q., Tang, G.: Single machine scheduling with discretely controllable processing times. Oper. Res. Lett. **21**(2), 69–76 (1997)
9. Cheng, T., Chen, Z., Li, C.L.: Parallel-machine scheduling with controllable processing times. IIE Trans. **28**(2), 177–180 (1996)
10. Cheng, T., Diamond, J., Lin, B.: Optimal scheduling in film production to minimize talent hold cost. J. Optim. Theor. Appl. **79**(3), 479–492 (1993)
11. Edis, E.B., Oguz, C., Ozkarahan, I.: Parallel machine scheduling with additional resources: notation, classification, models and solution methods. Eur. J. Oper. Res. **230**(3), 449–463 (2013)
12. Epasto, A., Lattanzi, S., Vassilvitskii, S., Zadimoghaddam, M.: Submodular optimization over sliding windows. In: Proceedings of the 26th International Conference on World Wide Web, pp. 421–430 (2017). International World Wide Web Conferences Steering Committee
13. Fink, A., Voß, S.: Applications of modern heuristic search methods to pattern sequencing problems. Comput. Oper. Res. **26**(1), 17–34 (1999)
14. Fokkink, R., Lidbetter, T., Végh, L.A.: On submodular search and machine scheduling. Math. Oper. Res. **44**(4), 1431–1449 (2019)
15. Graham, R.L., Lawler, E.L., Lenstra, J.K., Kan, A.R.: Optimization and approximation in deterministic sequencing and scheduling: a survey. In: Annals of Discrete Mathematics, vol. 5, pp. 287–326. Elsevier (1979)

16. Ishii, H., Martel, C., Masuda, T., Nishida, T.: A generalized uniform processor system. Oper. Res. **33**(2), 346–362 (1985)
17. Jegelka, S., Bach, F., Sra, S.: Reflection methods for user-friendly submodular optimization. In: Advances in Neural Information Processing Systems, pp. 1313–1321 (2013)
18. Kim, G., Xing, E.P., Fei-Fei, L., Kanade, T.: Distributed cosegmentation via submodular optimization on anisotropic diffusion. In: 2011 International Conference on Computer Vision, pp. 169–176. IEEE (2011)
19. Krause, A., Golovin, D.: Submodular function maximization (2014)
20. Masdari, M., Salehi, F., Jalali, M., Bidaki, M.: A survey of pso-based scheduling algorithms in cloud computing. J. Netw. Syst. Manag. **25**(1), 122–158 (2017)
21. Nemhauser, G.L., Wolsey, L.A., Fisher, M.L.: An analysis of approximations for maximizing submodular set functions–i. Math. Program. **14**(1), 265–294 (1978)
22. Nowicki, E., Zdrzałka, S.: A survey of results for sequencing problems with controllable processing times. Discret. Appl. Math. **26**(2–3), 271–287 (1990)
23. Ou, J., Zhong, X., Wang, G.: An improved heuristic for parallel machine scheduling with rejection. Eur. J. Oper. Res. **241**(3), 653–661 (2015)
24. Pei, J., Pardalos, P.M., Liu, X., Fan, W., Yang, S.: Serial batching scheduling of deteriorating jobs in a two-stage supply chain to minimize the makespan. Eur. J. Oper. Res. **244**(1), 13–25 (2015)
25. Shabtay, D., Kaspi, M.: Minimizing the total weighted flow time in a single machine with controllable processing times. Comput. Oper. Res. **31**(13), 2279–2289 (2004)
26. Shabtay, D., Steiner, G.: A survey of scheduling with controllable processing times. Discret. Appl. Math. **155**(13), 1643–1666 (2007)
27. Shakhlevich, N.V., Shioura, A., Strusevich, V.A.: Single machine scheduling with controllable processing times by submodular optimization. Int. J. Found. Comput. Sci. **20**(02), 247–269 (2009)
28. Shakhlevich, N.V., Strusevich, V.A.: Pre-emptive scheduling problems with controllable processing times. J. Sched. **8**(3), 233–253 (2005)
29. Shakhlevich, N.V., Strusevich, V.A.: Preemptive scheduling on uniform parallel machines with controllable job processing times. Algorithmica **51**(4), 451–473 (2008)
30. Shioura, A., Shakhlevich, N.V., Strusevich, V.A.: A submodular optimization approach to bicriteria scheduling problems with controllable processing times on parallel machines. SIAM J. Discret. Math. **27**(1), 186–204 (2013)
31. Shioura, A., Shakhlevich, N.V., Strusevich, V.A.: Decomposition algorithms for submodular optimization with applications to parallel machine scheduling with controllable processing times. Math. Program. **153**(2), 495–534 (2015)
32. Shioura, A., Shakhlevich, N.V., Strusevich, V.A.: Energy saving computational models with speed scaling via submodular optimization. In: Proceedings of Third International Conference on Green Computing, Technology and Innovation, pp. 7–18 (2015)
33. Shioura, A., Shakhlevich, N.V., Strusevich, V.A.: Application of submodular optimization to single machine scheduling with controllable processing times subject to release dates and deadlines. INFORMS J. Comput. **28**(1), 148–161 (2016)
34. Shioura, A., Shakhlevich, N.V., Strusevich, V.A.: Handling scheduling problems with controllable parameters by methods of submodular optimization. In: Kochetov, Y., Khachay, M., Beresnev, V., Nurminski, E., Pardalos, P. (eds.) DOOR 2016. LNCS, vol. 9869, pp. 74–90. Springer, Cham (2016). https://doi.org/10.1007/978-3-319-44914-2_7

35. Shioura, A., Shakhlevich, N.V., Strusevich, V.A.: Machine speed scaling by adapting methods for convex optimization with submodular constraints. INFORMS J. Comput. **29**(4), 724–736 (2017)
36. Shioura, A., Shakhlevich, N.V., Strusevich, V.A.: Preemptive models of scheduling with controllable processing times and of scheduling with imprecise computation: a review of solution approaches. Eur. J. Oper. Res. **266**(3), 795–818 (2018)
37. Shioura, A., Shakhlevich, N.V., Strusevich, V.A.: Scheduling problems with controllable processing times and a common deadline to minimize maximum compression cost. J. Glob. Optim., 1–20 (2018). https://doi.org/10.1007/s10898-018-0686-2
38. Vickson, R.: Two single machine sequencing problems involving controllable job processing times. AIIE Trans. **12**(3), 258–262 (1980)
39. Wang, J., Qiao, C., Yu, H.: On progressive network recovery after a major disruption. In: 2011 Proceedings IEEE INFOCOM, pp. 1925–1933. IEEE (2011)
40. Wang, S., Liu, M.: A branch and bound algorithm for single-machine production scheduling integrated with preventive maintenance planning. Int. J. Prod. Res. **51**(3), 847–868 (2013)
41. Wang, S., Liu, M., Chu, C.: A branch-and-bound algorithm for two-stage no-wait hybrid flow-shop scheduling. Int. J. Prod. Res. **53**(4), 1143–1167 (2015)
42. Wang, Y., Liu, Y., Kirschen, D.S.: Scenario reduction with submodular optimization. IEEE Trans. Power Syst. **32**(3), 2479–2480 (2016)
43. Wolsey, L.A., Nemhauser, G.L.: Integer and Combinatorial Optimization, vol. 55. Wiley, Hoboken (1999)
44. Yin, Y., Wang, Y., Cheng, T., Wang, D.J., Wu, C.C.: Two-agent single-machine scheduling to minimize the batch delivery cost. Comput. Ind. Eng. **92**, 16–30 (2016)
45. Zhang, H., Vorobeychik, Y.: Submodular optimization with routing constraints. In: Thirtieth AAAI Conference on Artificial Intelligence (2016)
46. Zhao, K., Lu, X.: Approximation schemes for two-agent scheduling on parallel machines. Theor. Comput. Sci. **468**, 114–121 (2013)

Edge Computing Integrated with Blockchain Technologies

Chuanwen Luo[1]([✉]), Liya Xu[2], Deying Li[1], and Weili Wu[3]

[1] School of Information, Renmin University of China,
Beijing 100872, People's Republic of China
chuanwen_luo@163.com, deyingli@ruc.edu.cn
[2] School of Information Science and Technology, Jiujiang University,
Jiujiang 332005, People's Republic of China
xuliya603@whu.edu.cn
[3] Department of Computer Science, University of Texas at Dallas,
Richardson, TX 75080, USA
weiliwu@utdallas.edu

Abstract. With the rapid increasing of the number of devices connected to the Internet of Things (IoTs), the traditional centralized cloud computing system is unable to satisfy the Quality of Service (QoS) for many applications, especially for areas with real-time, reliability and security. The edge computing as an extension of the cloud computing is introduced, which lies in its ability to transfer the sensitive data from cloud to the edge for increasing network security and to realize high frequency interaction and real-time transmission of data. However, that the edge servers maintain sensitive privacy information generates many important security issues for the edge computing network. Moreover, the data produced by IoT devices are separated into many parts and stored in different edges servers that are located in different locations, which is hard to guarantee data integrity due to data loss and incorrect data storage in edge servers. As the emergence of blockchain technologies, the various security problems and data integrity of the edge computing can be addressed by integrating blockchain technologies. In this paper, we present a comprehensive overview of edge computing integrated with blockchain technologies. Firstly, the blockchain technologies and the architecture of the edge computing are introduced. Secondly, the motivations and architecture of the edge computing integrated with blockchain are introduced. Thirdly, the related works about the edge computing integrated with blockchain that have been investigated are introduced. Finally, the research challenges are discussed.

Keywords: Edge computing · Blockchain · Internet of Things

1 Introduction

In the traditional cloud computing, all data produced by Internet of Thing (IoT) devices have to be uploaded to centralized servers, and the cloud servers provide services of storage and computing etc and send results back to the IoT

© Springer Nature Switzerland AG 2020
D.-Z. Du and J. Wang (Eds.): Ko Festschrift, LNCS 12000, pp. 268–288, 2020.
https://doi.org/10.1007/978-3-030-41672-0_17

devices [1], as shown in Fig. 1(a). The cloud computing can provide users the infinite computing and storage resource which are available on demand no matter where and when the users send request to cloud. Meanwhile, most users have no knowledge of where their data or application programming are stored or operated in cloud server. With rapid expanding of the number of IoT devices, a large huge volume of data produced by heterogeneous IoT devices is transmitted to cloud for computing and storage service, which require high performance for cloud platform and a large demand for network bandwidth and have potentially centralized risk [2]. Therefore, with techniques and IoT devices are getting more involved in human's life, the centralized cloud computing paradigm can hardly solve existing challenges such as security in centralized cloud, realtime data delivery and processing, and mobility support, etc.

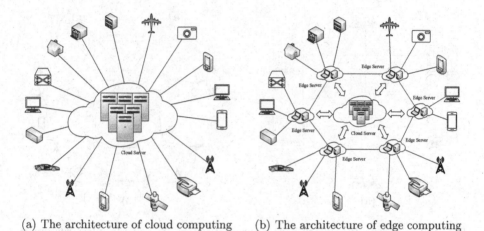

(a) The architecture of cloud computing   (b) The architecture of edge computing

Fig. 1. Compare the central cloud computing with the distributed edge computing.

To solve these problems, the edge computing (also called fog computing) as an evolving architecture that combines cloud computing and IoT is introduced, which are deployed between central cloud server and IoT devices [3,4], as shown in Fig. 1(b). The edge computing can push the frontier of computing applications, the privacy data storage and realtime data processing and analysis away from centralized cloud to the edge servers of the network, which can retain the core advantages of cloud computing and transfer the realtime control and sensitive data storage to the edge servers. Nonetheless, the security and privacy issues, e.g. authentication, intrusion detection, access control, etc, in edge computing architecture can hardly be resolved [5], since various software and applications are embedded in heterogeneous edge servers and the migration of services across edge servers is vulnerable. As the emergence of the blockchain technologies, the edge computing integrated with blockchain technologies is becoming a effective method to solve the above problems.

Blockchain as an underlying technology with digital cryptocurrency is first proposed by Nakamoto in 2008 and is implemented in 2009 for Bitcoin [6]. Blockchain can be defined as a distributed decentralized shared tamper-resistant database which can be maintained, shared, replicated and synchronized by multiple participants in the Peer-to-Peer (P2P) network, as shown in Fig. 2(a). It can facilitate establishing secure, trusted and decentralized intelligent system for solving the security and privacy problems in edge computing [7]. Consequently, the edge computing integrated with blockchain technologies can provide sensitive information hiding and reliable access and control of the network in distributed edge servers and cloud servers, as well as provide quick search in IoT devices, as shown in Fig. 2(b).

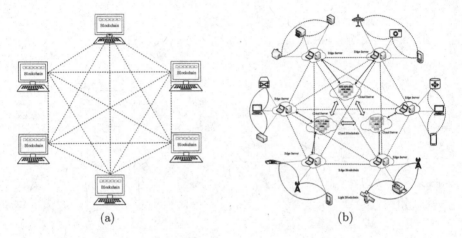

(a)   (b)

Fig. 2. The Peer-to-Peer blockchain network structure [8] and the infrastructure of edge computing integrated with blockchain.

So far, there are many related works on the application of edge computing integrated with blockchain to different areas, such as [9–11]. This survey aims to envision the potential contribution of blockchain technologies for revolutionizing the edge computing and current challenges for their integration. In this survey, we will review the blockchain technologies and the general architecture of the edge computing. Then the motivations on why to integrate the blockchain technologies into the edge computing and the integrated architecture of the edge computing are introduced. Then, we review the literature that has already been proposed about the edge computing integrated with blockchain technologies. Finally, we will give the challenges for the integration of edge computing and blockchain.

The rest of the paper is organized as follows. In Sect. 2, we introduce the technologies about blockchain. In Sect. 3, the structure of edge computing is introduced. In Sect. 4, we present the motivations and architecture, overview and challenges of edge computing integrated with blockchain. Section 5 concludes the paper.

2   Technologies of Blockchain

The blockchain is a distributed database that does not need third party verification and a central authority [6]. The distributed database transforms all transaction data of the network into associated strings stored in a block which is constructed in a certain period of time and points to the previous data block with hash pointer, and all blocks form a single and complete chain, as shown in Fig. 3. The distributed database is also verified by the public/private key pair of asymmetric encryption in cryptography to encrypt internal data and ensure data security.

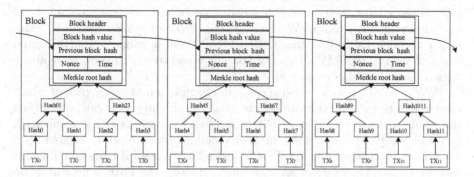

Fig. 3. The data structure of the blockchain [6].

2.1   The Key Technologies of Blockchain

The key components of blockchain are decentralized shared storage, consensus protocols, cryptograph and smart contracts. According to these technologies, we obtain that the characterizations of the blockchain are decentralized, synchronized and immutability.

2.1.1   Decentralized Shared Storage

In the traditional storage system, e.g. Google, Dropbox, there is a centralized cloud server to store, manage and process data, which may suffer from potential security threats. The blockchain which is used in the P2P network, provides a decentralized storage method to guarantee the decentralized control to solve the problems generated from the traditional cloud storage system [12]. In the network, all users own the same distributed ledger which is stored with blockchain structure, as shown in Fig. 2(a). The ledger records the transactions, such as exchange of data or assets, among the participants in the network, which can be shared, replicated and synchronized. All transactions are stored in blocks of the blockchain, and all users hold all transactions. A transaction in the blockchain network can be conducted between any two participants without authentication by the third party, which can reduce the server costs and mitigate the performance limitation of the central server [13].

2.1.2 Consensus Protocols

Blockchain is formed by a lot of blocks and each of them has a set of transactions, where each transaction can be seen as information, value or other data transferred between different entities by broadcasting in the network. Each block must contain a consensus protocol to be considered valid in the blockchain. In the P2P network, the end users who are equal to each other verify the same block in the blockchain, and they jointly maintain a consensus protocol to ensure that all users trust each other. The existing common consensus mechanisms in blockchain include Proof of Work (PoW) [6], Proof of Stake (PoS) [14], Practical Byzantine Fault Tolerance (PBFT) [15], Proof of Storage [16], Delegated Proof of Stake (DPoS) [17], Proof of Space [18] and so on. To know more about consensus protocols, one can refer to [19].

PoW is to define an expensive computer calculation of users, each of which completes a certain difficult problem to get a result, and others can easily check the result to verify whether or not it satisfies the corresponding problem. A group of trustless transactions bundled together into a block is created on a distributed blockchain when PoW is performed and a reward is given to the first user who solves the problem for each block. Because the PoW is dependent on the power energy consumption, which produces a lot of energy cost for the whole network. The PoS as an alternative mechanism for PoW is introduced to replace most PoW's work, which generates the next block based on users' shareholding in the network. For example, in PPcoin, the shareholding is currency amount times holding period, called coin age [16].

2.1.3 Cryptograph

Digital signature is a string of numbers that can only be generated by the sender of the information and cannot be forged by others [20]. Each user in the network has a pair of private and public keys that are generated randomly. The private key is saved by user and is used to sign transactions. The transactions that are signed with private key are broadcasted to the whole network, and they are verified by public key. For example in Fig. 4, when Alice wants to send transactions to Bob, she first executes hash function for transactions in block for obtaining hash abstract. Then she signs the hash abstract with her private key and sends the hash abstract and the transactions in block to Bob. After receiving the message, Bob decrypts the hash abstract with Alice's public key and computes the hash function from the original transactions. Then he verifies them to judge whether or not they are the same transactions.

In each block of the blockchain, hash pointer is a pointer which points to the previous block storage location and its location data, as shown in Fig. 3. The hash pointer can not only tell you where the block is stored, but also give you a way to verify whether or not the data has been tampered by invaders. For example,

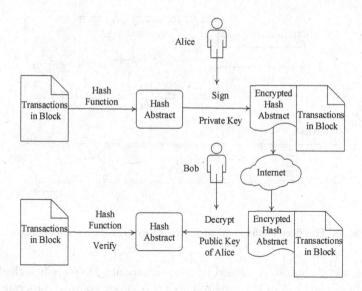

Fig. 4. Digital signature used in blockchain.

in Bitcoin, if the adversary tempered with the transactions in the k-th block in the blockchain, then the hash abstract in the $k + 1$-th block can not match the new hash abstract generating from the k-th updated block. Therefore, the adversary can not temper with any transaction information without undetected. The definition of the Merkle tree was first proposed by Merkle in [21]. As shown in Fig. 3, a Merkle tree is a tree structure that is used to store the transaction data, on which every leaf node is a transaction and every non-leaf node is the hash value of the two children. The Merkle tree structure can improve the storage and search efficiency. For example, if the adversary updated a transaction information, then the hash value in the upper layer can not match the original value. Even if he continues to modify the hash value of the upper layer, the updated value is finally passed to the root of the Merkle tree which is not tempered. Therefore, we can detect any modification of transactions only through storing the root of the Merkle tree. At the same time, any transaction information can be accessed in the running time of the depth of the Merkle tree.

2.1.4 Smart Contracts

The smart contracts also called digital contract or blockchain contract are first proposed by Szabo in [22], which are self-executing contracts with the terms of the agreement between participants in the P2P network and are written as program code stored on blockchain. Smart contracts allow transactions to be conducted in anonymous between involved participants without the need for a third party authority [23].

We illustrate the operation mechanisms of smart contracts when they are used in decentralized cryptocurrencies like Bitcoin [24], as shown in Fig. 5.

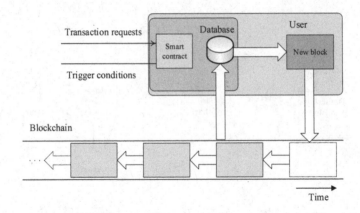

Fig. 5. The decentralized cryptocurrency system with smart contract.

After the smart contract is signed by all participants, they are attached to the blockchain in the form of the program codes. A smart contract program of user is executed when it received transaction requests or other trigger conditions.

3   The General Architecture of Edge Computing

The term of the edge computing is proposed for improving applications over Content Delivery Networks (CDNs) [25], which is to achieve massive scalability through taking advantage of proximity and resources of edge servers in CDNs. In the last decade, the edge computing has been widely used in IoT networks.

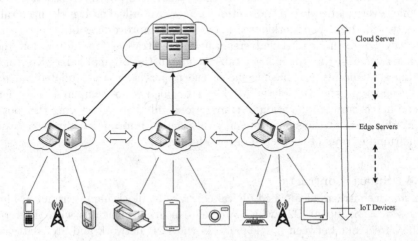

Fig. 6. The general architecture of the edge computing.

In the existing cloud-based IoT networks, the IoT devices have the limited resource and lower computation [26]. Therefore, the IoT devices can not bear the storage and processing of a large amount of data, since the IoT devices produce an amount of data exponentially with the growth of the technologies. The cloud computing for IoT applications helps enterprises and their end users to liberate from the specification uses of storage, computation constraint and network communication costs and other details. However, the cloud computing for IoT applications have the poor real-time dynamic and can hardly satisfy their requirements of location awareness, privacy security and mobility support [27]. Therefore, the edge computing is an emerging computing model that transfers part of cloud computing and storage to the edge services to solve the latency-sensitive applications and improve the QoS of mobility support, privacy security and location awareness for IoT networks [28].

The general architecture of the edge computing is depicted in Fig. 6, which is composed of three layers: IoT devices, edge servers and cloud servers. The IoT devices are any devices that allow people and things to be connected each other in the internet/networks [29], which are attached to one of the edge servers. The edge devices can be interconnected and each of them is linked to the cloud servers that can support many applications, such as real-time data processing and sensitive information storage. The cloud servers provide the high-performance computing, managing and storage for a large quantity of data produced by IoT devices but there is no strict requirement for realtime. Compared with the current cloud computing, the edge computing has many advantages: low latency, scalability and effective use of resources, privacy security [30].

Low Latency. In the traditional centralized cloud computing, the data produced by IoT devices should be uploaded to the cloud servers for acquiring services such as data processing, storage and management. Then the results computed by cloud servers are returned to the IoT devices. This interaction pattern can not meet many applications with high timeliness requirements, such as unmanned driving and realtime voice translation. In the edge computing, the edge servers are more closer to the IoT devices than the cloud servers, which can provide the efficient communication with low latency.

Scalability. With the massive growth of IoT devices, the performance of the cloud servers can not satisfy the requirements of data processing and storage of a large of number of data generated by IoT devices. Therefore, we need to deploy many distributed edge servers for dealing with a part of data produced by IoT devices, which can expand the access IoT devices and data processing capacity in the networks.

Versatility. The scalability of the edge computing also provides versatility for the edge computing. The enterprises can easily to invest in ideal markets through working with local edge data centers without spending on the expensive infrastructure. The edge servers enable them to serve for users with a small physical distance or latency. In addition, they have the flexibility to move to other markets when the economic environment changes.

Bandwidth Saving. There is a lot of data that can be computed locally at edge servers without having to be uploaded to the cloud servers. For example, the environment monitoring data produced by intelligent robots and road conditions generated by unmanned vehicles can be processed in the edge servers, which can also avoid the waste of bandwidth in the network.

Privacy Security. Since the edge servers have the capability of computing and storage, the sensitive privacy information about users can be stored in the edge servers without being uploaded in the centralized cloud together with the data generated by IoT devices, which can avoid the privacy information disclosure when the centralized cloud is invaded.

Fault Tolerance. With many edge servers which have capability of data storage and processing, it is difficult to shut down the server completely by any fault. The produced data can be rerouted in many ways to ensure that users maintain access to the requirements.

4   Integration of Edge Computing and Blockchain

In this section, we will introduce the motivations, general architecture, related works and challenges for the edge computing integrated with blockchain technologies. Firstly, we will introduce the motivations to show why the edge computing need to be integrated with blockchain technologies in Sect. 4.1. Secondly, the general architecture of the integration of edge computing and blockchain technologies is introduced in Sect. 4.2. Thirdly, we briefly review the literature about the edge computing integrated with blockchain technologies in Sect. 4.3. Finally, we give the challenges for their integration in Sect. 4.4.

4.1   Motivations

The edge computing is proposed to improve the performance of the existing centralized cloud computing. Although it can improve the security of the sensitive information compared with the cloud computing, there still exist many security problems in edge computing, such as authentication of IoT devices and edge servers, and malicious attacks [31,32].

  With the development of the blockchain techniques, the blockchain can solve the security problems with regard to the immutability and the general resistance to attacks and provide decentralized shared data storage scheme which allows the network to be permissionless and censorship-resistant in the edge computing. Every block in the chain contains hash abstract of the previous block, which enables blocks to be traced back all of the blocks in the blockchain. Therefore, it is impractical to modify transactions between IoT devices in a block once it is created. When the transactions between IoT devices are generated, they are remained within the blockchain throughout its life and are easy to be searched, which allows the blockchain to become the trust database [33]. The consensus mechanisms of blockchain can provide the new method for managing the distributed database and provide the fresh payment mode for the network services in the edge computing. Blockchain has been widely used in providing trustworthy and authorized identity registration and ownership tracking of assets, which can maintain the integrity of the transactions in the distributed edge computing, such as TrustChain [34]. For example, a malicious insider to the cloud provider maybe launch severe attacks for the cloud servers, such as data integrity breaking. The decentralized blockchain can avoid the data integrity breaking since all participants of cloud servers maintain the same ledger and each block points to the hash abstract of the previous block.

In the edge computing, one of the key challenges for IoT devices is to enable and control autonomous and self-organized machine-to-machine communication [35]. Besides, since the edge servers may be deployed in the places without rigorous surveillance and protection, the edge servers may encounter many malicious attacks, such as distributed Denial-of-Service (DoS) attacks, jamming attacks and sniffer attacks [1]. The communication security problems in the edge computing which are produced between edge servers and IoT devices or among edge servers are important challenges [32]. This is because that the edge servers may be deployed in the openness and heterogeneous environment, which produces many trustworthy problems. The digital signature of cryptograph used in the blockchain is core part to solve the problem of communication security, which can guarantee that the identifies of IoT devices are verifies and identified and that the messages are not tempered during transmission. From the IoT devices' perspective, a security and transparency mechanism is needed to achieve secure communication. The smart contract can operate transparently and distributed data securely [36]. For example, the secure update scheme for IoT devices that utilizes the smart contract to check a firmware version, validate the correctness of firmware, and download the latest firmware for the embedded devices. In the proposed scheme, the smart contract allows devices to store the hash of the latest firmware updated on the network and the devices can be found by the smart contract's address [37].

4.2   General Architecture

According to the structure of the edge computing in Fig. 6, the IoT is divided into three layers, i.e. IoT device layer, edge layer and cloud layer. Therefore, blockchain can be integrated in every layer to improve the performance of the network.

Figure 7 illustrates the basic architecture of the integration of edge computing and blockchain. In the architecture, each IoT device is connected to an edge server and communicate with the edge server. Each edge server together with IoT devices attached to it form a local network. Each edge server is a manager to control the IoT devices in the local network. The IoT devices are registered with certification authority by the edge server. Any pair of devices in the network communicate with each other supported by the edge servers. In the network, the basic communication primitive for exchanging information among IoT devices or IoT device to edge server or among servers, can be seen as a transaction. The edge server is a blockchain manager that is responsible for managing the blockchain, which includes creation, verification and storage of the individual transactions and blocks of transactions. Blockchain stored at cloud servers can form a decentralized storage system with blockchain incentives, which is an effective combination of blockchain and storage system. Compared with the existing centralized cloud computing, the distributed blockchain cloud has the following advantages: higher reliability, higher availability of services, higher fault tolerance and lower cost.

Since the IoT devices belong to the different local networks, the communication between IoT devices can be divided into two categories: Device-to-Device (D2D) communication in the same local network and D2D communication in the different local networks. In the first case, the transaction request from source is forward to its manager, where it is authenticated. Then they are broadcasted to the whole network by the internet. In the second case, as the devices are not registered with the same manager, the transaction of any two devices across the local network is authenticated by their respective manager. All of transactions among IoT devices are mined in a block of blockchain stored in edge servers.

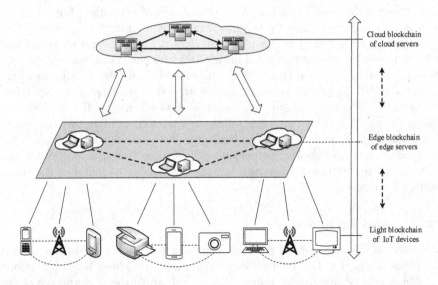

Fig. 7. The architecture of the edge computing integrated with blockchain.

In the network, the IoT devices produce a massive of data which is needed to transmit to the edge servers for processing and analysis. The edge servers quickly processes the data with high real-time requests and store the sensitive data with blockchain, where the requests can be as transactions stored in the blockchain held by edge servers. Then it forwards the pre-processed encrypted data with low real-time and security to the cloud servers for further processing and storage. Afterwards, the data is stored in the distributed cloud servers with blockchain. Therefore, based on the different the capabilities on computation power, storage space and source management for the devices in the different layers, three types of blockchain can be constructed in the network: light blockchain, edge blockchain and cloud blockchain.

Light Blockchain. Since the IoT devices has limited storage space and communication power, the consensus protocols that require high computing power or the large of storage space such as PoW and PoS are not suitable for the IoT device layer. Therefore, the edge server as a manager to control the transactions exchange and manage the blockchain which records the transactions of among the IoT devices. For the fundamental applications, e.g. transaction search, the simple light blockchain should be maintained by the IoT devices which only stores the header information of the blockchain stored in the edge servers [38].

Edge Blockchain. The edge blockchain that is maintained by all edge servers is a decentralized chain which stores the transactions among IoT devices and between IoT devices and edge servers which require unified, consistency and transparent. The edge servers have computational capabilities and storage space for correlative services. Each edge server has a management controller and storage pool. The management controller works as a third party to manage transactions from IoT devices. As the resources of IoT devices such as storage space, computation power and memory are constrained, a variety of transactions should be stored in edge servers. When the transactions are generated by IoT devices, then they are broadcasted to the internet supported by the edge servers. For security and privacy protection, the transactions produced by IoT devices are anonymous and encrypted, and attached with digital signatures of IoT devices. The edge servers working as blockchain managers will periodically integrate received transactions into a block with the consensus protocol (e.g. Proof of Storage [7]) and broadcast the block to other edge servers for verification. The edge servers with the most contribution in the network is rewarded over a period of time, which is an incentive to encourage edge servers to provide enough support for maintaining the blockchain.

Cloud Blockchain. The cloud blockchain is a decentralized distributed storage chain in which all data produced by IoT devices are stored in utilizing blockchain. All data blocks are instantiated and distributed to all cloud servers in the network, which allows us to store the data in an efficient, verifiable, and permanent way. The digital signature and hashes are used in blockchain to ensure the integrity of data. Once a data block has been inserted into the chain, the data in any block of chain can not be modified, since any block points to the hash value of its previous block, which can prevent the data in the block from being tampered with. For example, the healthcare data can be stored in the cloud in utilizing blockchain, which can provide security storage for the healthcare data [39].

4.3 State of the Art

In the following, we elaborate on the role of the edge computing integrated with blockchain in the following scenarios to demonstrate how the blockchain technologies are implemented in edge computing-based IoT networks: smart city, smart transportation, industrial IoTs, smart home and smart grid.

4.3.1   Smart City

With the development of blockchain, IoTs and other new generation technologies, smart city has become a predicable mode to solve the problems of urban development in the future [40]. It can improve the efficiency of urban management and operation, and promote the sustainable and leap forward development of the city by comprehensively and transparently perceiving information, quickly and safely transmitting information, intelligently and efficiently processing information. The fresh form of urban development makes the city automatically perceive, effectively decisions making and control [41].

In [42], Sharma et al. proposed a blockchain-based distributed cloud architecture with a Software Defined Networking (SDN) with blockchain technologies, which consists of three layers, IoT devices, edge servers and cloud servers, and can solve many problems of the traditional cloud computing such as real-time data transmission, scalability, security and high availability. Rahman et al. leveraged blockchain technologies and edge computing to design an infrastructure that meets the security and privacy requirement of smart contract in massive smart cities [43]. The geo-tagged multimedia transactions were handled by edge servers and the key information were extracted by artificial intelligence technology. Then the processing results were saved in blockchain and distributed cloud storage to support the services of sharing economy. Khan et al. presented an architecture of the edge computing based on the blockchain technologies in [44], which can achieve granular and security management of data in different administrative departments of districts. They applied a proof of concept on a well-designed case that allows citizen to take part in administrative measure by consensus. This use case emphasizes the necessity to retain and process citizen participation data at the local level by deploying regional chain codes, and to share consistent results only through permission chain codes. Some IoT devices generate and process sensitive personal data with security and privacy issues. In [45], Damianou et al. presented a hybrid blockchain-based architecture for the edge computing-based IoT networks to address the issues of IoT devices related to the limited storage capacity, security and privacy through combining edge computing with blockchain technologies. In [46], the authors proposed a BLockchain-ENabled Decentralized Microservices Architecture which is implemented in a hierarchical blockchain-based edge computing to protect data processing among different service providers and entities in a smart public safety system. In [47], the authors proposed a video surveillance system which uses blockchains, edge computing, InterPlanetary File System (IPFS) technology and convolution neural networks. The reliability and robustness of the system is improved by the blockchain technologies. The edge computing is used to gather information from the large-scale wireless sensors and to process and analyze the sensory data. IPFS and CNNs are respectively used to achieve massive video data storage and real-time monitoring. Kotobi and Sartipi [48] leveraged mobile Communications, Computing, and Caching (3C) systems to enhance the bandwidth and latency of wireless communications. They stated the huge data transmission and processing burden required for smart city applications, which will deplete current wireless infrastructure unless using edge computing and caching to address the issues. Then they designed a blockchain database to solve the problem of security from communication between the smart city and home devices and sensors.

4.3.2 Smart Transportation

The smart transportation system combining block chain and edge computing is a hot spot in the field of transportation research and development in the world [49]. It combines users and providers to establish a real-time, accurate and efficient comprehensive transportation management system that can play a role in a wide range and all-round way [50].

In [51], Li et al. proposed an efficient and privacy preserving carpooling scheme to support conditional privacy, one-to-many proximity matching, destination matching and data auditability by combining the vehicular fog computing and blockchain technologies, where the fog servers are RSUs that are used to match passengers with drivers and to construct a private blockchain. In [52], Liu et al. respectively proposed blockchain-inspired data coins and energy coins in the Electric Vehicle Cloud and Edge computing (EVCE) on the basis of the distributed consensus protocol that used the data contribution frequency and energy contribution amount to realize proof of work. In [53], Nguyen et al. proposed a blockchain-based Mobility-as-a-Service (MaaS) to improve transparency and trust between providers by eliminating the intermediate layer which are used to manage and control the relationships between transportation providers and passengers. In the proposed blockchain-based MaaS, the smart contracts are executed on the edge servers, which can directly connect travelers to providers in a more efficient way, and achieve many advantages including validation, confirmation and formation. In [54], Zhou et al. presented a secure and high efficiency Vehicle-to-Grid (V2G) energy trading mechanism by considering blockchain technologies, contract theory and edge computing, which is called consortium blockchain-based secure energy trading. Then an incentive mechanism was proposed by considering the situation of information asymmetry. Afterward, they presented an edge computing-based task offloading mechanism to improve the success probability of block creation and proposed an optimal pricing strategy for the edge computing services.

4.3.3 Industrial IoTs

The integration of the edge computing and blockchain provides a secure and unified platform for the current cloud computing, data processing, and access control, enabling the rapid distribution of computing services to edge servers, which greatly promotes the development of the Industrial IoTs.

Blockchain mining tasks require powerful computing capabilities, and deploying equipment that meets this computing power is costly. Therefore, offloading mining tasks to the edge servers to make full use of the limited computing power of IIoT for mining will be a promising solution. In order to solve the data processing and mining tasks in IIoT authorized by the blockchain, Chen et al. [55] proposed a multi-hop collaborative distributed computing offloading algorithm. Aimed at minimizing the cost of the IIoT equipment, they set the offloading problem as a game problem, and the IIoT equipment can independently decide to obtain the maximum benefit. In addition, they used message exchange between IIoT devices to propose an efficient distributed algorithm to achieve low complexity Nash equilibrium. In [56], the authors proposed an IIoT architecture based on edge intelligence and blockchain authorization to implement flexible and secure edge service management. In order to reduce the cost of edge services and improve service capabilities, they designed a cross-domain shared edge resource scheduling mechanism and a credit differential edge transaction approval mechanism. In order to solve the problem of edge computing security in IIoT, the authors [57] leveraged self-authenticated cryptography technology to achieve the registration and authentication of network entities, and designed a blockchain-based identity management and access control mechanism. They proposed a lightweight key agreement protocol based on self-certified public keys, which provide authentication, auditability, and confidentiality for IIoT. Gai et al. [58] combined edge computing, blockchain and IIoT to design a new model called Blockchain-based Internet-of-Edge (BIoE) model. This model makes full use of the advantages of edge computing and blockchain to propose a privacy protection mechanism for scalable and controllable IoT systems. In [59], the authors proposed the concept of the IIoT bazaar. This is a decentralized industrial edge application market that creates transparency for all stakeholders through blockchain technology. It can implement tracking applications installed on edge devices. Edge devices with limited computing power can be integrated into the IIoT Bazaar ecosystem through fog computing. Users can easily interact with edge devices through Augmented Reality (AR).

4.3.4   Smart Home

The researchers in literature [60] introduced a new security framework, which improves the integrity, confidentiality and availability by integrating blockchain technology into smart home. Tantidham and Aung [61] used Ethereum blockchain technology to design a smart home emergency service system. This system can decentralize process some untrusted services for example access control services between Home Service Providers (HSPs) and smart home IoT devices. Their SHS includes: (1) Smart Home Sensor Manager, Raspberry Pi (RPi) is used as an edge IoT gateway to collect data related to the environment; (2) HSP miners deploying Meteor and Ethereum platforms; (3) providing home users and HSP employees Web-based application. In [62], the authors designed a new architecture based on blockchain technology. They introduced edge computing and a new algorithm to improve the quality of data transmission and the detection of erroneous data. In [63], the authors had designed a framework based on blockchain technology. This framework allows medical institutions to collect some quality of life information from the home environment through smart sensors and share some safety information with other communities. In particular, this framework collects data that is beneficial to treatment by authorizing some sensors that can track the quality of life, such as physiological characteristics and environmental quality, and stores these data in secure and dedicated edge services.

4.3.5 Smart Grid

In [64], the authors proposed a model permissioned blockchain edge model, in which they leveraged blockchain techniques and edge computing to solve two key issues that is privacy security and energy security in smart grid. Wang et al. [65] leveraged edge computing to propose a blockchain through mutual authentication and key agreement protocol for smart grid. The benefits of this protocol was that can achieve conditional anonymity and key management without other complex cryptographic primitives. Aimed at addressing the tamper-resistant, reliable and distributed ledger issues, a smart-toy-edge-computing-oriented data exchange prototype was proposed by means of signing smart contract to solve the problem of distrust between participants [66]. This solution also helps to achieve P2P data exchange between isolated smart toy and other edge devices in IoT systems. In [67], the authors proposed a framework named SURVIVOR to achieve energy transaction in V2G scenario. The decisions of Energy transaction are made by electric vehicles closer to the edge nodes. The blockchain is used to ensure the security of energy transactions. It selected the approval node among all existing nodes and was responsible for verifying the transaction based on the utility function.

4.4 Challenges

Although the emergence of blockchain technology provides new ideas and research directions for edge computing, there are many significant research challenges to be addressed. In this section, we propose some of research challenges for the edge computing integrated with blockchain technologies.

Scalability. IoT devices have limited storage which are usually requirements for blockchain based networks. Although D2D transactions and data produced by IoT devices can be transferred to edge servers for processing and storage. However, the smart contract of blockchain requires every participants to maintain a same distributed ledger, and sometimes every transaction in the ledger should be traced. Therefore, with more and more devices connected to the network, a massive of transactions will be produced and stored every participant, which needs higher operation performance of IoT devices and servers. Therefore, it is difficult to optimize the overall efficiency of the network, which restricts the scalability of IoT networks. Therefore, how to extend the scalability of the network by using the current implementations is an important challenge.

Consensus Optimization. In the edge computing, many applications require quick response of processing results from edge servers to the IoT devices, such as accident warnings in smart transportation. However, the existing consensus mechanisms require verification of most participating nodes to complete mining, which will generate serious network latency. Meanwhile, the existing consensus mechanisms such as PoW and PoS can not be available for light servers with limited storage and computation power. Therefore, how to design new consensus mechanisms to balance the realtime of applications and latency produced by consensus mechanisms and to adapt to the light servers is an important challenge for the edge computing integrated with blockchain.

Interoperability and Cost Standardization. Since the edge computing is heterogeneous hierarchical, the protocols implemented at different layers need to interoperate by providing conversion mechanisms. As the IoT devices have limited resources, mining and transaction data storage must transfer to the edge servers. It means that any transaction between two IoT devices must be implemented by one or two intermediate edge servers. Therefore, the interaction patterns are needed for data communication and management effectively. Moreover, the edge computing incorporates the combination of various application platforms, which may need to take cooperation to complete a service. Thus, the interaction models for diverse applications running varying and heterogeneous platforms are needed. Moreover, with the huge amounts of data produced by IoT devices, the processing and storage capacity of edge servers can not satisfy the requirement. Therefore, they have to transfer precessed data to the cloud servers for further processing and storage. However, since the intelligent settlement of service fee generated by using smart contract, the cost distributed standards are needed to complete the intelligent service of the networks.

Security. Although the blockchain can provide the security approaches for the edge computing, the blockchain itself is vulnerable to attack, such as denial of service, man in the middel and Sybil. The data protection for the IoT devices and the edge servers without unauthorized access is a important challenge, since the security cryptographic software are required to integrate into IoT devices and edge servers and the limitation of their storage resource restrict the improvement of the security software. And there are some other security challenges, such as communication security and privacy protection of devices.

5   Conclusion

In this paper, we give a comprehensive review for the edge computing integrated with blockchain, including architectures, technologies and the related works. Particularly, the general architecture of edge computing integrated with blockchain are introduced, in which three categories of blockchain can be constructed based on the architecture of the edge computing and enabling technologies: light blockchain, edge blockchain and cloud blockchain. In addition, we summarize the related research papers about the edge computing integrated with blockchain technologies to show how the edge computing integrated with blockchain is to be implemented in real-work applications, which consists of smart cities, smart transportation, Industrial IoTs, smart home, and smart grid. Finally, we conclude the challenges about the integration of edge computing and blockchain.

References

1. Yu, W., et al.: A survey on the edge computing for the Internet of Things. IEEE Access **6**, 6900–6919 (2017)
2. Li, C., Zhang, L.-J.: A blockchain based new secure multi-layer network model for Internet of Things. In: 2017 IEEE International Congress on Internet of Things (ICIOT), pp. 33–41. IEEE (2017)
3. Garcia Lopez, P., et al.: Edge-centric computing: vision and challenges. ACM SIG-COMM Comput. Commun. Rev. **45**(5), 37–42 (2015)
4. Lin, J., Wei, Y., Zhang, N., Yang, X., Zhang, H., Zhao, W.: A survey on Internet of Things: architecture, enabling technologies, security and privacy, and applications. IEEE Internet Things J. **4**(5), 1125–1142 (2017)
5. Yang, R., Yu, F.R., Si, P., Yang, Z., Zhang, Y.: Integrated blockchain and edge computing systems: a survey, some research issues and challenges. IEEE Commun. Sur. Tutor. **21**(2), 1508–1532 (2019)
6. Nakamoto, S., et al.: Bitcoin: a peer-to-peer electronic cash system (2008)
7. Kang, J., et al.: Blockchain for secure and efficient data sharing in vehicular edge computing and networks. IEEE Internet Things J. **6**(3), 4660–4670 (2018)
8. Biswas, S., Sharif, K., Li, F., Nour, B., Wang, Y.: A scalable blockchain framework for secure transactions in IoT. IEEE Internet Things J. **6**(3), 4650–4659 (2018)
9. Sharma, P.K., Chen, M.-Y., Park, J.H.: A software defined fog node based distributed blockchain cloud architecture for IoT. IEEE Access **6**, 115–124 (2017)
10. Stanciu, A.: Blockchain based distributed control system for edge computing. In: 2017 21st International Conference on Control Systems and Computer Science (CSCS), pp. 667–671. IEEE (2017)
11. Xiong, Z., Feng, S., Niyato, D., Wang, P., Han, Z.: Optimal pricing-based edge computing resource management in mobile blockchain. In: 2018 IEEE International Conference on Communications (ICC), pp. 1–6. IEEE (2018)
12. Jiang, P., Guo, F., Liang, K., Lai, J., Wen, Q.: Searchain: blockchain-based private keyword search in decentralized storage. Futur. Gener. Comput. Syst. (2017)
13. Zheng, Z., Xie, S., Dai, H.-N., Chen, X., Wang, H.: Blockchain challenges and opportunities: a survey. Int. J. Web Grid Serv. **14**(4), 352–375 (2018)
14. King, S., Nadal, S.: PPCoin: peer-to-peer crypto-currency with proof-of-stake. Self-published paper, 19 August 2012
15. Castro, M., Liskov, B., et al.: Practical Byzantine fault tolerance. In: OSDI, vol. 99, pp. 173–186 (1999)
16. He, K., Chen, J., Ruiying, D., Qianhong, W., Xue, G., Zhang, X.: DeyPoS: deduplicatable dynamic proof of storage for multi-user environments. IEEE Trans. Comput. **65**(12), 3631–3645 (2016)
17. Larimer, D.: Delegated proof-of-stake (DPOS). Bitshare whitepaper (2014)
18. Dziembowski, S., Faust, S., Kolmogorov, V., Pietrzak, K.: Proofs of space. In: Gennaro, R., Robshaw, M. (eds.) CRYPTO 2015. LNCS, vol. 9216, pp. 585–605. Springer, Heidelberg (2015). https://doi.org/10.1007/978-3-662-48000-7_29
19. Chalaemwongwan, N., Kurutach, W.: State of the art and challenges facing consensus protocols on blockchain. In: 2018 International Conference on Information Networking (ICOIN), pp. 957–962. IEEE (2018)

20. Johnson, D., Menezes, A., Vanstone, S.: The elliptic curve digital signature algorithm (ECDSA). Int. J. Inf. Secur. **1**(1), 36–63 (2001)
21. Merkle, R.C.: Protocols for public key cryptosystems. In: 1980 IEEE Symposium on Security and Privacy, p. 122. IEEE (1980)
22. Szabo, N.: Smart contracts: building blocks for digital markets. EXTROPY: J. Transhumanist Thought (16), 18:2 (1996)
23. Wang, S., Yuan, Y., Wang, X., Li, J., Qin, R., Wang, F.-Y.: An overview of smart contract: architecture, applications, and future trends. In: 2018 IEEE Intelligent Vehicles Symposium (IV), pp. 108–113. IEEE (2018)
24. Delmolino, K., Arnett, M., Kosba, A., Miller, A., Shi, E.: Step by step towards creating a safe smart contract: lessons and insights from a cryptocurrency lab. In: Clark, J., Meiklejohn, S., Ryan, P.Y.A., Wallach, D., Brenner, M., Rohloff, K. (eds.) FC 2016. LNCS, vol. 9604, pp. 79–94. Springer, Heidelberg (2016). https://doi.org/10.1007/978-3-662-53357-4_6
25. Rabinovich, M., Xiao, Z., Aggarwal, A.: Computing on the edge: a platform for replicating internet applications. In: Douglis, F., Davison, B.D. (eds.) Web Content Caching and Distribution, pp. 57–77. Springer, Dordrecht (2004). https://doi.org/10.1007/1-4020-2258-1_4
26. Liono, J., Jayaraman, P.P., Qin, A.K., Nguyen, T., Salim, F.D.: QDaS: quality driven data summarisation for effective storage management in Internet of Things. J. Parallel Distrib. Comput. **127**, 196–208 (2019)
27. Vaquero, L.M., Rodero-Merino, L.: Finding your way in the fog: towards a comprehensive definition of fog computing. ACM SIGCOMM Comput. Commun. Rev. **44**(5), 27–32 (2014)
28. Hajibaba, M., Gorgin, S.: A review on modern distributed computing paradigms: cloud computing, jungle computing and fog computing. J. Comput. Inf. Technol. **22**(2), 69–84 (2014)
29. Kumar, J.S., Patel, D.R.: A survey on Internet of Things: security and privacy issues. Int. J. Comput. Appl. **90**(11) (2014)
30. Satyanarayanan, M.: The emergence of edge computing. Computer **50**(1), 30–39 (2017)
31. Stojmenovic, I., Wen, S., Huang, X., Luan, H.: An overview of fog computing and its security issues. Concurr. Comput. Pract. Exp. **28**(10), 2991–3005 (2016)
32. Mukherjee, M., et al.: Security and privacy in fog computing: challenges. IEEE Access **5**, 19293–19304 (2017)
33. Veena, P., Panikkar, S., Nair, S., Brody, P.: Empowering the edge-practical insights on a decentralized Internet of Things. IBM Institute for Business Value (2015)
34. Otte, P., de Vos, M., Pouwelse, J.: Trustchain: a sybil-resistant scalable blockchain. Futur. Gener. Comput. Syst. (2017)
35. Restuccia, F., Kanhere, S.D., Melodia, T., Das, S.K.: Blockchain for the Internet of Things: present and future. arXiv preprint arXiv:1903.07448 (2019)
36. Christidis, K., Devetsikiotis, M.: Blockchains and smart contracts for the Internet of Things. IEEE Access **4**, 2292–2303 (2016)
37. Lee, B., Lee, J.-H.: Blockchain-based secure firmware update for embedded devices in an Internet of Things environment. J. Supercomput. **73**(3), 1152–1167 (2017)
38. Dorri, A., Kanhere, S.S., Jurdak, R., Gauravaram, P.: LSB: a lightweight scalable blockchain for IoT security and anonymity. J. Parallel Distrib. Comput. **134**, 180–197 (2019)
39. Esposito, C., De Santis, A., Tortora, G., Chang, H., Choo, K.-K.R.: Blockchain: a panacea for healthcare cloud-based data security and privacy? IEEE Cloud Comput. **5**(1), 31–37 (2018)

40. Gaur, A., Scotney, B., Parr, G., McClean, S.: Smart city architecture and its applications based on IoT. Procedia Comput. Sci. **52**, 1089–1094 (2015)
41. Tang, B., Chen, Z., Hefferman, G., Wei, T., He, H., Yang, Q.: A hierarchical distributed fog computing architecture for big data analysis in smart cities. In: 2015 Proceedings of the ASE BigData & SocialInformatics, p. 28. ACM (2015)
42. Sharma, P.K., Park, J.H.: Blockchain based hybrid network architecture for the smart city. Futur. Gener. Comput. Syst. **86**, 650–655 (2018)
43. Rahman, M.A., Rashid, M.M., Hossain, M.S., Hassanain, E., Alhamid, M.F., Guizani, M.: Blockchain and IoT-based cognitive edge framework for sharing economy services in a smart city. IEEE Access **7**, 18611–18621 (2019)
44. Khan, Z., Abbasi, A.G., Pervez, Z.: Blockchain and edge computing-based architecture for participatory smart city applications. Concurr. Comput. Pract. Exp., e5566 (2019)
45. Damianou, A., Angelopoulos, C.M., Katos, V.: An architecture for blockchain over edge-enabled IoT for smart circular cities. In: 2019 15th International Conference on Distributed Computing in Sensor Systems (DCOSS), pp. 465–472. IEEE (2019)
46. Xu, R., Nikouei, S.Y., Chen, Y., Blasch, E., Aved, A.: BlendMAS: a blockchain-enabled decentralized microservices architecture for smart public safety. arXiv preprint arXiv:1902.10567 (2019)
47. Wang, R., Tsai, W.-T., He, J., Liu, C., Li, Q., Deng, E.: A video surveillance system based on permissioned blockchains and edge computing. In: 2019 IEEE International Conference on Big Data and Smart Computing (BigComp), pp. 1–6. IEEE (2019)
48. Kotobi, K., Sartipi, M.: Efficient and secure communications in smart cities using edge, caching, and blockchain. In: 2018 IEEE International Smart Cities Conference (ISC2), pp. 1–6. IEEE (2018)
49. Sharma, P.K., Moon, S.Y., Park, J.H.: Block-VN: a distributed blockchain based vehicular network architecture in smart city. JIPS **13**(1), 184–195 (2017)
50. Sherly, J., Somasundareswari, D.: Internet of Things based smart transportation systems. Int. Res. J. Eng. Technol. **2**(7), 1207–1210 (2015)
51. Li, M., Zhu, L., Lin, X.: Efficient and privacy-preserving carpooling using blockchain-assisted vehicular fog computing. IEEE Internet Things J. **6**(3), 4573–4584 (2019)
52. Liu, H., Zhang, Y., Yang, T.: Blockchain-enabled security in electric vehicles cloud and edge computing. IEEE Netw. **32**(3), 78–83 (2018)
53. Nguyen, T.H., Partala, J., Pirttikangas, S.: Blockchain-based mobility-as-a-service. In: 2019 28th International Conference on Computer Communication and Networks (ICCCN), pp. 1–6. IEEE (2019)
54. Zhou, Z., Wang, B., Dong, M., Ota, K.: Secure and efficient vehicle-to-grid energy trading in cyber physical systems: integration of blockchain and edge computing. IEEE Trans. Syst. Man Cybern. Syst. **50**(1), 43–57 (2019)
55. Chen, W., et al.: Cooperative and distributed computation offloading for blockchain-empowered industrial Internet of Things. IEEE Internet Things J. **6**(5), 4833–8446 (2019)
56. Zhang, K., Zhu, Y., Maharjan, S., Zhang, Y.: Edge intelligence and blockchain empowered 5G beyond for the industrial Internet of Things. IEEE Netw. **33**(5), 12–19 (2019)
57. Ren, Y., Zhu, F., Qi, J., Wang, J., Sangaiah, A.K.: Identity management and access control based on blockchain under edge computing for the industrial Internet of Things. Appl. Sci. **9**(10) (2019). https://doi.org/10.3390/app9102058

58. Gai, K., Wu, Y., Zhu, L., Zhang, Z., Qiu, M.: Differential privacy-based blockchain for industrial Internet of Things. IEEE Trans. Ind. Inform. (2019)
59. Seitz, A., Henze, D., Miehle, D., Bruegge, B., Nickles, J., Sauer, M.: Fog computing as enabler for blockchain-based IIoT app marketplaces-a case study. In: 2018 Fifth International Conference on Internet of Things: Systems, Management and Security, pp. 182–188. IEEE (2018)
60. Dorri, A., Kanhere, S.S., Jurdak, R., Gauravaram, P.: Blockchain for IoT security and privacy: the case study of a smart home. In: 2017 IEEE International Conference on Pervasive Computing and Communications Workshops (PerCom Workshops), pp. 618–623. IEEE (2017)
61. Tantidham, T., Aung, Y.N.: Emergency service for smart home system using Ethereum blockchain: system and architecture. In: 2019 IEEE International Conference on Pervasive Computing and Communications Workshops (PerCom Workshops), pp. 888–893. IEEE (2019)
62. Casado-Vara, R., de la Prieta, F., Prieto, J., Corchado, J.M.: Blockchain framework for IoT data quality via edge computing. In: Proceedings of the 1st Workshop on Blockchain-Enabled Networked Sensor Systems, pp. 19–24. ACM (2018)
63. Rahman, M.A., Rashid, M., Barnes, S., Hossain, M.S., Hassanain, E., Guizani, M.: An IoT and blockchain-based multi-sensory in-home quality of life framework for cancer patients. In: 2019 15th International Wireless Communications & Mobile Computing Conference (IWCMC), pp. 2116–2121. IEEE (2019)
64. Gai, K., Wu, Y., Zhu, L., Xu, L., Zhang, Y.: Permissioned blockchain and edge computing empowered privacy-preserving smart grid networks. IEEE Internet Things J. 6(5), 7992–8004 (2019)
65. Wang, J., Wu, L., Choo, K.-K.R., He, D.: Blockchain based anonymous authentication with key management for smart grid edge computing infrastructure. IEEE Trans. Ind. Inform. 16(3), 1984–1992 (2019)
66. Yang, J., Zhihui, L., Jie, W.: Smart-toy-edge-computing-oriented data exchange based on blockchain. J. Syst. Arch. 87, 36–48 (2018)
67. Jindal, A., Aujla, G.S., Kumar, N.: SURVIVOR: a blockchain based edge-as-a-service framework for secure energy trading in SDN-enabled vehicle-to-grid environment. Comput. Netw. 153, 36–48 (2019)

Author Index

Printed in the United States
By Bookmasters